INTRODUCTIO.
INVERSE PROBLEMS
IN IMAGING

Related Titles

Inverse Problems in Scattering and Imaging
Edited by M Bertero, Università di Genova, and E R Pike, King's College, London
(ISBN: 0 7503 0143 0)

Inverse Problems in Astronomy: A Guide to Inversion Strategies for Remotely Sensed Data
I J D Craig, University of Waikato, New Zealand, and J C Brown, University of Glasgow
(ISBN: 0 85274 369 6)

Basic Methods of Tomography and Inverse Problems
Edited by P C Sabatier, Université des Sciences et Techniques du Languedoc
(ISBN: 0 85274 475 7)

INTRODUCTION TO INVERSE PROBLEMS IN IMAGING

Mario Bertero

Department of Computer and Information Sciences University of Genova

Patrizia Boccacci

Department of Physics University of Genova

Institute of Physics Publishing
Bristol and Philadelphia

British Library Cataloguing-in-Publication Data

A catalogue record for this book is available from the British Library.

ISBN 0 7503 0439 1 (hbk)
0 7503 0435 9 (pbk)

Library of Congress Cataloging-in-Publication Data

Bertero, Mario
Introduction to inverse problems in imaging / Mario Bertero, Patrizia Boccacci
p. cm.
Includes bibliographical references and index.
ISBN 0-7503-0439-1 (akl. paper). – 0 7503 0435 9 (pbk. : pbk. alk. paper)
1. Image processing–Mathematics. 2. Inverse problems (Differential equations) I. Boccacci, Patrizia. II. Title
TA1637.B47 1998
621.36'7'0151535–dc21 98-14461
CIP

First published 1998
Reprinted 2002 (pbk)

Published by Institute of Physics Publishing, wholly owned by The Institute of Physics, London

Institute of Physics Publishing, Dirac House, Temple Back, Bristol BS1 6BE, UK

US Office: Institute of Physics Publishing, The Public Ledger Building, Suite 929, 150 South Independence Mall West, Philadelphia, PA 19106, USA

Typeset in TEX using the IOP Bookmaker Macros
Printed in the UK by Bookcraft Ltd, Bath

'Most people, if you describe a train of events to them, will tell you what the result would be. They can put those events together in their minds, and argue from them that something will come to pass. There are few people, however, who, if you told them a result, would be able to evolve from their own inner consciousness what the steps were which led up to that result. This power is what I mean when I talk of reasoning backwards, or analytically.' ... '... the grand thing is to be able to reason backwards. That is a very useful accomplishment, and a very easy one, but people do not practise it much. In the every-day affairs of life it is more useful to reason forwards, and so the other comes to be neglected. There are fifty who can reason synthetically for one who can reason analytically.'
(Sherlock Holmes in 'A Study in Scarlet')

Contents

Preface

This book arises from a series of lectures in image restoration and tomography that we teach to students in computer science and physics at the University of Genova. It is a course of approximately fifty one-hour lectures, supplemented by about fifty hours of training with the computer devoted to the implementation and validation of the principal methods presented in the lectures.

The mathematical background common to these students consists of first courses in mathematical analysis, geometry, linear algebra, probability theory and Fourier analysis. This is also the mathematical background we assume for our readers.

The selection of the methods and applications is partly subjective, since it is determined by our research experience, and partly objective, since it is dictated by the need to present efficient methods which are also easily implementable. Therefore, not only does the book not pretend to be complete but it is in fact based on a strategy of incompleteness. In spite of this choice, which also concerns references, we hope that the book will provide the reader with the appropriate background for a clear understanding of the essence of inverse problems (ill-posedness and its cure) and, consequently, for an intelligent assessment of the rapidly, and sometimes chaotically, growing literature on these problems.

We wish to thank our friends and colleagues Paola Brianzi, Piero Calvini and Carlo Pontiggia, who have read and commented on preliminary drafts of the book. In particular we are indebted to Christine De Mol, from the University of Brussels, for an accurate and critical reading of the manuscript and for several suggestions intended to improve the clarity of the presentation. We also express our gratitude to Laura Opisso for her help during the preparation of the LaTeX file.

For illustrating some of the methods described in the book, we used images (both synthetic and real) from the Hubble Space Telescope. These images were produced by the Space Telescope Science Institute, with support operated by the Association of Universities for Research in Astronomy, Inc., from NASA contract NAS5-26555, and were reproduced with permission from AURA/STScI.

The manuscript of the book was completed while M. Bertero was with the Department of Physics of the University of Genova.

M. Bertero and P. Boccacci
Genova, November 1997

Chapter 1

Introduction

The first clinical machine for the detection of head tumours, based on a new technique called X-ray *computed tomography* (CT), was installed in 1971 at the Atkinson Morley's Hospital, Wimbledon, Great Britain.

The announcement of this machine, by G.H. Hounsfield at the 1972 British Institute of Radiology annual conference [1], has been considered the greatest achievement in radiology since the discovery of X-rays by W.C. Röngten in 1895. In 1979 G.H. Hounsfield shared with A. Cormack the Nobel Prize for Physiology and Medicine.

The CT computer-generated images provided the first example of images obtained by solving a mathematical problem which belong to the class of the so-called inverse and ill-posed problems. The first workshop completely devoted to these problems was organized in 1971, that first year of tomography, by L. Colin at Ames Research Center, Moffet Field, California [2]. Among the topics considered were inversion methods applied to passive and active atmospheric sounding, ionospheric sounding, particle scattering, electromagnetic scattering and seismology.

In recent years important advances, certainly driven by the success of tomography, were made both in the theory and in the practice of inverse problems. A typical mathematical property of these problems, the so-called ill-posedness, has been understood and methods for overcoming the difficulties due to this property have been developed. Applications to imaging have been especially impressive; in particular the applications to diagnostic medicine have contributed to the development of medical imaging, a stimulating, important and still expanding discipline.

1.1 What is an inverse problem?

From the point of view of a mathematician the concept of an inverse problem has a certain degree of ambiguity which is well illustrated by a frequently quoted statement of J.B. Keller [3]: 'We call two problems *inverses* of one another if the formulation of each involves all or part of the solution of the other. Often, for historical reasons, one of the two problems has been studied extensively for some time, while the other has never been studied and is not so well understood. In such cases, the former is called the *direct problem*, while the latter is the *inverse problem*'.

In any domain of mathematical physics one finds problems satisfying the requirements stated by Keller. In general these problems are related by a sort of duality in the sense that one problem can be obtained from the other by exchanging the role of the data and that of the unknowns: the data of one problem are the unknowns of the other and conversely. As a consequence of this duality it may seem arbitrary to decide what is the direct and what is the inverse problem.

For a physicist, however, the situation is quite different because the two problems are not on the same level: one of them, and precisely that called the direct problem, is considered to be more fundamental than the other and, for this reason, is also much more investigated. In other words, the historical reasons mentioned by Keller are basically physical reasons.

With reference to physics, one can say that a direct problem is a problem oriented along a cause–effect sequence or, also, a problem which consists of computing the consequences of given causes; then, the corresponding inverse problem is associated with the reversal of the cause–effect sequence and consists of finding the unknown causes of known consequences [4]. It follows that the definition of a direct–inverse pair must be based on well-established physical laws, which specify what are the causes and what are the effects and provide the equations relating the effects to the causes. It also follows that for each domain of physics (mechanics, astronomy, wave propagation, heat conduction, geophysics, etc) it is necessary to specify the direct problems typical of that domain as well as the corresponding inverse problems. A few examples may clarify these statements.

In classical mechanics a direct problem is, for instance, the computation of the trajectories of particles from the knowledge of the forces. Then the inverse problem is the determination of the forces from the knowledge of the trajectories. From this point of view Newton not only stated the basic laws of mechanics, and therefore the basic equations of the direct problem, but also solved the first inverse problem when he determined the gravitation force from the Kepler laws describing the trajectories of the planets.

Other examples, however, are more appropriate for the modern applications of inverse methods. In scattering and diffraction theory, the direct problem is the computation of the scattered (or diffracted) waves from the knowledge of the sources and obstacles, while the inverse problem consists of the determination of the obstacles from the knowledge of the sources and of the scattered (or diffracted) waves. Inverse problems of this kind are fundamental for various methods of non-destructive evaluation (including medical imaging) which consist of sounding an object by means of a suitable radiation source.

Another example of a direct problem in wave-propagation theory is the computation of the field radiated by a given source, for instance the radiation pattern of a given antenna; then the inverse problem is the determination of the source from the knowledge of the radiated field (in the previous case, the determination of the current distribution in the antenna from the knowledge of the radiation pattern). Analogously in potential theory, which is basic in geodesy, the direct problem is the determination of the potential generated by a known mass distribution, while the inverse problem is the determination of the mass distribution from the values of the potential, and so on.

Other examples come from instrumental physics, i.e. the physics of instruments such as electronic devices, imaging systems, etc. Here the direct problem is the computation of the output of the instrument (the image) being given the input (the object) and the characteristics of the instrument (impulse response function, etc). Then the inverse problem is the identification of the input of a given instrument from the knowledge of the output. The first part of this book is devoted to a particular important case of this inverse problem.

As we stated above, a direct problem is a problem oriented along a cause–effect sequence; it is also a problem directed towards a loss of information: its solution defines a transition from a physical quantity with a certain information content to another quantity with a smaller information content. This property, which is common to most direct problems, will be reformulated in a more precise way in the next section. In general it implies that the solution is much smoother than the data: the image provided by a bandlimited system is smoother than the corresponding object, the scattered wave due to an obstacle is smooth even if the obstacle is rough, and so on. Here we briefly discuss a simple example where this property has a nice physical interpretation.

In the case of heat propagation, let us consider the direct problem of computing the temperature distribution at a time $t > 0$, given the temperature distribution at $t = 0$ (plus additional boundary conditions). A simplified version of the problem is the following: to solve in the domain $\mathcal{D} = \{0 \leq x \leq a; t > 0\}$ the one-dimensional equation of the heat conduction

$$\frac{\partial^2 u}{\partial x^2} = \frac{1}{D}\frac{\partial u}{\partial t}, \tag{1.1}$$

(the positive constant D measures the thermal conductivity) given the following initial and boundary conditions

$$u(x, 0) = f(x) \quad ; \quad u(0, t) = u(a, t) = 0. \tag{1.2}$$

This is our direct problem, the cause the given temperature distribution $f(x)$ at $t = 0$ and the corresponding effect being the temperature distribution $u(x, t)$ at the time $t > 0$.

The problem can be easily solved by means of Fourier series expansions.

If we expand the data function as follows

$$f(x) = \sum_{n=1}^{\infty} f_n \sin\left(\pi n \frac{x}{a}\right) \tag{1.3}$$

with

$$f_n = \frac{2}{a} \int_0^a f(x) \sin\left(\pi n \frac{x}{a}\right) dx, \tag{1.4}$$

then the solution of the problem is given by

$$u(x, t) = \sum_{n=1}^{\infty} f_n e^{-D(\pi n/a)^2 t} \sin\left(\pi n \frac{x}{a}\right). \tag{1.5}$$

Let us assume that the initial state is known with a certain precision ε, in the sense that only the Fourier coefficients f_n, such that $|f_n| > \varepsilon$, are known. Since the Fourier coefficients f_n tend to zero when $n \to \infty$, it follows that only a finite number of Fourier coefficients, let us say N_ε, is known. This is the information content of the data of the direct problem, or also of the initial state of the physical system. Then, if the final state is the temperature distribution at the time $t = T$, i.e. $u(x, T)$, its number of Fourier coefficients greater than ε is much smaller than N_ε as an effect of the decaying factor $\exp[-D(\pi n/a)^2 T]$. We conclude that the information content of the solution is much smaller than the information content of the data. This result is related to the well-known fact that the entropy of the system increases for increasing time.

The corresponding inverse problem is the problem of determining the temperature distribution at $t = 0$, being given the temperature distribution at the time $t = T > 0$. Therefore the data function is now given by $u(x, T)$, while the unknown function is $f(x)$. From the previous discussion of the direct problem it follows that, if we know the data with precision ε, then it will be difficult (maybe impossible) to obtain $f(x)$ with the same precision because some information has been lost in the natural evolution of the system. In other words, there exist many functions $f(x)$ which correspond, within the precision ε, to the given $u(x, T)$.

This conceptual difficulty is common to most inverse problems because, by solving these problems, we would like to accomplish a transformation which should correspond to a gain of information. It provides the explanation of a typical mathematical property of inverse problems which is known as ill-posedness. This point will be considered in the next section. Here we only observe as an interesting fact that the different physical interpretations of the two problems, direct and inverse, are associated with different mathematical properties.

The example previously discussed is an example of a linear problem and, in this book, we will consider only linear problems. The reason is twofold: first linear problems, eventually deriving from the linearization of nonlinear ones, are, at the moment, the most important for the applications; secondly, well-developed mathematical methods and efficient numerical algorithms for their solution are already available.

Undoubtedly most inverse problems are basically nonlinear and very interesting from the mathematical point of view. They will be ready for applications for the future because the mathematical research in this area is rapidly growing and many results are already available. We would also like to remember that the first inverse problems, which were the object of elegant and deep mathematical investigations, were nonlinear, namely, the *inverse Sturm–Liouville problem* and the *inverse scattering problem in quantum theory*.

The first dates back to Lord Rayleigh who, in describing the vibrations of strings of variable density, discussed the possibility of deriving the density distribution from the frequencies of vibration. Important contributions to this problem have been made by outstanding mathematicians, such as Levinson, Marchenko and Krein. An easy-to-read survey of these results is given in a famous paper by M. Kac [5].

As concerns the second problem, it was originated by an attempt to determine the unknown nuclear forces from the measured scattering data. Basic results in this problem are due, for instance, to Levinson, Jost, Kohn, Gel'fand, Levitan, Marchenko and many others. A valuable and complete presentation of these results is the book of Chadan and Sabatier [6].

Among books devoted to specific problems we would also like to mention the book of Colton and Kress [7] on inverse problems in acoustic and electromagnetism, the book of Craig and Brown [8] on inverse problems in astronomy, the book of Tarantola [9] on inverse problem in geophysics and the book of Herman *et al* [10] on tomography and related problems.

1.2 What is an ill-posed problem?

In the previous section we mentioned that a typical property of inverse problems is ill-posedness, a property which is opposite to that of well-posedness. We comment now on these concepts.

The basic concept of a *well-posed problem* was introduced by the French mathematician Jacques Hadamard in a paper published in 1902 on boundary-value problems for partial differential equations and their physical interpretation [11]. In this first formulation, a problem is called well-posed when its solution is unique and exists for arbitrary data. In subsequent work Hadamard emphasizes the requirement of continuous dependence of the solution on the data [12], claiming that a solution which varies considerably for a small variation of the data is not really a solution in the physical sense. Indeed, since physical data are never known exactly, this should imply that the solution is not known at all.

From an analysis of several cases Hadamard concludes that only problems motivated by physical reality are well-posed. An example is provided by the initial value problem, also called the Cauchy problem, for the D'Alembert equation which is basic in the description of wave propagation

$$\frac{\partial^2 u}{\partial x^2} - \frac{1}{c^2}\frac{\partial^2 u}{\partial t^2} = 0 \tag{1.6}$$

where c is the wave velocity. If we consider, for instance, the following initial data at $t = 0$

$$u(x, 0) = f(x), \quad \frac{\partial u}{\partial t}(x, 0) = 0 \tag{1.7}$$

then there exists a unique solution given by

$$u(x, t) = \frac{1}{2}\left[f(x - ct) + f(x + ct)\right]. \tag{1.8}$$

This solution exists for any continuous function $f(x)$. Moreover it is obvious that a small variation of $f(x)$ produces a small variation of $u(x, t)$, as given by equation (1.8).

Another example of a well-posed problem is provided by the forward problem of the heat equation, briefly analyzed in the previous section. The solution, as given by equation (1.5), clearly exists, is unique, and depends continuously on the data function $f(x)$.

The previous problems are well-posed and, of course, are basic in the description of physical phenomena. They are examples of direct problems. An impressive example of an ill-posed problem and, in particular, of non-continuous dependence on the data, was also provided by Hadamard [12]. This problem which, at that time, was deprived of physical motivations, is the Cauchy problem of the Laplace equation in two variables

$$\frac{\partial^2 u}{\partial x^2} + \frac{\partial^2 u}{\partial y^2} = 0. \tag{1.9}$$

If we consider the following Cauchy data at $y = 0$

$$u(x, 0) = \frac{1}{n}\cos(nx), \quad \frac{\partial u}{\partial y}(x, 0) = 0 \tag{1.10}$$

then the unique solution of equation (1.9), satisfying the conditions (1.10), is given by

$$u(x, y) = \frac{1}{n}\cos(nx)\cosh(ny). \tag{1.11}$$

The factor $\cos(nx)$ produces an oscillation of the surface representing the solution of the problem. When n is sufficiently large, this oscillation is imperceptible near $y = 0$ but becomes enormous at any given finite distance from the x-axis. More precisely, when $n \to \infty$, the data of the problem tend to zero but, for any finite value of y, the solution tends to infinity.

This is now a classical example illustrating the effects produced by a non-continuous dependence of the solution on the data. If the oscillating function

(1.10) describes the experimental errors affecting the data of the problem then the error propagation from the data to the solution is described by equation (1.11) and its effect is so dramatic that the solution corresponding to real data is deprived of physical meaning. Moreover it is also possible to show that the solution does not exist for arbitrary data but only for data with some specific analyticity property. In any case it is known now that the basic inverse problem of electrocardiography [13], i.e. the reconstruction of the epicardial potential from body surface maps, can be formulated just as a Cauchy problem for an elliptic equation, i.e. a generalization of the Laplace equation.

Another example of non-continuous dependence of the solution on the data is provided by the backward problem of the heat equation, i.e. the inverse problem discussed in the previous section. If we consider the following data at the time $t = T$

$$u(x, T) = \frac{1}{n} \sin\left(\pi n \frac{x}{a}\right) \tag{1.12}$$

then, the solution of the heat equation (1.1) is given by

$$u(x, t) = \frac{1}{n} \sin\left(\pi n \frac{x}{a}\right) e^{D(\pi n/a)^2 (T-t)}. \tag{1.13}$$

For $t < T$ and, in particular, for $t = 0$ we find again that, when $n \to \infty$, the data function tends to zero while the solution tends to infinity. Since this pathology is clearly due to the exponential factor which, as discussed in the previous section, is responsible for the loss of information in the solution of the direct problem, we find a link between this loss of information and the lack of continuity in the solution of the inverse problem.

A problem satisfying the requirements of uniqueness, existence and continuity is now called *well-posed* (in the sense of Hadamard), even if the complete formulation in terms of the three requirements is first given by R. Courant [14]. The problems which are not well-posed are called *ill-posed* or also *incorrectly posed* or *improperly posed*. Therefore an ill-posed problem is a problem whose solution is not unique or does not exist for arbitrary data or does not depend continuously on the data.

The conviction of Hadamard that problems motivated by physical reality must be well-posed is essentially generated by the physics of the nineteenth century. The requirements of existence, uniqueness and continuity of the solution are deeply inherent in the ideal of a unique, complete and stable determination of the physical events. As a consequence of this point of view, ill-posed problems were considered, for many years, as mathematical anomalies and were not seriously investigated. The discovery of the ill-posedness of inverse problems has completely modified this conception.

The previous observations and considerations can justify now the following general statement: a direct problem, i.e. a problem oriented along a cause–effect sequence, is well-posed while the corresponding inverse problem, which implies a reversal of the cause–effect sequence, is ill-posed. This statement, however, is meaningful only if we provide a suitable mathematical setting for the description of direct and inverse problems. To this purpose we take into account that we are mainly considering problems of imaging and therefore we use a language appropriate to these problems.

The first point is to define the class of the objects to be imaged, which will be described by suitable functions with certain properties. In this class we also need a *distance*, in order to establish when two objects are close and when they are not. In such a way our class of objects takes the structure of a *metric space* of functions. We denote this space by $\mathcal{X}$ and we call it the *object space*.

The second point is to solve the direct problem, i.e. to compute, for each object, the corresponding image which can be called the computed image or the *noise-free image*. Since the direct problem is well-posed, to each object we associate one, and only one, image. As we already remarked, this image may be rather smooth as a consequence of the fact that its information content is smaller than the information content of the corresponding object. This property of smoothness, however, may not be true for the measured images, also called *noisy images*, because they correspond to some noise-free image corrupted by the noise affecting the measurement process.

Therefore the third point is to define the class of the images in such a way that it contains both the noise-free and the noisy images. It is convenient to introduce a distance also in this class. We denote the corresponding function space by $\mathcal{Y}$ and we call it the *image space*.

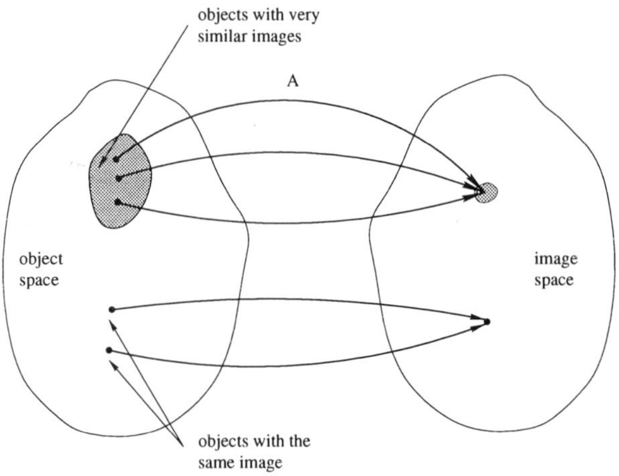

Figure 1.1. Schematic representation of the relationship between objects and images. The shaded subsets in $\mathcal{X}$ and $\mathcal{Y}$ illustrate the loss of information due to the imaging process.

In conclusion, the solution of the direct problem defines a mapping (operator), denoted by A, which transforms any object of the space $\mathcal{X}$ into a noise-free image of the space $\mathcal{Y}$. This operator is continuous, i.e. the images of two close objects are also close, because the direct problem is well-posed. The set of the noise-free images is usually called, in mathematics, the *range* of the operator A, and, as follows from our previous remark, this range does not coincide with the image space $\mathcal{Y}$ because this space contains also the noisy images.

By means of this mathematical scheme it is possible to describe the loss of information which, as we said, is typical in the solution of the direct problem. It has two consequences which are described in a different way from the mathematical point of view, even if their practical effects are very close. First, it may be possible that two, or even more, objects have exactly the same image. In the case of a linear operator this is related to the existence of objects whose image is exactly zero. These objects will be called *invisible objects*. Then, given any object of the space $\mathcal{X}$, if we add to it an invisible object, we obtain a new object which has exactly the same image. Secondly, and this fact is much more general than the previous one, it may be possible that two very distant objects have images which are very close. In other words there exist very broad sets of distinct objects such that the corresponding sets of images are very small. All these properties are illustrated in figure 1.1.

If we consider now the inverse problem, i.e. the problem of determining the object corresponding to a given image, we find that this problem is ill-posed as a consequence of the loss of information intrinsic to the solution of the the direct one. Indeed, if we have an image corresponding to two distinct objects, the solution of the inverse problem is not unique. If we have a noisy image, which is not in the range of the operator A, then the solution of the inverse problem does not exist. If we have two neighbouring images such that the corresponding objects are very distant, then the solution of the inverse problem does not depend continuously on the data.

1.3 How to cure ill-posedness

The property of non-continuous dependence of the solution on the data strictly applies only to ill-posed problems formulated in infinite dimensional spaces. In practice one has discrete data and one has to solve discrete problems. These, however, are obtained by discretizing problems with very bad mathematical properties. What happens in these cases?

If we consider a linear inverse problem, its discrete version is a linear algebraic system, apparently a rather simple mathematical problem. Many methods exist for solving numerically this problem. However the solution does not work. A lively description of the first attempts at data inversions is given by S. Twomey in the preface of his book [15]: 'The crux of the difficulty was that numerical inversions were producing results which were physically unacceptable but were mathematically acceptable (in the sense that *had* they existed they should have given measured values identical or almost identical with what was measured)'. These results were 'rejected as impossible or ridiculous by the recipient of the computer's answer. And yet the computer was often blamed, even though it had done all that had been asked of it'. ... 'Were it possible for computers to have ulcers or neuroses there is little doubt that most of those of which early numerical inversion attempts were made would have required both afflictions'.

The explanation can be found having in mind the examples discussed in the previous section, where small oscillating data produce large oscillating solutions. In any inverse problem data are always affected by noise which can be viewed as a small randomly oscillating function. Therefore the solution method amplifies the noise producing a large and wildly oscillating function which completely hides the physical solution corresponding to the noise-free data. This property holds true also for the discrete version of the ill-posed problem. Then one says that the corresponding linear algebraic system is *ill-conditioned*: even if the solution exists and is unique, it is completely corrupted by a small error on the data.

In conclusion, we have the following situation: we can compute one, and only one, solution of our algebraic system but this solution is unacceptable for the reasons indicated above; the physically acceptable solution we are looking for is not a solution of the problem but only an approximate solution in the sense that it does reproduce the data not exactly but only within the experimental errors. However, if we look for approximate solutions, we find that they constitute a set which is extremely broad and contains completely different functions, a consequence of the loss of information in the direct problem, as we discussed in the previous sections. Then the question arises: how can we choose the good ones?

We can state now what we consider the 'golden rule' for solving inverse problems which are ill-posed: search for approximate solutions satisfying additional constraints coming from the physics of the problem. This point can be clarified in the framework of the mathematical model introduced in the previous section.

The set of the approximate solutions corresponding to the same data function is just the set of the objects with images close to the measured one (the two sets are the shaded sets represented in figure 1.1). The set of the objects is too broad, as a consequence of the loss of information due to the imaging process. Therefore we need some additional information to compensate this loss. This information, which is also called *a priori* or *prior* information, is additional in the sense that it cannot be derived from the image or from the properties of the mapping A which describes the imaging process but expresses some expected physical properties of the object. Its role is to reduce the set of the objects compatible with the given image or also to discriminate between interesting objects and spurious objects, generated by uncontrolled propagation of the noise affecting the image.

The most simple form of additional information is that the object cannot be too large. This implies a constraint which consists of an upper bound on the object itself, or its intensity, or its energy, etc. Another kind of additional information may be that the object is smooth so that, for instance, its derivatives must be smaller than a certain quantity. Moreover it may be known that the object must be non-negative or that it must be different from zero only inside a given bounded region. A quite different kind of additional information may consist of statistical properties of the objects. In this case one assumes that the objects to be restored are realizations of a random process with known probability distribution. This can be a way for expressing our previous experience about the possible objects to be restored. Even if a complete knowledge of the probability distribution is not available, a partial knowledge of statistical properties consisting, for instance, in the knowledge of the expectation values and covariance matrices may be useful.

The idea of using prescribed bounds to produce approximate and stable solutions was introduced by C. Pucci in the case of the Cauchy problem for the Laplace equation [16], i.e. the first example of an ill-posed problem discussed by Hadamard. The constraint of positivity was used by F. John in the solution of the heat equation for preceding time [17], i.e. the problem previously considered of determining the temperature distribution at the time $t = 0$ from the knowledge of the temperature distribution at the time $t = T > 0$. A general version of similar ideas was formulated independently by V.K. Ivanov [18]. His method and the method of D.L. Phillips for Fredholm integral equations of the first kind [19] were the first examples of *regularization methods* for the solution of ill-posed problems. The theory of these methods was formulated by A.N. Tikhonov one year later [20].

The principle of the regularization methods is to use the additional information explicitly, at the start, to construct families of approximate solutions, i.e. of objects compatible with the given image. These methods are now one of the most powerful tools for the solution of inverse problem, another one being provided by the so-called Bayesian methods, where the additional information used is of statistical nature.

1.4 An outline of the book

The book of Tikhonov and Arsenin [21], published in 1977 (the Russian edition appeared in 1974 and the French translation in 1976) is the first book on regularization theory. Subequently a dozen books, written by mathematicians for mathematicians, have appeared on the same subject. We only mention the book of Groetsch [22], which is not too technical and contains a commented list of references, and the very recent book of Engl, Hanke and Neubauer [23], which provides a valuable and well-organized presentation of the most important mathematical results on regularization theory.

The main purpose of our book, which is not addressed to mathematicians (but we believe that it may be useful also to mathematicians) is to introduce the reader to the basic ideas and methods for the solution of inverse and ill-posed problems even if he/she does not master the mathematics of the functional treatment of operator equations. To attain this purpose we found it useful to focus on a few significant examples, mainly image deconvolution and tomography, because the basic mathematical tool which is needed in these cases is common to anyone working in applied science: Fourier analysis.

The book consists of two main parts. The first one is completely devoted to the problem of image deconvolution which is a particular example of the more general problem of image restoration (other, more or less, equivalent names: image deblurring, image enhancement, image reconstruction, etc). This problem has all the conceptual difficulties of any inverse problem but, as we already observed, its treatment only requires the knowledge of Fourier analysis. Moreover in this framework it is possible to introduce in a natural and easy way the most useful methods for the solution of inverse problems and to describe these methods in terms of concepts derived from the theory of linear systems.

The first part contains six chapters. Chapter 2 is a short description of the mathematical methods used for the analysis and the numerical treatment of image deconvolution, namely the Fourier transform and the discrete Fourier transform. If these tools are familiar to the reader then he/she can skip this chapter and use it only to understand the notations we use.

Chapter 3 contains the description of several examples of space-invariant imaging systems. In such a case the mapping A, introduced in section 1.2, is a convolution operator and the information loss due to the imaging process can be described in terms of the transfer function of the system. It is due both to the possible zeros of the transfer function and to its behaviour at high frequencies. Indeed, in all examples discussed in this chapter the transfer function tends to zero at high frequencies so that the imaging system acts essentially as a Fourier filter (low-pass filter).

The subsequent four chapters constitute the central part of the book. In chapter 4 the ill-posedness of the problem of image deconvolution is investigated as well as the relationship between the ill-posedness of the problem formulated in function spaces and the ill-conditioning of the corresponding discrete problem. Moreover the concept of least-squares solution and generalized solution is introduced and the relationship with the method known as inverse filtering is indicated. Examples where this method can work are also discussed.

The concept of constrained least-squares solution is used to introduce in chapter 5 the Tikhonov regularization method. The behaviour of the regularized solutions as a function of the regularization parameter is investigated and the property called *semiconvergence* is emphasized. Thanks to this property, the user knows that an optimal value of the regularization parameter exists, even if its determination may be difficult. Optimal here means that among all regularized solutions, that corresponding to this value of the regularization parameter provides the best approximation of the unknown object.

Next the general concept of regularization method or regularization algorithm is illustrated by giving other examples of these methods. The concept of global PSF is introduced as a tool for comparing the performances of different linear regularization algorithms. Finally a short description of various methods for the choice of the regularization parameter is given.

An interesting property of several iterative methods for the solution of a linear problem is that they can be viewed as regularization methods if the problem is ill-posed. In these cases the role of the regularization parameter is played by the number of iterations and the semiconvergence property, which holds true again, implies the existence of an optimal value, in the sense specified above, of this number of iterations. These iterative methods are investigated in chapter 6. With the exception of the so-called Landweber method, all methods are nonlinear and therefore they provide examples of nonlinear regularization methods for linear inverse problems. In particular the projected Landweber method is discussed as a method which approximates the solutions of constrained least-squares problems in the case of convex constraints (positivity etc). In particular cases this method can provide very accurate solutions even if it is not very efficient from the computational point of view.

The last chapter of the first part is devoted to statistical methods, i.e. methods which are based on the knowledge of statistical properties of the noise and, possibly, of the object. The methods where only statistical properties of the noise are used are the so-called maximum likelihood methods. They are closely related to the least-squares method and have, in general, similar mathematical difficulties, i.e. they still lead to ill-posed problems. In the case of Poisson noise, maximum likelihood provides the foundation of an iterative method known as the expectation–maximization or Lucy–Richardson method. Numerical simulations indicate that this method may have the semiconvergence property even if this result has not been proved. On the other hand Bayesian methods are based on *a priori* information about the probability distribution of the object and, according to our 'golden rule', they may lead to well-posed problems. It is shown that this is, indeed, the case for Gaussian processes where the Bayes method leads to the Wiener filter. The strict analogy between the Wiener filter and the Tikhonov regularization method (with a particular choice of the regularization parameter) is also shown.

The second part of the book, which consists of four chapters, is devoted to linear imaging systems which are not described by convolution operators but by linear operators whose basic property is that they can be represented by means of the singular value decomposition. This representation is essentially a generalization of the well-known representation of a symmetric matrix in terms of its eigenvalues and eigenvectors (spectral representation), not only to arbitrary matrices but also to some particular classes of linear operators. In chapter 8 examples of the imaging systems under consideration are provided with particular attention to tomographic systems (both transmission and emission tomography are described). Chapter 9 is devoted to the singular value decomposition. It is derived in the case of a matrix and in the case of a semi-discrete imaging operator. The case of integral operators is also discussed and a rather simple derivation of the singular value decomposition of the Radon transform, the basic operator in tomography, is also given.

The inversion methods discussed in the case of image deconvolution are reproposed in chapter 10 in the case of imaging systems described by a singular value decomposition. Since the application of this technique may be difficult in the case of large images, due to the large number of singular values and singular functions to be computed, in chapter 11 we discuss two problems where Fourier based methods can still be used, namely tomography and super-resolution in image deconvolution.

In spite of our attempt to avoid the technicalities of functional analysis a few basic concepts of this beautiful mathematical theory are needed for a better understanding of the methods presented in the book. Therefore we attempt an elementary presentation of these concepts in the third part of the book (mathematical appendices). Here we also discuss a few questions of linear algebra which cannot be easily found in books on numerical analysis.

References

[1] Hounsfield G N 1973 *Br. J. Radiol.* **46** 1016

[2] Colin L ed 1972 *Mathematics of Profile Inversion*, Proc. of a workshop held at Ames Research Center, Moffet Field, California, July 12-16, 1971 (NASA Technical Memorandum, TMX-62-150).

[3] Keller J B 1976 *Am. Math. Monthly* **83** 107

[4] Turchin V F, Kozlov V P and Malkevich M S 1971 *Soviet. Phys. Usp.* **13** 681

[5] Kac M 1966 *Am. Math. Monthly* **73** (4) Part II 1

[6] Chadan K and Sabatier P C 1989 *Inverse Problems in Quantum Scattering Theory* 2nd edn (Berlin: Springer)

[7] Colton D and Kress R 1992 *Inverse Acoustic and Electromagnetic Scattering Theory* (Berlin: Springer)

[8] Craig I J D and Brown J C 1986 *Inverse Problems in Astronomy* (Bristol: Adam Hilger)

[9] Tarantola A 1987 *Inverse Problem Theory* (Amsterdam: Elsevier)

[10] Herman G T, Tuy H K, Langenberg K J and Sabatier P C 1987 *Basic Methods of Tomography and Inverse Problems* (Bristol: Adam Hilger)

[11] Hadamard J 1902 *Bull. Univ. Princeton* **13** 49

[12] Hadamard J 1923 *Lectures on Cauchy's Problem in Linear Partial Differential Equations* (New Haven, CT: Yale University Press)

[13] Colli Franzone P, Taccardi B and Viganotti C 1977 *Adv. Cardiol.* **21** 167

[14] Courant R and Hilbert D 1962 *Methods of Mathematical Physics*, vol. II (New York: Interscience) 227

[15] Twomey S 1977 *Introduction to the Mathematics of Inversion in Remote Sensing and Indirect Measurements* (New York: Elsevier)

[16] Pucci C 1955 *Atti Acc. Naz. Lincei* **18** 473

[17] John F 1955 *Ann. Mat. Pura Appl.* **40** 129

[18] Ivanov V K 1962 *Soviet Math. Dokl.* **3** 981
[19] Phillips D L 1962 *J. Assoc. Comput. Mach.* **9** 84
[20] Tikhonov A N 1963 *Soviet Math. Dokl.* **4** 1035
[21] Tikhonov A N and Arsenin V Y 1977 *Solutions of Ill-posed Problems* (Washington: Winston/Wiley)
[22] Groetsch C W 1993 *Inverse Problems in the Mathematical Sciences* (Braunschweig: Vieweg)
[23] Engl H W, Hanke M and Neubauer A *Regularization of Inverse Problems* (Dordrecht: Kluwer)

PART 1

IMAGE DECONVOLUTION

Chapter 2

Some mathematical tools

In this chapter we summarize the mathematical methods which are basic both for a theoretical analysis and a practical solution of the problem of image deconvolution. These methods are essentially the Fourier Transform (FT) and the Discrete Fourier Transform (DFT). The DFT is fundamental for the numerical treatment of the problem because a very efficient algorithm for its computation, the so-called Fast Fourier Transform (FFT), is available. We also discuss convolution operators both in the continuous and in the discrete case.

2.1 The Fourier Transform (FT)

The Fourier Transform is discussed in so many textbooks that we need spend only a few words just to establish the notations we use and recalling the properties we need. For a review of the fundamental mathematical concepts, see the books by Papoulis [1] and Bracewell [2].

We consider a function $f(\boldsymbol{x})$ of the q-dimensional variable $\boldsymbol{x} = \{x_1, x_2, \ldots, x_q\}$ (in the problems of image restoration and data deconvolution we have only $q = 1, 2, 3$); then the *Fourier Transform* (FT) of f is the function defined by the following integral, whenever it is convergent

$$\hat{f}(\boldsymbol{\omega}) = \int e^{-i\boldsymbol{\omega}\cdot\boldsymbol{x}} f(\boldsymbol{x}) d\boldsymbol{x}. \tag{2.1}$$

Here $d\boldsymbol{x} = dx_1 \ldots dx_q$, $\boldsymbol{\omega} \cdot \boldsymbol{x} = \omega_1 x_1 + \omega_2 x_2 + \ldots + \omega_q x_q$ and the integral is extended to all variables from $-\infty$ to $+\infty$. The following inversion formula holds true, if also in this case the integral is convergent

$$f(\boldsymbol{x}) = \frac{1}{(2\pi)^q} \int e^{i\boldsymbol{x}\cdot\boldsymbol{\omega}} \hat{f}(\boldsymbol{\omega}) d\boldsymbol{\omega}. \tag{2.2}$$

If the components of the vector $\boldsymbol{x}$ have the physical meaning of space variables, then the components of the vector $\boldsymbol{\omega} = \{\omega_1, \omega_2, \cdots, \omega_q\}$ are called space frequencies. The set of all vectors ω is also called the frequency domain or Fourier domain.

Two classes of functions will be considered: (Lebesgue) integrable and square-integrable functions. A function f is *integrable* if

$$\int |f(\boldsymbol{x})| d\boldsymbol{x} \quad < \quad +\infty \tag{2.3}$$

where the integral is intended in the sense of Lebesgue.

The Fourier transform of an integrable function has the following properties (*Riemann–Lebesgue theorem*) [1]: the function $\hat{f}(\omega)$ is continuous and bounded

$$|\hat{f}(\omega)| < \int |f(\boldsymbol{x})| d\boldsymbol{x}; \tag{2.4}$$

moreover $|\hat{f}(\omega)|$ tends to zero when $|\omega| \to \infty$.

Remark 2.1. Integrability of the Fourier transform. *Even if* $|\hat{f}(\omega)| \to 0$ *at infinity, the FFT of an integrable function in general is not integrable. Examples of integrable functions whose FT is not integrable are provided by the characteristic functions. The characteristic function of a set* $\mathcal{D}$ *is defined as the function which is* 1 *over the set and zero elsewhere and is denoted by* $\chi_{\mathcal{D}}(\boldsymbol{x})$. *In the case of one variable, if* $\mathcal{D}$ *is the interval* $[-\pi, \pi]$, *then*

$$\hat{\chi}_{\mathcal{D}}(\omega) = 2\pi \operatorname{sinc}(\omega) \tag{2.5}$$

the function $\operatorname{sinc}(\xi)$ *being defined by*

$$\operatorname{sinc}(\xi) = \frac{sin(\pi\xi)}{\pi\xi}. \tag{2.6}$$

The function $\operatorname{sinc}(\xi)$ *is not integrable because the integral of its modulus does not converge. Analogously, in two variables, if* $\chi_{\mathcal{D}}(\boldsymbol{x})$ *is the characteristic function of the disc of radius* π, *then its Fourier transform is given by*

$$\hat{\chi}_{\mathcal{D}}(\omega) = 2\pi^2 \frac{J_1(\pi|\omega|)}{|\omega|} \tag{2.7}$$

where $J_1(\xi)$ *is the Bessel function of order 1. Also in this case the Fourier transform is not integrable.*

The characteristic functions considered in this remark are square-integrable and their FT have the same property. Indeed, a function f is *square-integrable* if

$$\int |f(\boldsymbol{x})|^2 d\boldsymbol{x} \quad < \quad \infty. \tag{2.8}$$

For these functions the FT, in general, is not defined by means of the integral (2.1). A rigorous theory requires Lebesgue theory of measure and integration. Roughly speaking, one defines the FT by restricting the integral (2.1) to a ball of radius R and then taking the limit when $R \to \infty$.

It can be proved that the FT defined in such a way is also square-integrable and that the following *Parseval equality* holds true

$$\int |f(x)|^2 dx = \frac{1}{(2\pi)^q} \int |\hat{f}(\omega)|^2 d\omega, \tag{2.9}$$

as well the *generalized Parseval equality*

$$\int f(x)h^*(x)dx = \frac{1}{(2\pi)^q} \int \hat{f}(\omega)\hat{h}^*(\omega)d\omega \tag{2.10}$$

where $*$ denotes complex conjugation.

Before concluding this section, we briefly mention the geometrical structure of the space of all square-integrable functions, the so-called *L^2-space*. We do not give a precise definition of this space because it also requires Lebesgue theory of measure and integration. An element of this space, for instance, is not a function but a class of functions which are equal almost everywhere (see appendix A); however, for the sake of simplicity, we call such an element a square-integrable function.

The L^2-space is linear because a linear combination of square-integrable functions is still a square-integrable function. Moreover, given two functions f, h of this space, it is possible to define their *scalar product* as follows

$$(f, h) = \int f(x)h^*(x)dx. \tag{2.11}$$

A linear space equipped with a scalar product is called a *Euclidean space*. Then two functions f, h are said to be *orthogonal* if their scalar product is zero. From the generalized Parseval equality it follows that, if two functions f, h are orthogonal, their FT, $\hat{f}, \hat{h}$ are also orthogonal.

To each function f one can associate a *norm* defined as follows

$$\|f\| = \left(\int |f(x)|^2 dx \right)^{1/2} \tag{2.12}$$

which is related to the scalar product (2.11) by $\|f\| = (f, f)^{1/2}$. A function f is said to be *normalized* if $\|f\| = 1$. A set of functions $f_1, f_2, \ldots, f_n$ is said to be an *orthonormal set* if they satisfy the conditions

$$(f_j, f_k) = \delta_{jk} \tag{2.13}$$

where δ_{jk} is the Kronecker symbol, equal to 1 when $j = k$ and equal to 0 when $j \neq k$.

It is also possible to introduce the *distance* of f from h, defined as follows

$$d(f, h) = \|f - h\|. \tag{2.14}$$

This distance is the *mean square error* we commit if we take h as an approximation of f. The ball of centre f and radius ϵ is the set of all functions h satisfying the condition $\|f - h\| \leq \epsilon$ and therefore is the set of all functions which approximate f with a mean square error no greater than ϵ. A few basic properties of Euclidean spaces of functions and Hilbert spaces are summarized in appendix A.

2.2 Bandlimited functions and sampling theorems

We first introduce the concept of support of a function. A precise definition of this concept in the general case is rather involved, even if the support is essentially the set of points where the function is different from zero. Here we give a definition which applies to the case of continuous or piecewise continuous functions.

Consider the set of the points $\boldsymbol{x}$ where $f(\boldsymbol{x}) \neq 0$; the *support* of f is the closure of this set. We explain this point by means of a few examples, mainly in the case of functions of one variable. The first is the function $f_1(x)$ defined by $f_1(x) = 1 - |x|$, when $|x| < 1$, and $f_1(x) = 0$ when $|x| \geq 1$. Then the set of the points where $f_1(x) \neq 0$ is the open interval $(-1, 1)$ while the support of $f_1(x)$ is the closed interval $[-1, 1]$. This is obtained by adding to $(-1, 1)$ the limit points ± 1 which do not belong to this interval. The second example is the function $f_2(x)$ defined by $f_2(x) = \mathrm{sinc}(x)$ (see equation (2.6)). This function is different from zero everywhere except at the points $x_n = n$, $(n = \pm 1, \pm 2, \ldots)$. Therefore its support is the whole axis $(-\infty, +\infty)$. As a third example we can consider the function defined in equation (2.7), which is a function of two variables, vanishing over a countable set of circles centred at the origin. The closure of the set where the function is different from zero is the whole 2D plane.

A function f whose support is a bounded set is said to be *spacelimited*, while a function f whose FT $\hat{f}$ has a bounded support is said to be *bandlimited*. In such a case the support of $\hat{f}$ is also called the *band* of f. For instance, in the previous examples the function $f_1(x)$ is spacelimited while the function $f_2(x)$ is bandlimited because it is the inverse Fourier transform of the characteristic function of the interval $[-\pi, \pi]$ (which is the band of $f_2(x)$).

In the treatment of imaging problems we will need both spacelimited and bandlimited functions. In general, the physical objects to be imaged are spacelimited while the images provided, for instance, by an optical instrument are bandlimited. Then, it is important to point out that a *spacelimited function is never bandlimited* as well as *a bandlimited function is never spacelimited*. The first part of this statement follows from the fact that the FT of a spacelimited function is an entire function, i.e a function which is analytic everywhere. But an analytic function, which vanishes on some interval, does also vanish everywhere. As a consequence, if $f(x)$ is spacelimited and not identically zero, then the support of $\hat{f}(\omega)$ must be the whole axis $(-\infty, +\infty)$. The second part of the statement follows from the symmetry between the FT and the inverse FT. Moreover the statement can be generalized to functions of two and more variables.

A basic property of a bandlimited function is the possibility of representing such a function, without any loss of information, by means of its samples taken at equidistant points. This property is expressed by the so-called *sampling theorem* or *Wittaker–Shannon theorem* [3]. Thanks to the symmetry mentioned above, it also applies to the FT of a spacelimited function which, therefore, can be represented by means of samples taken at equidistant points in the Fourier domain. Here we give the formulae in the case of bandlimited functions and we only indicate the straightforward modifications which are required in the case of the FT of a spacelimited function.

Let $f(x)$ be a bandlimited and square-integrable function of the space variable x, with a band interior to the interval $[-\Omega, \Omega]$. In this case we call Ω the *bandwidth* of f and we also say that f is Ω-bandlimited. Then, according to the sampling theorem, f can be represented as follows

$$f(x) = \sum_{n=-\infty}^{+\infty} f\left(n\frac{\pi}{\Omega}\right) \operatorname{sinc}\left[\frac{\Omega}{\pi}\left(x - n\frac{\pi}{\Omega}\right)\right] \tag{2.15}$$

the sinc function being defined in equation (2.6). The points $x_n = n\pi/\Omega$ are called the *sampling points* and the distance π/Ω is called the *sampling distance* or also the *Nyquist distance*. We stress the fact that the sampling distance is inversely proportional to the bandwidth of f. Moreover, since the sampling functions

$$S_\Omega(x - x_n) = \operatorname{sinc}\left[\frac{\Omega}{\pi}\left(x - n\frac{\pi}{\Omega}\right)\right] \tag{2.16}$$

are smooth between adjacent sampling points, it is clear that a bandlimited function with bandwidth Ω cannot change very rapidly over distances much smaller than π/Ω.

Analogously, if $f(x)$ is spacelimited, with support interior to the interval $[-X, X]$ (in such a case we also say that f is X-spacelimited), then its FT, $\hat{f}(\omega)$, can be represented by the expansion (2.15) with x replaced by ω and Ω replaced by X. Therefore the sampling distance in Fourier domain is given by π/X and the sampling functions are given by $S_X(\omega - \omega_n) = \operatorname{sinc}\left[(\omega - n\pi/X)X/\pi\right]$.

The sampling functions defined in equation (2.16) have very interesting properties. The function $S_\Omega(x - x_n)$, associated with the sampling point x_n, is equal to 1 at the point x_n and is zero at the other sampling points x_m, $m \neq n$ (see figure 2.1). Moreover the sampling functions associated with different sampling

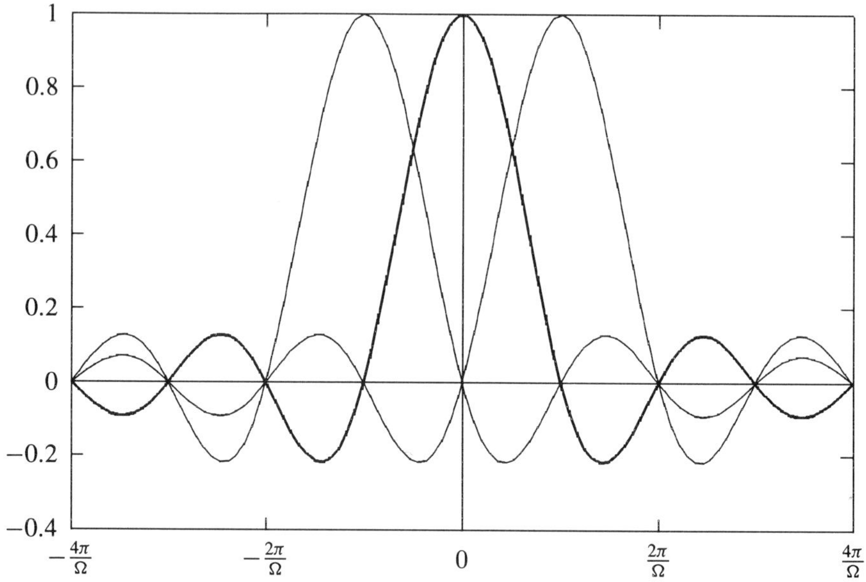

Figure 2.1. The sampling functions corresponding to the sampling points $x_0 = 0$ and $x_{\pm 1} = \pm\frac{\pi}{\Omega}$.

points are orthogonal, i.e.

$$\int_{-\infty}^{+\infty} S_\Omega(x - x_m) S_\Omega(x - x_n) dx = \frac{\pi}{\Omega}\delta_{mn}. \tag{2.17}$$

These relations imply also the following formula for the scalar product of two bandlimited functions having the same bandwidth Ω

$$\int_{-\infty}^{+\infty} f(x) h^*(x) dx = \sum_{n=-\infty}^{+\infty} f\left(n\frac{\pi}{\Omega}\right) h^*\left(n\frac{\pi}{\Omega}\right)\frac{\pi}{\Omega} \tag{2.18}$$

which shows that the scalar product can be computed from the samples of $f(x)$ and $h(x)$ using the trapezoidal rule.

As follows from the general results discussed in appendix A, the sampling functions (2.16) constitute an orthogonal basis in the subspace of all square-integrable and Ω-bandlimited functions. Then equation (2.18) is a particular example of the generalized Parseval equality given in equation (A.14).

The sampling expansion (2.15) can be extended to functions of many variables. For simplicity we consider only the case of two variables. If the function $f(\boldsymbol{x}) = f(x_1, x_2)$ is bandlimited and its band is a set $\mathcal{B}$ interior to the square $|\omega_1| \leq \Omega$, $|\omega_2| \leq \Omega$ (see figure 2.2) then it can be represented by means

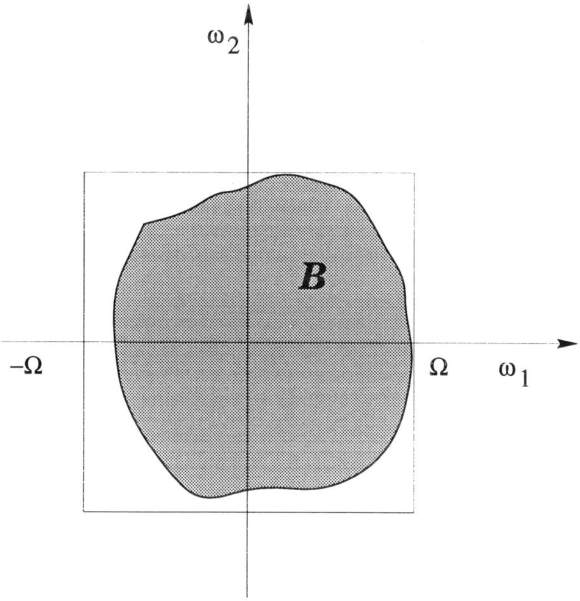

Figure 2.2. The smallest square with side 2Ω containing the band $\mathcal{B}$

of the following expansion

$$f(x_1, x_2) = \sum_{n_1, n_2 = -\infty}^{+\infty} f(x_{n_1}, x_{n_2}) S_\Omega(x_1 - x_{n_1}) S_\Omega(x_2 - x_{n_2}) \tag{2.19}$$

where $x_{n_1} = n_1\pi/\Omega$, $x_{n_2} = n_2\pi/\Omega$ and the sampling function $S_\Omega(x)$ is defined in equation (2.16). The function $S_\Omega(\boldsymbol{x}) = S_\Omega(x_1)S_\Omega(x_2)$ is the inverse FT of the characteristic function of the square with centre at the origin and side 2Ω. A picture of the modulus of this function is given in figure 2.3.

In general, however, the sampling expansion (2.19) is not the most efficient one, in the sense that it does not require the minimum number of sampling points per unit area. More general sampling expansions can be used [4] with sampling points forming a non-rectangular lattice. The optimum choice depends on the form of the band. For instance, if the band is a disc with centre at the origin and radius Ω, the optimum sampling lattice is the 120° rhombic [4], generated by the vectors (see figure 2.4)

$$\boldsymbol{v}_1 = \left\{ \frac{\pi}{\sqrt{3}\Omega}, -\frac{\pi}{\Omega} \right\}, \ \boldsymbol{v}_2 = \left\{ \frac{\pi}{\sqrt{3}\Omega}, \frac{\pi}{\Omega} \right\}. \tag{2.20}$$

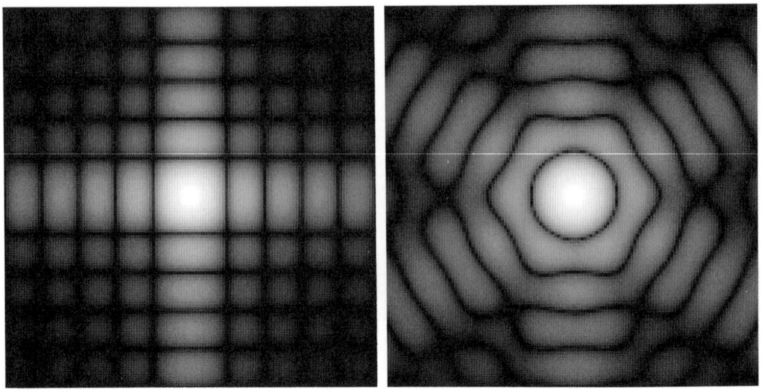

Figure 2.3. Left: picture of the modulus of the sampling function in the expansion (2.19); Right: picture of the modulus of the sampling function in the expansion (2.22). The two pictures correspond to the same value of Ω.

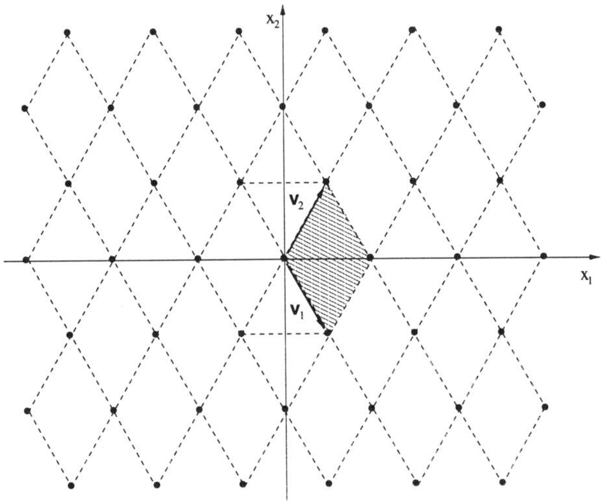

Figure 2.4. Sampling lattice for functions with circular band.

The sampling points are given by

$$x_{n_1,n_2} = n_1 v_1 + n_2 v_2 \tag{2.21}$$

and the corresponding sampling expansion is

$$f(x) = \sum_{n_1,n_2=-\infty}^{+\infty} f(x_{n_1,n_2}) S_\Omega(x - x_{n_1,n_2}). \tag{2.22}$$

Here the sampling function $S_\Omega(x)$ is the inverse FT of the characteristic function of the regular hexagon $\mathcal{H}_\Omega$ of side $2\Omega/\sqrt{3}$ (this is the smallest regular hexagon $\mathcal{H}_\Omega$ inscribing the circle of radius Ω) [4]:

$$S_\Omega(x) = \frac{C_\Omega}{(2\pi)^2} \int \chi_{\mathcal{H}_\Omega}(\omega) e^{ix\cdot\omega} d\omega. \tag{2.23}$$

A picture of the modulus of this function is given in figure 2.3. The constant C_Ω is the area of the elementary cell of the lattice, i.e. of the rhombus defined by the vectors v_1, v_2 (shaded region in figure 2.2): $C_\Omega = (2/\sqrt{3})(\pi/\Omega)^2$.

The sampling functions of equation (2.22) satisfy an orthogonality condition similar to (2.17)

$$\int S_\Omega(x - x_{m_1,m_2}) S_\Omega(x - x_{n_1,n_2}) dx = C_\Omega \delta_{m_1,n_1} \delta_{m_2,n_2} \tag{2.24}$$

the integral being extended to all variables from $-\infty$ to $+\infty$. As a consequence we get a relation similar to (2.18)

$$\int f(x) h^*(x) dx = \sum_{n_1,n_2=-\infty}^{+\infty} f(x_{n_1,n_2}) h^*(x_{n_1,n_2}) C_\Omega \tag{2.25}$$

which is also a sort of trapezoidal rule for computing the 2D integral: the value of the integrand at each one of the points of the sampling lattice of figure 2.4 is multiplied by the area of the cell on its right-hand side and the sum of all these discrete contributions is the value of the integral. Again this formula is a particular case of the generalized Parseval equality given in appendix A.

2.3 Convolution operators

The *convolution product* of two functions $f(x)$ and $K(x)$ is the function $g(x)$ defined as follows

$$g(x) = \int K(x - x') f(x') dx' \tag{2.26}$$

whenever the integral exists. The following notation is rather usual

$$g = K * f. \tag{2.27}$$

One of the basic properties of the Fourier transform is that it transforms the convolution product of two functions into the product of their Fourier transforms (*convolution theorem*) [1]

$$\hat{g}(\omega) = \hat{K}(\omega)\hat{f}(\omega). \tag{2.28}$$

We will not state here general conditions under which this equation is valid. For our purposes it is sufficient to know that it holds true when $\hat{K}(\omega)$ is bounded and $\hat{f}(\omega)$ is square-integrable. In such a case $\hat{g}(\omega)$ is square-integrable and therefore the function $g = K * f$ is also square-integrable, as follows from the Parseval equality, equation (2.9).

- *Definition of a convolution operator*

Let us consider a fixed function $K(\boldsymbol{x})$ whose FT $\hat{K}(\omega)$ is bounded, and let us also consider the convolution product of this function with all square-integrable functions. The results are also square-integrable and therefore, in this way, we define an operator in L^2, the space of all square-integrable functions, which will be called a *convolution operator* and denoted by A

$$Af = K * f. \tag{2.29}$$

Such an operator is a particular case of more general integral operators which will be considered in the second part of the book.

- *Spectral representation of a convolution operator*

From equation (2.28) and the inversion formula (2.2) of the FT, we obtain the following *spectral representation* of a convolution operator

$$(Af)(\boldsymbol{x}) = \frac{1}{(2\pi)^q} \int \hat{K}(\omega)\hat{f}(\omega)e^{i\boldsymbol{x}\cdot\omega}d\omega. \tag{2.30}$$

- *Properties of a convolution operator*

The operator A is *linear*, i.e. it satisfies the following condition

$$A(\alpha_1 f^{(1)} + \alpha_2 f^{(2)}) = \alpha_1 A f^{(1)} + \alpha_2 A f^{(2)} \tag{2.31}$$

where α_1, α_2 are arbitrary real or complex numbers. Moreover it is *bounded* in the following sense. If we denote by $\hat{K}_{max}$ the maximum value of $\hat{K}(\omega)$

$$\hat{K}_{max} = \max_{\omega} |\hat{K}(\omega)| \tag{2.32}$$

then from equation (2.28), from the Parseval equality (2.9) and from the definition (2.12) of the norm we find that

$$\|Af\| \le \hat{K}_{max}\|f\|. \tag{2.33}$$

From this inequality we see that the operator A is also *continuous*, in the sense that, if $\|f\|$ is small (tends to zero) then $\|Af\|$ is also small (tends to zero). More details on linear operators in Euclidean spaces of functions are given in appendix B.

- *The adjoint of a convolution operator*

For future use it is also necessary to introduce the *adjoint operator* A^*. This is a generalization of the hermitian conjugate of a matrix, i.e. of the matrix whose elements are the complex conjugate of the elements of the transposed matrix. Therefore, in the following, the $*$, when applied to an operator, will mean its adjoint while, when applied to a complex number, will mean its complex conjugate.

The adjoint operator A^* is the unique operator such that, for any pair of square-integrable functions f, g

$$(Af, g) = (f, A^*g) \tag{2.34}$$

the scalar product being defined in equation (2.11). Since, from the generalized Parseval equality (2.10) we have

$$\begin{aligned}(Af, g) &= \frac{1}{(2\pi)^q}\int \hat{K}(\boldsymbol{\omega})\hat{f}(\boldsymbol{\omega})\hat{g}^*(\boldsymbol{\omega})d\boldsymbol{\omega} \\ &= \frac{1}{(2\pi)^q}\int \hat{f}(\boldsymbol{\omega})\left[\hat{K}^*(\boldsymbol{\omega})\hat{g}(\boldsymbol{\omega})\right]^* d\boldsymbol{\omega} \\ &= (f, A^*g)\end{aligned} \tag{2.35}$$

we see that A^* is given by

$$(A^*g)(\boldsymbol{x}) = \frac{1}{(2\pi)^q}\int \hat{K}^*(\boldsymbol{\omega})\hat{g}(\boldsymbol{\omega})e^{i\boldsymbol{x}\cdot\boldsymbol{\omega}}d\boldsymbol{\omega}. \tag{2.36}$$

This is the spectral representation of the adjoint operator which is also a convolution operator given by

$$(A^*g)(\boldsymbol{x}) = \int K^*(\boldsymbol{x}' - \boldsymbol{x})g(\boldsymbol{x}')d\boldsymbol{x}'. \tag{2.37}$$

Here the convolution theorem has been used as well the fact that the inverse Fourier transform of $\hat{K}^*(\boldsymbol{\omega})$ is $K^*(-\boldsymbol{x})$.

- *The inverse of a convolution operator*

When $\hat{K}(\omega)$ is different from zero everywhere or has isolated zeros, more precisely when the support of $\hat{K}(\omega)$ coincides with the whole frequency space, then one can define the inverse operator A^{-1} as follows

$$(A^{-1}g)(x) = \frac{1}{(2\pi)^q} \int \frac{\hat{g}(\omega)}{\hat{K}(\omega)} e^{ix\cdot\omega} d\omega. \tag{2.38}$$

Since the function $1/\hat{K}(\omega)$ is not bounded, because $\hat{K}(\omega)$ has isolated zeros and tends to zero at infinity (as a consequence, for instance, of the Riemann–Lebesgue theorem), the operator A^{-1} is not bounded and the integral (2.38) is not defined for every $\hat{g}(\omega)$, but only for those $\hat{g}(\omega)$ such that the function $\hat{g}(\omega)/\hat{K}(\omega)$ is square-integrable.

- *The null space and the range of a convolution operator*

The inverse operator does not exist when the function $K(x)$ is bandlimited, with a band $\mathcal{B}$. In such a case there exist functions $f \neq 0$ such that $Af = 0$. Indeed, if a function f is such that $\hat{f}(\omega) = 0$ when ω is in $\mathcal{B}$ and $\hat{f}(\omega) \neq 0$ when ω is not in $\mathcal{B}$, then $\hat{K}(\omega)\hat{f}(\omega) = 0$ everywhere and therefore equation (2.28) implies that $Af = 0$. An example is provided in figure 2.5. The set of all functions f having this property is called the *null space* of the operator A. If A represents an imaging system then an object f in the null space of A has an image which is zero and, for this reason, it can be called an *invisible object* (see the discussion of section 1.2). In such a case, the null space will also be called the *space of the invisible objects.*

Another important concept is that of *range* of the operator A. This is the set of all functions g which have the form $g = Af$, i.e. are images of some object f, if A describes an imaging system. The range of A is also the space of the *noise-free images* which have already been introduced in section 1.2 and will be defined more precisely in the next chapter. From equation (2.28) it follows that if K is bandlimited all these functions are also bandlimited with a band which is contained in the band $\mathcal{B}$ of $K(x)$.

We conclude this section by observing that, if f is a function in the null space of A and g is a function in the range of A, then these functions are orthogonal because the product $\hat{f}(\omega)\hat{g}(\omega)$ is always zero. In other words all invisible objects are orthogonal to all noise-free images.

2.4 The Discrete Fourier Transform (DFT)

Let $f = \{f_0, f_1, \ldots, f_{N-1}\}$ be an N-dimensional vector. Then its *Discrete Fourier Transform* (DFT) is the N-dimensional vector $\hat{f} = \{\hat{f}_0, \hat{f}_1, \ldots, \hat{f}_{N-1}\}$

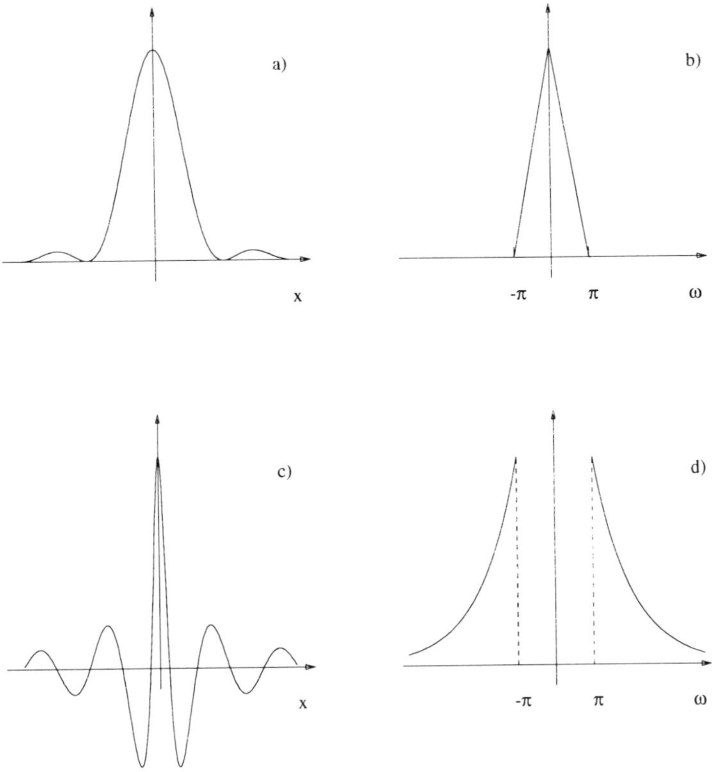

Figure 2.5. An example of invisible object: (a) the function $K(x)$ defining the convolution operator; (b) the Fourier transform of $K(x)$; (c) a function $f(x)$ such that $(K * f)(x) = 0.$; (d) the Fourier transform of $f(x)$.

defined by

$$\hat{f}_m = \sum_{n=0}^{N-1} f_n \exp\left(-i\frac{2\pi}{N}mn\right). \tag{2.39}$$

A complete discussion of the properties of the DFT can be found in [5].

The DFT has an inversion formula analogous to that of the FFT

$$f_n = \frac{1}{N}\sum_{m=0}^{N-1} \hat{f}_m \exp\left(i\frac{2\pi}{N}mn\right). \tag{2.40}$$

In the following it will be assumed, when necessary, that N *is a power of* 2, $N = 2^p$, because in such a case one can use the Fast Fourier Transform (FFT) algorithm for the computation of the DFT and of the inverse DFT on a binary computer. The description of this algorithm as well as the proof of equation (2.40) are given in appendix D.

The following *Parseval equality*

$$\sum_{n=0}^{N-1} |f_n|^2 = \frac{1}{N} \sum_{m=0}^{N-1} |\hat{f}_m|^2 \tag{2.41}$$

and the following *generalized Parseval equality*

$$\sum_{n=0}^{N-1} f_n h_n^* = \frac{1}{N} \sum_{m=0}^{N-1} \hat{f}_m \hat{h}_m^* \tag{2.42}$$

hold true. Their proof is also given in appendix D.

If we consider now the set of all N-dimensional vectors, this is a linear space where we can introduce the canonical scalar product

$$(\boldsymbol{f} \cdot \boldsymbol{h}) = \sum_{n=0}^{N-1} f_n h_n^* \tag{2.43}$$

with the associated norm of a vector $\|\boldsymbol{f}\| = (\boldsymbol{f} \cdot \boldsymbol{f})^{1/2}$, i.e.

$$\|\boldsymbol{f}\| = \left(\sum_{n=0}^{N-1} |f_n|^2 \right)^{1/2}, \tag{2.44}$$

and the associated distance of a vector $\boldsymbol{f}$ from a vector $\boldsymbol{h}$

$$d(\boldsymbol{f}, \boldsymbol{h}) = \|\boldsymbol{f} - \boldsymbol{h}\|. \tag{2.45}$$

From equation (2.42) we see that, if two vectors are orthogonal, i.e. $(\boldsymbol{f} \cdot \boldsymbol{h}) = 0$, then also the corresponding DFT are orthogonal: $(\hat{\boldsymbol{f}} \cdot \hat{\boldsymbol{h}}) = 0$.

It is possible to represent the DFT as the decomposition of a vector with respect to an orthonormal basis. Let us introduce the vectors $\boldsymbol{v}_0, \boldsymbol{v}_1, \ldots, \boldsymbol{v}_{N-1}$ defined as follows ($(\boldsymbol{v}_m)_n$ denotes the nth component of the vector $\boldsymbol{v}_m$):

$$(\boldsymbol{v}_m)_n = \frac{1}{\sqrt{N}} \exp\left(i \frac{2\pi}{N} mn \right). \tag{2.46}$$

From equation (D.4) of appendix D it follows that they satisfy the following orthonormality conditions

$$(\boldsymbol{v}_m \cdot \boldsymbol{v}_l) = \delta_{ml}. \tag{2.47}$$

Then equations (2.39) and (2.40) can be written as follows

$$\hat{f}_m = \sqrt{N} (\boldsymbol{f} \cdot \boldsymbol{v}_m) \tag{2.48}$$

$$\boldsymbol{f} = \frac{1}{\sqrt{N}} \sum_{m=0}^{N-1} \hat{f}_m \boldsymbol{v}_m . \tag{2.49}$$

We see that the values of the DFT of $\boldsymbol{f}$ are (except for a factor $\sqrt{N}$) the components of $\boldsymbol{f}$ with respect to the orthonormal basis $\boldsymbol{v}_0, \boldsymbol{v}_1, \ldots, \boldsymbol{v}_{N-1}$. The inversion formula is simply the representation of $\boldsymbol{f}$ in terms of these components.

Remark 2.2. Oscillating behaviour of the basis vectors. *All components of the vector $\boldsymbol{v}_0$ have the same sign, since $(\boldsymbol{v}_0)_n = N^{-1/2}$, while the real and imaginary parts of the components of the other vectors $\boldsymbol{v}_m$ have an oscillating behaviour. The frequency of oscillation increases with m up to $m = N/2$. In this case $(\boldsymbol{v}_{N/2})_n = N^{-1/2}(-1)^n$. Then the oscillation frequency decreases, as follows from the relation*

$$\boldsymbol{v}_{N-m} = \boldsymbol{v}_m^* \quad ; m = 1, 2, \ldots, \frac{N}{2} - 1. \tag{2.50}$$

In other words these vectors are the discrete versions of the exponentials $\exp(i\omega x)$ *and* $\exp(-i\omega x)$.

In view of the applications which will be investigated in the next sections it is convenient to consider the DFT not only as a transformation of vectors but also as a transformation of periodic sequences with period N. Given the vector $\boldsymbol{f}$, one can consider the periodic sequence generated by $\boldsymbol{f}$ and such that $f_{n+N} = f_n$. Then equation (2.39) generates also a periodic sequence $\hat{f}_m$, defined for any integer m, and such that $\hat{f}_{m+N} = \hat{f}_m$.

When we have a periodic sequence we can consider translations of this sequence and investigate the effect of these translations on the DFT. The following formula is proved in appendix D

$$\sum_{n=0}^{N-1} f_{n+p} \exp\left(-i\frac{2\pi}{N}mn\right) = \exp\left(i\frac{2\pi}{N}pm\right)\hat{f}_m \tag{2.51}$$

showing that a translation of the sequence is equivalent to a multiplication of the DFT by a phase factor. This property is the discrete version of a well known property of the Fourier transform.

A translation of the sequence is equivalent to a cyclic permutation of the components of the vector (see figure 2.6). Since the translation by $N/2$ is important in the discretization both of the Fourier transform and of the convolution product, it is convenient to give a name to the result of this operation. The vector obtained by the $N/2$-translation of the periodic sequence generated by $\boldsymbol{f}$ (see figure 2.6) will be called the *shifted vector* and denoted by $\boldsymbol{f}_s$. Equation (2.51) implies a very simple relationship between the DFT of $\boldsymbol{f}$ and the DFT of $\boldsymbol{f}_s$. The precise definition and relationship are the following:

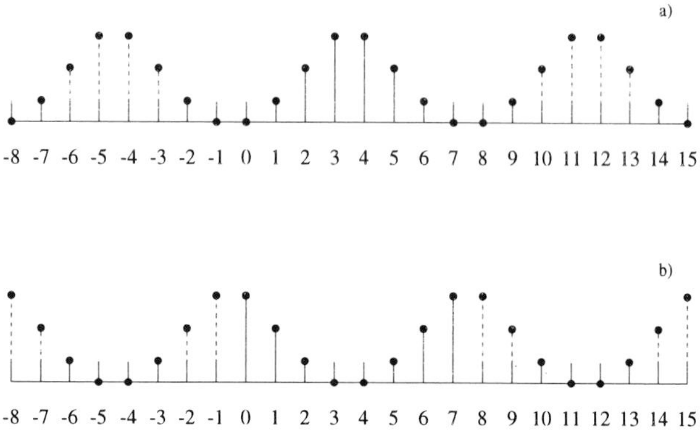

Figure 2.6. Illustrating the equivalence between translation of the periodic sequence and cyclic permutation of the components of the vector: (a) the original periodic sequence generated by the vector corresponding to the full vertical bars; (b) the same sequence translated by $N/2$, the vector generating the sequence being indicated again by full vertical bars.

- the components of the shifted vector $\boldsymbol{f}_s$ are related to the components of $\boldsymbol{f}$ by: $(\boldsymbol{f}_s)_n = f_{n+N/2}$ when $n = 0, 1, \cdots, N/2 - 1$ and $(\boldsymbol{f}_s)_n = f_{n-N/2}$ when $n = N/2, N/2 + 1, \cdots, N - 1$. It can be easily shown by the reader that the index rearrangement can also be written in terms of a shift matrix $\mathbf{R}$, which coincides with the permutation matrix $\mathbf{R}_p$, defined in appendix D, with $p = N/2$.

- the DFT of $\boldsymbol{f}_s$ is related to the DFT of $\boldsymbol{f}$ by $(\boldsymbol{f}_s)^{\wedge}_m = (-1)^m \hat{f}_m$. This is a particular case of equation (2.51) with $p = N/2$. The same relationship holds true between the inverse DFT of $\boldsymbol{f}$ and $\boldsymbol{f}_s$.

The previous analysis can be extended to the case of multidimensional images. For simplicity we give formulae only for 2D images (which are the most frequent in the applications). The extension to 3D images is obvious.

Let us suppose that we have a square $(N \times N)$ array of data. The more general case of rectangular arrays is discussed in [5]. An element $f_{m,n}$ of the array is characterized by two indices $\{m, n\}$, both taking the values

$0, 1, \ldots, N-1$. Then the DFT of the array $\boldsymbol{f}$ is defined as follows

$$\hat{f}_{k,l} = \sum_{m,n=0}^{N-1} f_{m,n} \exp\left[-i\frac{2\pi}{N}(km + ln)\right]. \tag{2.52}$$

All formulae for the one-dimensional DFT can be extended to the 2D case if the summation is replaced by a double summation with respect to the two indices. Therefore the inversion formula is given by

$$f_{m,n} = \frac{1}{N^2} \sum_{k,l=0}^{N-1} \hat{f}_{k,l} \exp\left[i\frac{2\pi}{N}(km + ln)\right] \tag{2.53}$$

while the *Parseval equality* and the *generalized Parseval equality* are given by

$$\sum_{m,n=0}^{N-1} |f_{m,n}|^2 = \frac{1}{N^2} \sum_{k,l=0}^{N-1} |\hat{f}_{k,l}|^2 \tag{2.54}$$

$$\sum_{m,n=0}^{N-1} f_{m,n} h^*_{m,n} = \frac{1}{N^2} \sum_{k,l=0}^{N-1} \hat{f}_{k,l} \hat{h}^*_{k,l}. \tag{2.55}$$

It is possible to give a Euclidean structure to the space of all square arrays by introducing the scalar product

$$(\boldsymbol{f} \cdot \boldsymbol{h}) = \sum_{m,n=0}^{N-1} f_{m,n} h^*_{m,n}. \tag{2.56}$$

In a similar way one can define the norm of an array as $\|\boldsymbol{f}\| = (\boldsymbol{f} \cdot \boldsymbol{f})^{1/2}$ and the distance between two arrays, $d(\boldsymbol{f} - \boldsymbol{h}) = \|\boldsymbol{f} - \boldsymbol{h}\|$. Then it is possible to introduce a basis consisting of the orthonormal arrays

$$(\boldsymbol{v}_{m,n})_{k,l} = \frac{1}{N} \exp\left[i\frac{2\pi}{N}(km + ln)\right], \tag{2.57}$$

where $(\boldsymbol{v}_{m,n})_{k,l}$ denotes the $\{k, l\}$ element of the array $\boldsymbol{v}_{m,n}$, and the equations (2.47),(2.48) and (2.49) can be easily extended by taking into account the definition (2.56) of the scalar product. Since the array (2.57) can be written as the tensor product of the vectors $\boldsymbol{v}_m$ and $\boldsymbol{v}_n$

$$\boldsymbol{v}_{m,n} = \boldsymbol{v}_m \otimes \boldsymbol{v}_n, \tag{2.58}$$

the symmetry properties of the arrays $\boldsymbol{v}_{m,n}$ can be easily derived from equation (2.50).

Finally the array $f_{m,n}$ can be extended to a periodic 2D lattice such that $f_{m+N,n+N} = f_{m,n}$ and the extension of equation (2.51) can be easily obtained. Very important for the applications is also the *shifted array* which is now defined by:

- $(f_s)_{m,n} = f_{m+N/2,n+N/2}$ for $m, n = 0, 1, \ldots, N/2 - 1$;
- $(f_s)_{m,n} = f_{m-N/2,n+N/2}$ for $m = N/2, N/2 + 1, \ldots, N - 1$ and $n = 0, 1, \ldots, N/2 - 1$;
- $(f_s)_{m,n} = f_{m+N/2,n-N/2}$ for $m = 0, 1, \ldots, N/2 - 1$ and $n = N/2, N/2 + 1, \ldots, N - 1$;
- $(f_s)_{m,n} = f_{m-N/2,n-N/2}$ for $m, n = N/2, N/2 + 1, \ldots, N - 1$.

In other words the shift operation is applied to both indices m, n. An example of the result of this operation will be given in the next chapter, section 3.1.

2.5 Cyclic matrices

Let us consider two vectors of length N, $\boldsymbol{K}$ and $\boldsymbol{f}$, with the corresponding periodic sequences of period N; then their *cyclic-convolution product* is the vector of length N, $\boldsymbol{g}$, with the corresponding sequence of period N, defined by

$$g_m = \sum_{n=0}^{N-1} K_{m-n} f_n. \tag{2.59}$$

Indeed it is easy to verify, using the periodicity of K_n, that this equation, considered for any m, defines a periodic sequence. We will use the following notation which is similar to that used for the convolution product (2.26)

$$\boldsymbol{g} = \boldsymbol{K} * \boldsymbol{f}. \tag{2.60}$$

By taking the DFT of both sides of equation (2.59) one can easily prove the following *cyclic-convolution theorem* (see appendix D)

$$\hat{g}_m = \hat{K}_m \hat{f}_m. \tag{2.61}$$

• *Definition of a cyclic matrix*

If we fix a vector $\boldsymbol{K}$ and we compute its cyclic-convolution product with all vectors $\boldsymbol{f}$, we thus define a linear operator in the N-dimensional Euclidean space. This operator can be expressed in terms of a matrix $\mathbf{A}$

$$\mathbf{A}\boldsymbol{f} = \boldsymbol{K} * \boldsymbol{f}, \tag{2.62}$$

with matrix elements $(\mathbf{A})_{m,n} = K_{m-n}$. By taking into account the periodicity of the sequence K_m, we have

$$\mathbf{A} = \begin{pmatrix} K_0 & K_{N-1} & K_{N-2} & \cdots & K_1 \\ K_1 & K_0 & K_{N-1} & \cdots & K_2 \\ K_2 & K_1 & K_0 & \cdots & K_3 \\ \cdots & \cdots & \cdots & \cdots & \cdots \\ K_{N-1} & K_{N-2} & K_{N-3} & \cdots & K_0 \end{pmatrix}. \tag{2.63}$$

The rows are just the cyclic permutations of the components of the vector $\boldsymbol{K}$ and any matrix having this structure is called a *cyclic matrix* (or circulant matrix). It is a particular case of a *Toeplitz matrix*, i.e. a matrix whose elements $\{n, n+p\}$, with p fixed, are constant.

• *Diagonalization of a cyclic matrix*

Any cyclic matrix is diagonalized by the DFT. Indeed the vectors $\boldsymbol{v}_m$ defined in equation (2.46) are the eigenvectors of any cyclic matrix $\mathbf{A}$ (see appendix D for the proof) and the components of the DFT of the vector $\boldsymbol{K}$, which generates $\mathbf{A}$, are the corresponding eigenvalues; i.e.

$$\mathbf{A}\boldsymbol{v}_m = \hat{K}_m \boldsymbol{v}_m. \tag{2.64}$$

• *Spectral representation of a cyclic matrix*

From equations (2.48),(2.49) and (2.64) we obtain the following *spectral representation* of a cyclic matrix

$$\mathbf{A}\boldsymbol{f} = \sum_{m=0}^{N-1} \hat{K}_m (\boldsymbol{f} \cdot \boldsymbol{v}_m) \boldsymbol{v}_m \tag{2.65}$$

which is the discrete version of equation (2.30).

If we put

$$\hat{K}_{max} = \max_m |\hat{K}_m| \tag{2.66}$$

then we can derive from equation (2.65) a bound similar to the bound (2.33) for a convolution operator

$$\|\mathbf{A}\boldsymbol{f}\| \le \hat{K}_{max} \|\boldsymbol{f}\|, \tag{2.67}$$

the norm being now that defined in equation (2.44).

• *The adjoint of a cyclic matrix*

As concerns the *adjoint matrix* (i.e. the hermitian conjugate) $\mathbf{A}^*$, it is the unique matrix such that, for any vectors $\boldsymbol{f}, \boldsymbol{g}$ one has

$$(\mathbf{A}\boldsymbol{f} \cdot \boldsymbol{g}) = (\boldsymbol{f} \cdot \mathbf{A}^*\boldsymbol{g}) \tag{2.68}$$

and its matrix elements are given by

$$(\mathbf{A}^*)_{m,n} = (\mathbf{A})^*_{n,m}. \tag{2.69}$$

Its spectral representation, which can be obtained from equation (2.65) and (2.68), is the following

$$\mathbf{A}^* g = \sum_{m=0}^{N-1} \hat{K}_m^* (g \cdot v_m) v_m. \tag{2.70}$$

This is the discrete version of equation (2.36).

- *The inverse of a cyclic matrix*

Equation (2.65) implies that the matrix $\mathbf{A}$ is non-singular if and only if $\hat{K}_m \neq 0$ for any m. In such a case the inverse matrix $\mathbf{A}^{-1}$ is given by

$$\mathbf{A}^{-1} g = \sum_{m=0}^{N-1} \frac{1}{\hat{K}_m} (g \cdot v_m) v_m \tag{2.71}$$

and, if we put

$$\hat{K}_{min} = \min_m |\hat{K}_m| \tag{2.72}$$

we have the bound

$$\|\mathbf{A}^{-1} g\| \leq \frac{1}{\hat{K}_{min}} \|g\|. \tag{2.73}$$

- *The null space and the range of a cyclic matrix*

The inverse matrix does not exist when at least one of the $\hat{K}_m$ is zero. In such a case the *null space* of the matrix is the subspace spanned by the vectors v_m corresponding to values of m such that $\hat{K}_m = 0$; the *invisible objects* are all possible linear combinations of these v_m. Moreover, as in the case of a convolution operator, we define the range of $\mathbf{A}$ as the set of all vectors g having the form $g = \mathbf{A} f$. It follows that all invisible objects are orthogonal to the range of $\mathbf{A}$.

The extension to the case of 2D images is straightforward, since the convolution product is defined by

$$g_{m,n} = \sum_{k,l=0}^{N-1} K_{m-k,n-l} f_{k,l}. \tag{2.74}$$

Using the definition (2.52) of the DFT it is easy to extend to the present case all equations from equation (2.60) to equation (2.73).

2.6 Relationship between FT and DFT

Since a fast algorithm for the computation of the DFT is available, it is important to clarify the relationship between the FT of a function and the DFT of a vector formed by means of samples of this function. In this section we investigate this problem. We consider only the case of functions of one variable. The extension to two and three variables is easy.

An exact relationship between the Fourier transform and the discrete Fourier transform was established by Cooley, Lewis and Welch [6]: let $f(x)$, $\hat{f}(\omega)$ be an FT pair and let $f_X(x)$, $\hat{f}_\Omega(\omega)$ be their periodized (or aliased) versions with period $2X$ and 2Ω respectively; then, if one takes N equispaced samples of these functions in the intervals $[-X, X]$ and $[-\Omega, \Omega]$, these samples form a DFT pair provided that $\Omega X = \pi N/2$. This theorem is very useful for understanding the errors committed in the computation of the Fourier transform of a function by means of the DFT of its samples. For instance, error bounds in the case of square-integrable functions have been established [7]. Here we adopt a more intuitive approach which provides in a very elementary way the basic relationship between the samplings of $f(x)$ and $\hat{f}(\omega)$.

Let us assume that the function $f(x)$ is small (or zero) outside the interval $[-X, X]$ so that we can approximate its Fourier transform by restricting the integral to the interval $[-X, X]$. Then we approximate this integral by the trapezoidal rule using the following sampling points in $[-X, X]$:

$$x_n = -X + n\delta_x \ ; \quad n = 0, 1, \cdots, N-1 \ ; \quad \delta_x = \frac{2X}{N}. \tag{2.75}$$

We find the following approximation for $\hat{f}(\omega)$

$$\hat{f}_{X,N}(\omega) = \sum_{n=0}^{N-1} f(x_n) \exp(-i\omega x_n)\delta_x. \tag{2.76}$$

Assume now that we intend to compute $\hat{f}_{X,N}(\omega)$ in N equispaced points ω_m interior to some interval $[-\Omega, \Omega]$ using the DFT. This requirement will fix the value of Ω.

Indeed, if we consider the points

$$\omega_m = -\Omega + m\delta_\omega \ ; \quad m = 0, 1, \cdots, N-1 \ ; \quad \delta_\omega = \frac{2\Omega}{N} \tag{2.77}$$

we have

$$\omega_m x_n = \Omega X - (X\delta_\omega m + \Omega\delta_x n) + \delta_x\delta_\omega mn. \tag{2.78}$$

The exponential $\exp(-i\omega_m x_n)$ contains as a factor the exponential of the DFT, $\exp(-i2\pi mn/N)$, if and only if

$$\delta_x\delta_\omega = \frac{2\pi}{N}. \tag{2.79}$$

Since δ_x has been fixed, this equation gives the value of δ_ω and therefore also the value of Ω. It is easy to find that the following relations hold true

$$\delta_\omega = \frac{\pi}{X} \ , \quad \delta_x = \frac{\pi}{\Omega} \ , \quad \frac{2}{\pi}\Omega X = N. \tag{2.80}$$

The first is precisely the sampling distance of the FT of a function which is zero outside the interval $[-X, X]$, while the second is the sampling distance of a function whose FT is zero outside the interval $[-\Omega, \Omega]$ (see section 2.2). The third relation is that provided by the theorem proved in [6]. We also point out that these equations show the relationship between the samplings of $f(x)$ and $\hat{f}(\omega)$. Indeed, these samplings depend on three parameters N, X, Ω. Choosing two of them (for instance N and X or N and Ω) the third one is fixed by the requirement of using the DFT for the computation of $\hat{f}_{X,N}(\omega_m)$.

By substituting the relations (2.80) in equation (2.78) we find that

$$\omega_m x_n = \frac{\pi N}{2} - \pi(m+n) + \frac{2\pi}{N}mn. \tag{2.81}$$

In the case $N = 2^p$, with $p \geq 2$, the term $\pi N/2$ is a multiple of 2π, and therefore this term does not contribute to $\exp(-i\omega_m x_n)$. Moreover the terms πm and πn contribute by factors $(-1)^m$ and $(-1)^n$ respectively, so that from equations (2.76) and (2.81) we get

$$\hat{f}_{X,N}(\omega_m) = (-1)^m \sum_{n=0}^{N-1} (-1)^n f(x_n)\delta_x \exp\left(-i\frac{2\pi}{N}mn\right). \tag{2.82}$$

We obtain the following result: in order to compute $\hat{f}_{X,N}(\omega_m)$ we must compute the DFT of the vector $(-1)^n f(x_n)\delta_x$ and then multiply the result by $(-1)^m$. Since $(-1)^m \hat{f}_{X,N}(\omega_m)$ is the DFT of $(-1)^n f(x_n)\delta_x$, from the inversion formula of the DFT and the equations (2.80)–(2.81) we get

$$\begin{aligned} f(x_n) &= \frac{(-1)^n}{N\delta_x} \sum_{m=0}^{N-1} (-1)^m \hat{f}_{X,N}(\omega_m) \exp\left(i\frac{2\pi}{N}nm\right) \\ &= \frac{1}{2\pi} \sum_{m=0}^{N-1} \hat{f}_{X,N}(\omega_m) \exp(ix_n\omega_m)\delta_\omega. \end{aligned} \tag{2.83}$$

It follows that the sampling values $f(x_n)$ can be exactly recovered by discretizing the inversion formula of the FFT and by writing $\hat{f}_{X,N}(\omega_m)$ in place of $\hat{f}(\omega_m)$.

However, in order to reduce the number of multiplications, very often one does not compute equation (2.82) but the DFT $\hat{f}_m$ of the vector $f_n = f(x_n)$. This can be written in the following way

$$\hat{f}_m = \sum_{n=0}^{N-1} f_n \exp\left(-i\frac{2\pi}{N}mn\right)$$

$$= \frac{1}{\delta_x} \sum_{n=0}^{N-1} (-1)^n \delta_x f(x_n) \exp\left[-i\frac{2\pi}{N} n \left(m - \frac{N}{2}\right)\right] \tag{2.84}$$

and a comparison with equation (2.82) shows that

$$\hat{f}_{X,N}(\omega_m) = (-1)^m \delta_x \hat{f}_{m+N/2}, \tag{2.85}$$

where the periodic extension of $\hat{f}_m$ is used. This relation is illustrated in figure 2.7.

In equation (2.85), which is equivalent to equation (2.82), the values $\hat{f}_{m+N/2}$ can be replaced by the components of the shifted vector $\hat{\boldsymbol{f}}_s$ defined in section 2.4. Then, if we remember that the DFT of the shifted vector $\boldsymbol{f}_s$ is the DFT of the vector $\boldsymbol{f}$ multiplied by $(-1)^m$, we find that in order to obtain $\hat{f}_{X,N}(\omega_m)$ from the vector $\boldsymbol{f}$ whose components are given by $f_n = f(x_n)$, the following procedure is equivalent to equation (2.82):

- apply the shift operation to $\boldsymbol{f}$ and obtain $\boldsymbol{f}_s$
- compute the DFT of the shifted vector $\boldsymbol{f}_s$
- apply the shift operation to the result of the previous step
- multiply the result by δ_x.

The extension to the case of two or three variables is obvious.

Remark 2.3. Symmetry properties of the DFT. *It is well known that the Fourier transform of a function $f(x)$ has the following properties:*

- *if $f(x)$ is real valued, then $\hat{f}^*(\omega) = \hat{f}(-\omega)$;*
- *if $f(x)$ is even or odd, then $\hat{f}(\omega)$ is also even or odd.*

Similar properties hold true for the discrete Fourier transform. If we consider the periodic sequence associated with a vector of length N, then this sequence is even or odd if $f_{-n} = f_n$ or $f_{-n} = -f_n$. Thanks to the periodicity of the sequence these relations also imply $f_{N-n} = f_n$ in the even case and $f_{N-n} = -f_n$ in the odd case. Therefore, in terms of vectors defined only for values of the index between 0 and $N - 1$, these symmetries show up in relationships between the values at n and $N - n$.

Then, from the definition (2.39) one easily derives that:

- *if the vector f_n is real valued then $\hat{f}_m$ satisfies the condition $\hat{f}_m^* = \hat{f}_{N-m}$;*
- *if the vector f_n is even or odd, then the vector $\hat{f}_m$ is also even or odd.*

If the function $f(x)$ is real valued and even (odd), then the vector f_n, formed with its samples at the points (2.75), is also real valued and even (odd). As a consequence of the previous results the DFT $\hat{f}_m$ has the same properties of the samples of $\hat{f}(\omega)$.

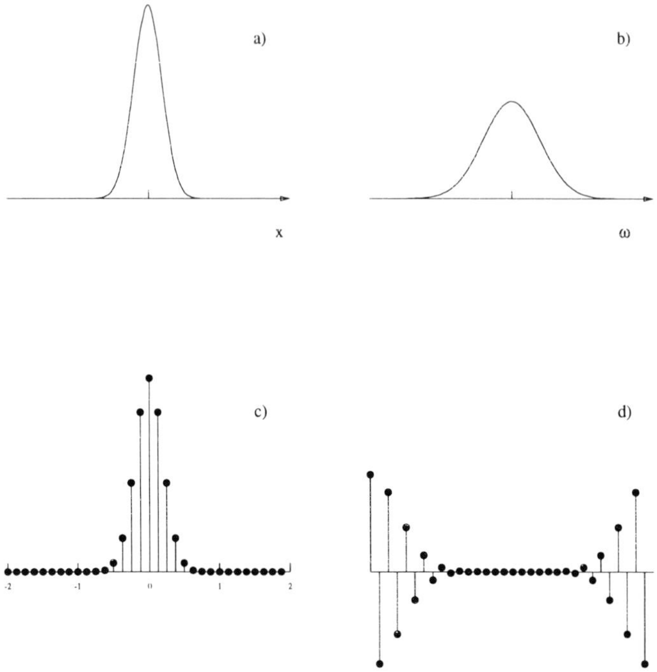

Figure 2.7. Illustrating the relationship between the FT of a function and the DFT of its samples: (a) a function $f(x)$ with a Gaussian shape; (b) the FT of the function $f(x)$; (c) the samples of $f(x)$; (d) the DFT of the samples of $f(x)$.

2.7 Discretization of the convolution product

We consider now the problem of approximating the convolution product of two functions by means of the cyclic-convolution product of their samples. Again we mainly investigate the case of functions of one variable and, in order to simplify our analysis, we assume that both $f(x)$ and $K(x)$ are zero outside the interval $[-X, X]$ and at the boundary points $\pm X$. Then, if $f(x)$ is sampled at the points x_n, equation (2.75), we can approximate the convolution product by means of the trapezoidal rule and we get the following approximation $g_N(x)$ for

$g(x) = (K * f)(x)$

$$g_N(x) = \sum_{n=0}^{N-1} K(x - x_n) f(x_n) \delta_x. \tag{2.86}$$

Consider now the problem of computing $g_N(x)$ precisely at the sampling points (2.75). We get

$$g_N(x_m) = \sum_{n=0}^{N-1} K(x_m - x_n) f(x_n) \delta_x \tag{2.87}$$

and therefore we need the values of $K(x)$ at the points $x_m - x_n = (m - n)\delta_x$. By adding and subtracting X and by taking into account that $X = N\delta_x/2$, we can write $x_m - x_n$ as follows

$$x_m - x_n = -X + \left(m - n + \frac{N}{2}\right) \delta_x = x_{m-n+N/2}. \tag{2.88}$$

Since both m and n take N values from 0 to $N - 1$, the difference $m - n$ takes $2N - 1$ values from $-(N - 1)$ up to $N - 1$, so that the function $K(x)$ must be sampled at equispaced points, with spacing δ_x, ranging from $-2X + \delta_x$ up to $2X - \delta_x$.

If we introduce the vectors $g_m = g_N(x_m)$ and $f_n = f(x_n)$, both with N components, and the vector $\overline{K}_n = K(x_n)\delta_x$, with $2N - 1$ components, $(n = -N/2 + 1, -N/2 + 2, \cdots, 3N/2 - 1)$, equation (2.87) can also be written as follows

$$g_m = \sum_{n=0}^{N-1} \overline{K}_{m-n+N/2} f_n, \tag{2.89}$$

i.e. the computation of the vector $\boldsymbol{g}$ implies a matrix-vector multiplication $\boldsymbol{g} = \mathbf{A}\boldsymbol{f}$, where $(\mathbf{A})_{m,n} = \overline{K}_{m-n+N/2}$. The matrix $\mathbf{A}$ is Toeplitz but it is not cyclic: its matrix elements are obtained from a vector of length $2N - 1$, while the matrix elements of a cyclic matrix are obtained from a vector of length N. This is an initial difficulty because, in the case of a Toeplitz matrix of general type, the product $\mathbf{A}\boldsymbol{f}$ can still be computed by means of the FFT algorithm [8] but the procedure is not so simple as in the case of a cyclic convolution. A second difficulty is that the product $\mathbf{A}\boldsymbol{f}$ provides approximate values of $g(x)$ only inside the interval $[-X, X]$ while the support of $g(x)$ in general coincides with the interval $[-2X, 2X]$ since both the support of $f(x)$ and the support of $K(x)$ are contained in $[-X, X]$.

In figure 2.8 we give an example of the approximation which can be obtained from equation (2.89) in the case where $f(x)$ is the characteristic function of the open interval $(-1, 1)$ (so that $f(-1) = f(1) = 0$) while $K(x)$ is the triangular function defined by $K(x) = 1 - |x|$, for $|x| < 1$; $K(x) = 0$, for $|x| \geq 1$. The function $g(x) = (K * f)(x)$ is bell-shaped, its support being the interval $[-2, 2]$. The approximation provided by equation (2.89), using 16

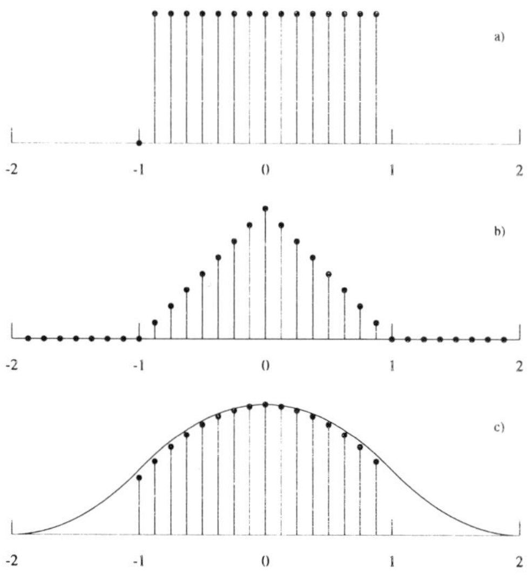

Figure 2.8. (a) Samples of the function $f(x) = 1$ for $|x| < 1$, $f(x) = 0$ for $|x| \geq 1$ (16 points); (b) samples of the function $K(x) = 1 - |x|$ for $|x| < 1$, $K(x) = 0$ for $|x| \geq 1$ (31 points); (c) approximation of $g = K * f$ obtained by means of equation (2.89). The full line indicates the function g.

points in the interval $[-1, 1]$ for $f(x)$ and 31 points in the interval $[-2, 2]$ for $K(x)$ is satisfactory at the points interior to $[-1, 1]$.

A simple way for overcoming the two difficulties mentioned above consists of adding zeros in order to have vectors of length $2N$ (*padding by zeros*) and then computing the cyclic convolution of these vectors. More precisely the procedure is the following. We consider, in the interval $[-2X, 2X]$, the sampling points

$$x_n = -2X + n\delta_x \quad ; \quad n = 0, 1, \cdots, 2N - 1 \tag{2.90}$$

with the same δ_x of equation (2.75) so that, if we assume again that N is even, the sampling points in $[-X, X]$ correspond to $n = N/2, N/2+1, \cdots, 3N/2-1$ and coincide with the sampling points (2.75). It follows that $f(x_n) = 0$ for $n = 0, 1, \cdots, N/2-1$ and $n = 3N/2, 3N/2, \cdots, 2N-1$ (recall that $f(X) = 0$, by assumption).

Equation (2.87) is now replaced by

$$g_{2N}(x_m) = \sum_{n=0}^{2N-1} K(x_m - x_n) f(x_n) \delta_x \tag{2.91}$$

and, for $m = N/2, N/2+1, \cdots, 3N/2-1$ we get exactly the values of $g_N(x_m)$ given by equation (2.87) for $m = 0, 1, \cdots, N-1$. In addition equation (2.91) provides approximate values of $g(x)$ outside $[-X, X]$.

If we proceed as in the derivation of equation (2.88) we obtain

$$x_m - x_n = -2X + (m - n + N)\delta_x = x_{m-n+N} \tag{2.92}$$

and therefore, if we introduce vectors of length $2N$, $g_m = g(x_m)$, $f_n = f(x_n)$ and a vector of length $4N-1$, $\overline{K}_n = K(x_n)\delta_x$ ($n = -N+1, -N+2, \cdots, 3N-1$), equation (2.91) can also be written as follows

$$g_m = \sum_{n=0}^{2N-1} \overline{K}_{m-n+N} f_n. \tag{2.93}$$

This relation implies again a matrix-vector multiplication with a Toeplitz matrix $2N \times 2N$. However let us consider the periodic continuation (with period $2N$) of the values of $\overline{K}_n$ for $n = 0, 1, \cdots, 2N-1$ (i.e. the samples of $K(x)\delta_x$ inside $[-2X, 2X]$). This periodic continuation, which will be denoted by $\overline{\overline{K}}_n$, is zero for $n = -N/2, -N/2+1, \cdots, -1$ and for $n = 2N, 2N+1, \cdots, 5N/2-1$ and therefore we have $\overline{\overline{K}}_n = \overline{K}_n$ for all values of the index from $n = -N/2$ up to $n = 5N/2-1$. In figure 2.9 this relationship follows from a comparison of the panels (b) and (c). This figure can also be used for understanding the next step.

Equation (2.93) requires a rearrangement of the components of $\overline{K}_n$. We can perform the same operation on the components of $\overline{\overline{K}}_n$ and this is equivalent to an N-translation of the periodic sequence generated by $\overline{\overline{K}}_n$. The result is represented in panel (d) of figure 2.9. For $n = 0, 1, \cdots, 2N-1$ we obtain precisely the components of the shifted vector of $\overline{\overline{K}}_n$, as defined in section 2.4. We put $\boldsymbol{K} = \overline{\overline{\boldsymbol{K}}}_s$ and we consider the cyclic convolution of the vectors $\boldsymbol{K}$ and $\boldsymbol{f}$. The result can be easily understood by looking again at figure 2.9 where the vector $\boldsymbol{f}$ is represented in panel (a) and the periodic sequence associated with $\boldsymbol{K}$ in panel (d). Since this sequence is symmetric with respect to $n = 0$, we do not need a reflection of this sequence for computing the cyclic-convolution product. Then the component $m = 0$ of $\boldsymbol{K} * \boldsymbol{f}$ is obtained by summing the products of the corresponding components of $\boldsymbol{f}$ and $\boldsymbol{K}$. The result is zero and this is true in general because we have assumed that both $f(x)$ and $K(x)$ are zero at the points $\pm X$. The next component is obtained by the same operations after a shift of one unit in the index of the periodic sequence of $\boldsymbol{K}$ and so on. Since the components of $\overline{\overline{\boldsymbol{K}}}$ not coinciding with the corresponding components of $\overline{\boldsymbol{K}}$ are always multiplied by components of $\boldsymbol{f}$ which are zero, it follows that the computation of the cyclic convolution of $\boldsymbol{K}$ and $\boldsymbol{f}$ is equivalent to the computation of equation (2.91).

In conclusion, the procedure is the following:

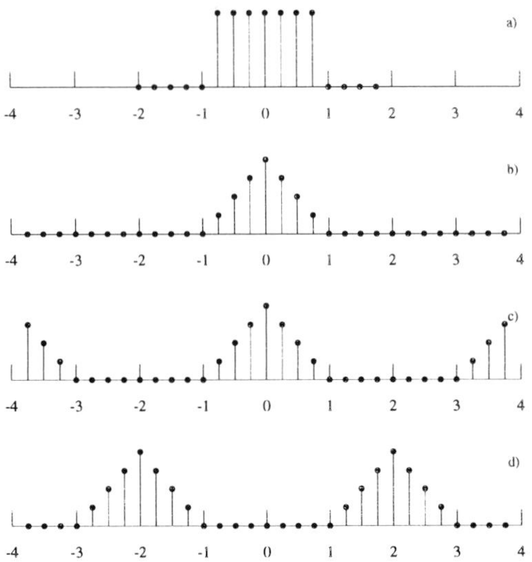

Figure 2.9. (a) Samples of the function $f(x) = 1$ for $|x| < 1$, $f(x) = 0$ for $|x| \geq 1$ in the interval $[-2, 2]$ (16 points); b) samples of the function $K(x) = 1 - |x|$ for $|x| < 1$, $K(x) = 0$ for $|x| \geq 1$ in the interval [-4,4] (31 points); (c) periodic continuation of the samples of $K(x)$ in the interval [-2,2]; (d) N-translation (with $N = 8$) of the sequence of (c).

- extend the vectors of length N, formed by the samples of $f(x)$ and $K(x)\delta_x$ in $[-X, X]$, to vectors $\boldsymbol{f}$ and $\overline{\boldsymbol{K}}$ of length $2N$ by adding N zeros to the left and to the right (padding by zeros);
- generate the shifted vector $\overline{\boldsymbol{K}}_s$ and denote it by $\boldsymbol{K}$;
- compute the cyclic convolution of the vectors $\boldsymbol{K}$ and $\boldsymbol{f}$.

Remark 2.4. The need for introducing the shifted vector, i.e. for considering the N-translation of the periodic sequence associated with the samples of $K(x)\delta_x$, is related to the fact that the component $m = 0$ of the vector does not correspond to the value of the function at $x = 0$ but to the value of the function at $x = -2X$. It is easy to recognize that a $2X$-translation in the x-variable correspond to an N-translation in the index of the samples.

The cyclic convolution in the last step can be computed by computing the DFT of the two vectors, by multiplying the two DFT component by component and then by taking the inverse DFT of this product. Since these operations can be performed by means of the FFT algorithm it is clear that the number of operations is of the order of $N \log_2 N$. In figure 2.10 we show the result of the application of this procedure to the example of figure 2.8. The result is satisfactory also for the values of $g(x)$ outside the interval $[-1, 1]$.

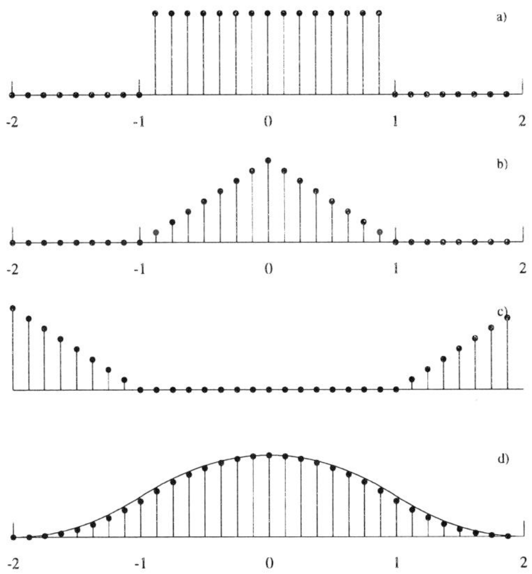

Figure 2.10. (a) Samples of the function $f(x) = 1$ for $|x| < 1$, $f(x) = 0$ for $|x| \geq 1$ (32 points); (b) samples of the function $K(x) = 1 - |x|$ for $|x| < 1$, $K(x) = 0$ for $|x| \geq 1$ (32 points); (c) the shifted vector obtained from the samples in (b); (d) result obtained by computing the cyclic convolution of the vectors in (a) and (c). The full line indicates the function $g = K * f$.

A typical situation in the case of the imaging problems which will be considered in the next chapter is the following: the domain where the function $K(x)$ is significantly different from zero is smaller and, in general, much smaller than the support of the function $f(x)$. We can take advantage of this situation. If the support of $f(x)$ is $[-X, X]$ while the support of $K(x)$ is $[-X', X']$ with $X' < X$, we can apply to the interval $[-(X + X'), (X + X')]$ the procedure outlined above in the case of the interval $[-2X, 2X]$.

However, when the support of $K(x)$ is considerably smaller than the support of $f(x)$, a more simple, even if less accurate, procedure is the following one:

- introduce vectors of length N defined by $f_n = f(x_n)$ and $\overline{K}_n = K(x_n)\delta_x$, the sampling points x_n being again those defined in equation (2.75);

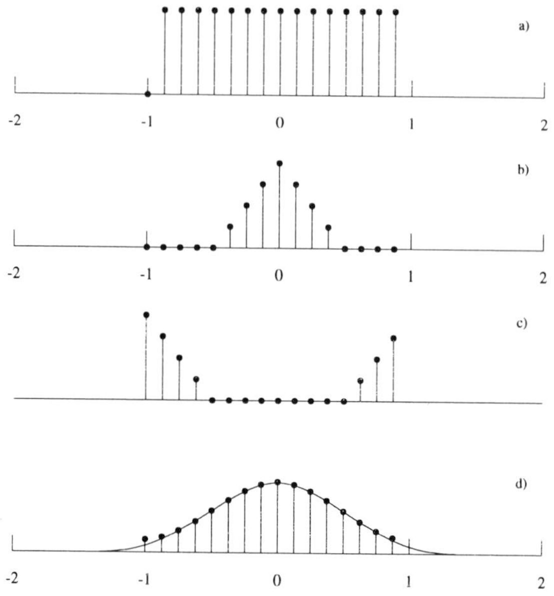

Figure 2.11. (a) Samples of the function $f(x) = 1$ for $|x| < 1$, $f(x) = 0$ for $|x| \geq 1$ (16 points); (b) samples of the function $K(x) = 2(1 - 2|x|)$ for $|x| < 1/2$, $K(x) = 0$ for $|x| \geq 1/2$ (16 points); (c) the shifted vector obtained from the samples in (b); (d) result obtained by computing the cyclic convolution of the vectors in (a) and (c). The full line indicates the function $g = K * f$.

- generate the shifted vector $\overline{\boldsymbol{K}}_s$ (see equation (2.88)) and denote it by $\boldsymbol{K}$;
- compute the cyclic convolution of the vectors $\boldsymbol{f}$ and $\boldsymbol{K}$.

It should be clear that this procedure provides the same values of the previous one except at points in neighbourhoods of $-X$ and X whose size is of the order of the support of $K(x)$. In figure 2.11 we show the result of the application of this procedure to a modification of the example of figure 2.8: the function $f(x)$ is the same while the function $K(x)$ is a triangular function with support $[-1/2, 1/2]$ instead of $[-1, 1]$. The effect of the periodicity of the cyclic convolution at the edges of the interval is evident even if it is small.

Remark 2.5. The components $\hat{g}_m$ of the DFT of the vector $\boldsymbol{g} = \boldsymbol{K} * \boldsymbol{g}$, computed by means of the procedure indicated above, are related to the samples of $\hat{g}(\omega) = \hat{K}(\omega)\hat{f}(\omega)$ by a relationship analogous to that of equation (2.85). This *means, in particular, that the values of* $\hat{g}(\omega)$ *in a neighbourhood of* $\omega = 0$ *are related to the components* $\hat{g}_m$ *for* m *close to* 0 *and* $N - 1$*. This relationship must be taken into account when one performs operations on the Fourier transform of* $g(x)$ *such as a low-pass filtering.*

The extension to the case of functions of two or three variables is easy.

References

[1] Papoulis A 1962 *The Fourier Integral and its Applications* (New York: McGraw-Hill)
[2] Bracewell R N 1965 *The Fourier Transform and its Applications* (New York: McGraw-Hill)
[3] Jerri A J 1977 *Proc. IEEE* **65** 1565
[4] Petersen D P and Middleton D 1962 *Inf. Control* **5** 279
[5] Briggs W L and Henson V E 1995 *The DFT* (Philadelphia, PA: SIAM)
[6] Cooley J W, Lewis P A and Welch P D 1967 *IEEE Trans Audio Electroacoustics* **AU15** 79
[7] Auslander L and Grünbaum F A 1989 *Inverse Problems* **5** 149
[8] Bitmead R R and Anderson B D O 1984 *Linear Algebra and Appl.* **34** 103

Chapter 3

Examples of image blurring

An image is a signal carrying information about a physical object which is not directly observable. In general the information consists of a degraded representation of the original object and one can roughly distinguish two sources of degradation: the process of image formation and the process of image recording.

The degradation due to the process of image formation is usually denoted by *blurring* and is a sort of bandlimiting of the object. In the case of aerial photographs, for instance, the blurring is due to relative motion between the camera and the ground, to aberrations of the optical components of the camera and, finally, to atmospheric turbulence. All these different kinds of blurring will be described in this chapter.

The degradation introduced by the recording process is usually denoted by *noise* and is due to measurement errors, counting errors, etc.

Blurring is a deterministic process and, in most cases, one has a sufficiently accurate mathematical model for its description. On the other hand, noise is a statistical process so that the noise affecting a particular image is not known. One can, at most, assume a knowledge of the statistical properties of the process.

In this chapter we first discuss blurring and noise in general terms and then we provide several examples of blurring. These examples are given both to show the large number of applications of the deconvolution problem and to provide models that are useful for testing the methods discussed in this book.

3.1 Blurring and noise

We will use, for convenience, the language of optical image formation, even if some of the examples discussed in the following are not derived from optics.

In optical image formation, an unknown spatial radiance distribution, which will be called the *object* and denoted by $f^{(0)}(\boldsymbol{x})$ (most frequently a function of two variables, $\boldsymbol{x} = \{x_1, x_2\}$, which can be the 2D mapping of a 3D scene), produces a radiance distribution in the image domain of the optical system (most frequently a plane). This distribution will be called, for a reason which will become clear shortly, the *noise-free image* of $f^{(0)}(\boldsymbol{x})$ and denoted by $g^{(0)}(\boldsymbol{x})$. If the image formation can be modelled as a linear process, then $g^{(0)}(\boldsymbol{x})$ is a linear superposition of the values of $f^{(0)}(\boldsymbol{x})$ and therefore it is given mathematically by

$$g^{(0)}(\boldsymbol{x}) = \int K(\boldsymbol{x}, \boldsymbol{x}') f^{(0)}(\boldsymbol{x}')d\boldsymbol{x}' \tag{3.1}$$

where $K(\boldsymbol{x}, \boldsymbol{x}')$ is the *impulse response function*, also called *point spread function* (PSF) of the linear imaging system. The effect of the PSF is called *blurring* and the noise-free image $g^{(0)}(\boldsymbol{x})$ is called a blurred version of the object $f^{(0)}(\boldsymbol{x})$.

The term point spread function for $K(\boldsymbol{x}, \boldsymbol{x}')$ derives from the fact that it is the image of a point source located at the point $\boldsymbol{x}'$. Indeed, if the point source is given by $f^{(0)}(\boldsymbol{x}'') = \delta(\boldsymbol{x}'' - \boldsymbol{x}')$ (where $\delta(\boldsymbol{x})$ is the Dirac delta distribution) then, from equation (3.1) with integration variable denoted now by $\boldsymbol{x}''$, one obtains $g^{(0)}(\boldsymbol{x}) = K(\boldsymbol{x}, \boldsymbol{x}')$. A schematic representation of this process is given in figure 3.1.

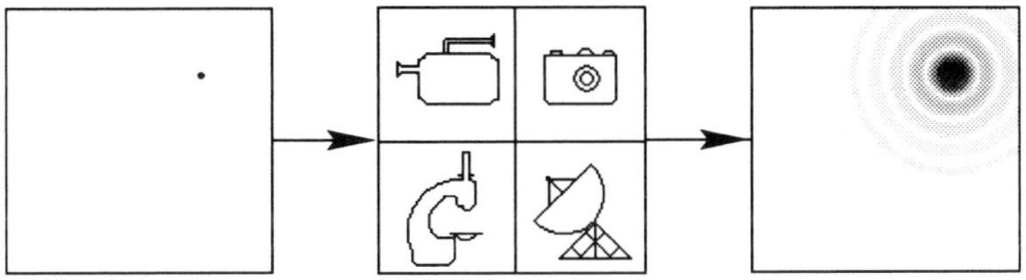

Point source $\delta(\boldsymbol{x} - \boldsymbol{x}')$ | Imaging system | Image of the point source $K(\boldsymbol{x}, \boldsymbol{x}')$

Figure 3.1. Schematic representation of the point spread function.

In principle, the PSF can be obtained by solving the direct problem associated with the imaging process. In the case of an optical system, for instance, this means computing the propagation of light from a point source in the object domain to a point of the image domain through the elements of the system (lenses, mirrors etc). If no exact or approximate solution of the direct problem is available, the PSF can eventually be measured by generating a point source, by moving the point source in the object domain and by detecting the images produced by the system for the various positions of the source. Once the PSF is known, equation (3.1) follows from the linearity of the system if one writes $f^{(0)}(\boldsymbol{x})$ as a linear superposition of point sources, i.e.

$$f^{(0)}(\boldsymbol{x}) = \int \delta(\boldsymbol{x} - \boldsymbol{x}') f^{(0)}(\boldsymbol{x}')d\boldsymbol{x}'. \tag{3.2}$$

In order to measure the blurred image of $f^{(0)}(\boldsymbol{x})$, a recording system is placed in the image domain (for instance an array of detectors). For simplicity we assume that the output of this system is proportional to $g^{(0)}(\boldsymbol{x})$ (and we will put the constant equal to 1 in the following). However the effect of the recording process is the addition of a noise contribution, so that the recorded image, denoted by $g(\boldsymbol{x})$ and called the *noisy image*, is given by

$$g(\boldsymbol{x}) = \int K(\boldsymbol{x}, \boldsymbol{x}')f^{(0)}(\boldsymbol{x}')d\boldsymbol{x}' + w(\boldsymbol{x}). \tag{3.3}$$

In figure 3.2 we give a schematic representation where the two processes, imaging and detection, are clearly separated.

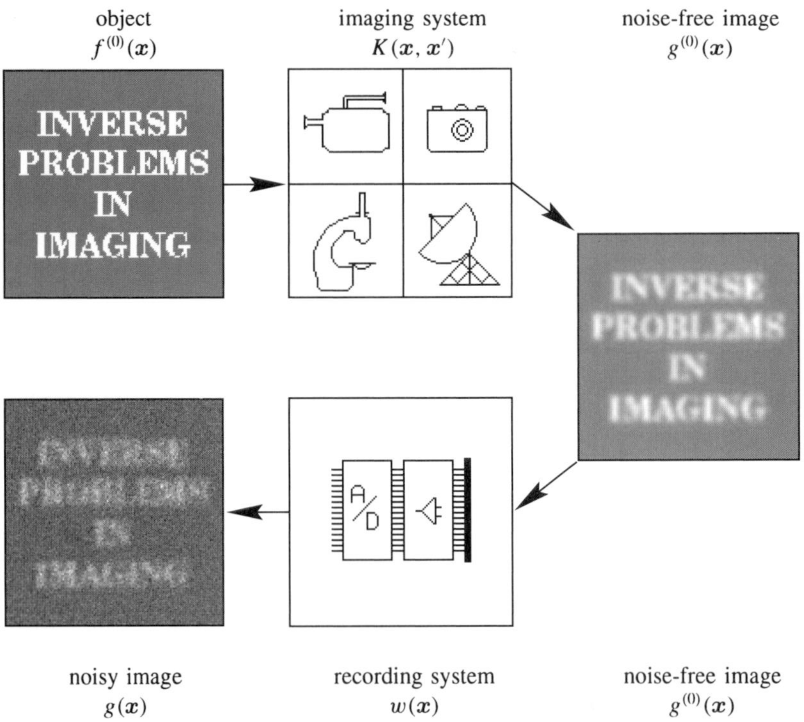

Figure 3.2. Schematic representation of the formation of the noisy image g.

The noise term w in equation (3.3) is in general a realization of a random process. This realization is not known in practice. One knows, at most, statistical properties of the random process, such as mean value, variance, etc. We can also know, of course, if it is additive or multiplicative, correlated or uncorrelated,

Gaussian or Poissonian, etc. These properties, when known, should be used in the treatment of the problem.

The amount of noise present in a discrete image is measured by the so-called *signal-to-noise ratio* (SNR) which is defined by

$$SNR = 10\log_{10}\left(\frac{\text{variance of the image}}{\text{variance of the noise}}\right)\ (\text{dB}) \tag{3.4}$$

and can be estimated by

$$SNR = 20\log_{10}\left(\frac{\|g\|}{\|\boldsymbol{w}\|}\right)\ (\text{dB}), \tag{3.5}$$

the norm being the Euclidean one associated with the scalar product (2.56). When the SNR is of the order of 40 or 50 dB, the noise is not visible in the image and the effect of the blur is dominant.

In many imaging systems the PSF is invariant with respect to translations in the following sense: the image $K(\boldsymbol{x}, \boldsymbol{x}_0)$ of a point source located at $\boldsymbol{x}_0$ is translated by $\boldsymbol{x}_0$ of the image $K(\boldsymbol{x}, \mathbf{0})$ of a point source located at the origin of the object plane, i.e. $K(\boldsymbol{x}, \boldsymbol{x}_0) = K(\boldsymbol{x} - \boldsymbol{x}_0, \mathbf{0})$. It follows that $K(\boldsymbol{x}, \boldsymbol{x}')$ is a function of the difference $\boldsymbol{x} - \boldsymbol{x}'$ and we write $K(\boldsymbol{x} - \boldsymbol{x}')$ instead of $K(\boldsymbol{x}, \boldsymbol{x}')$. Such an imaging system is called *space invariant* and the corresponding PSF is also called *space invariant*. If the imaging system is not space invariant, its PSF is said to be *space variant*. A space-invariant PSF can be determined by detecting the image of a single point source, for instance located in the centre of the object domain.

In the case of a space-invariant system, equation (3.3) becomes

$$g(\boldsymbol{x}) = \int K(\boldsymbol{x} - \boldsymbol{x}')f^{(0)}(\boldsymbol{x}')d\boldsymbol{x}' + w(\boldsymbol{x}). \tag{3.6}$$

or also

$$g = K * f^{(0)} + w. \tag{3.7}$$

The function $K(\boldsymbol{x})$ is also called the *point spread function* (PSF) and its Fourier transform $\hat{K}(\boldsymbol{\omega})$ is called the *transfer function* (TF) of the imaging system. These functions play an important role in the knowledge of the behaviour of the system. The PSF provides the response of the system to any point source wherever it is located. On the other hand, the TF tells us how a signal of a fixed frequency is propagated through the linear system, so that the blurring can be viewed as a sort of frequency filtering. Indeed, in terms of the Fourier transforms, equation (3.6) becomes

$$\hat{g}(\boldsymbol{\omega}) = \hat{K}(\boldsymbol{\omega})\hat{f}^{(0)}(\boldsymbol{\omega}) + \hat{w}(\boldsymbol{\omega}). \tag{3.8}$$

We will say that an imaging system is *bandlimited* if its PSF is bandlimited. The band $\mathcal{B}$ of the PSF will also be called the band of the imaging system.

In the following we will give examples of blurring which can be modelled by means of a convolution integral as in equation (3.6). Some of them will be examples of bandlimited blurring. In the other cases, the TF tends to zero for large space frequencies so that the corresponding imaging systems are approximately bandlimited.

Real images, however, are discrete and therefore a function $g(\boldsymbol{x})$ of two continuous variables is replaced by an $N \times N$ array of values of $g(\boldsymbol{x})$ taken at the points of a regular lattice. These values will be denoted by $g_{m,n}$. Then the original object $f^{(0)}(\boldsymbol{x})$ can also be replaced by an $N \times N$ array of values, denoted by $f^{(0)}_{k,l}$. Finally the convolution integral can be replaced by a cyclic convolution, as discussed in section 2.7. This approximation is reasonable if the PSF is significantly different from zero over a domain much smaller than the image domain so that the periodicity effects are negligible. By discretizing equation (3.6) along these lines, in the case of 2D images we obtain the following equation

$$g_{m,n} = \sum_{k,l=0}^{N-1} K_{m-k,n-l} f^{(0)}_{k,l} + w_{m,n} \tag{3.9}$$

or also

$$\boldsymbol{g} = \boldsymbol{K} * \boldsymbol{f}^{(0)} + \boldsymbol{w}. \tag{3.10}$$

We recall the relationship between the PSF of equation (3.6) and the *cyclic PSF* of equation (3.9). If we discretize $K(\boldsymbol{x})$ by using, for instance, equispaced sampling points, then the cyclic PSF, $\boldsymbol{K}$, of equation (3.10) is obtained by applying the shift operation to the array formed by the samples of $K(\boldsymbol{x})$. The result of this procedure is illustrated in figure 3.3.

3.2 Linear motion blur

The simplest example of blurring is that due to relative motion, during exposure, between the camera and the object being photographed. A rather simple case is that of a visible object moving across a uniform background (corresponding, for instance, to zero photographic density) in such a way that no part of the object goes out of the image domain during the motion. Examples arise, for instance, when photographing the Earth, Moon and planets using aerial vehicles such as aircraft and spacecraft. Examples can also be found in forensic science when potentially useful photographs are sometimes of extremely bad quality due to motion blur.

Let $f^{(0)}(\boldsymbol{x})$ be the illuminance, from a scene or object being photographed, which would result in the image plane of the camera in the absence of relative motion between the camera and the object. Nonlinear behaviour of the photographic material is neglected so that we can assume that it reacts to successive exposures just by linearly adding their effects. The relative motion can be a translation or a rotation or even a more general one. The blurring is

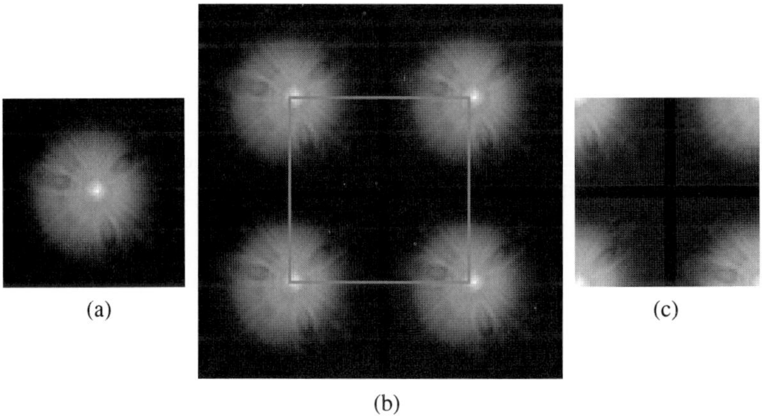

Figure 3.3. Relationship between the space-invariant PSF and the cyclic PSF: (a) discretization of the space-invariant PSF (60×60); (b) piece (120×120) of the translated periodic continuation of the discretized PSF; the square indicates the region of the points corresponding to the shifted array; (c) cyclic PSF (60×60), corresponding to the square in (b), to be used in the discrete equation (3.10).

space-invariant only in the case where the object and image planes are parallel and the motion is a translation.

If we assume that these conditions are satisfied, then the translation motion is described by a time-dependent vector

$$\boldsymbol{a}(t) = \{a_1(t), a_2(t)\}, \quad 0 \leq t \leq T \tag{3.11}$$

where T is the exposure interval. This vector defines the path, with respect to a fixed coordinated system in the image plane, which is described by the origin of a coordinate system fixed with respect to the moving object. One can assume that, at time $t = 0$, the origins of the two systems coincide. Then the trajectory described by an arbitrary point $\boldsymbol{x}'$ of the moving object is given by $\boldsymbol{x}' + \boldsymbol{a}(t)$ so that, if $\boldsymbol{x}$ is a given point in the fixed image plane, this point is reached, at time t, by the point of the moving object which, at time $t = 0$, is given by $\boldsymbol{x}' = \boldsymbol{x} - \boldsymbol{a}(t)$. From the linearity of the photographic material, it follows that the noise-free image $g^{(0)}(\boldsymbol{x})$ at the point $\boldsymbol{x}$ is the result of the addition of the contributions of all points $\boldsymbol{x}' = \boldsymbol{x} - \boldsymbol{a}(t)$ passing through $\boldsymbol{x}$ in the interval $[0, T]$, i.e.

$$g^{(0)}(\boldsymbol{x}) = \frac{1}{T} \int_0^T f^{(0)}[\boldsymbol{x} - \boldsymbol{a}(t)] dt \tag{3.12}$$

(the factor $1/T$ is introduced as a normalization factor).

This relationship can be written in the form of a convolution product by introducing the PSF

$$K(\boldsymbol{x}) = \frac{1}{T}\int_0^T \delta[\boldsymbol{x} - \boldsymbol{a}(t)]dt, \tag{3.13}$$

where $\delta(\boldsymbol{x})$ is the 2D delta distribution. Then, by means of straightforward computations, one finds that the TF is given by

$$\hat{K}(\boldsymbol{\omega}) = \frac{1}{T}\int_0^T e^{-i\boldsymbol{\omega}\cdot\boldsymbol{a}(t)}dt. \tag{3.14}$$

A very simple form of the TF is obtained in the case of linear motion, i.e.

$$\boldsymbol{a}(t) = t\boldsymbol{c}, \tag{3.15}$$

where $\boldsymbol{c}$ is the constant velocity of the motion. In this case the integral (3.14) can be computed and the result is

$$\hat{K}(\boldsymbol{\omega}) = \exp\left(-i\frac{T}{2}\boldsymbol{c}\cdot\boldsymbol{\omega}\right)\,\mathrm{sinc}\left(\frac{T}{2\pi}\boldsymbol{c}\cdot\boldsymbol{\omega}\right) \tag{3.16}$$

which can also be written in terms of the displacement vector $\boldsymbol{s} = T\boldsymbol{c}$ as follows

$$\hat{K}(\boldsymbol{\omega}) = \exp\left(-\frac{i}{2}\boldsymbol{s}\cdot\boldsymbol{\omega}\right)\,\mathrm{sinc}\left(\frac{1}{2\pi}\boldsymbol{s}\cdot\boldsymbol{\omega}\right). \tag{3.17}$$

The sinc-function is defined in equation (2.6).

As seen from equation (3.17) the TF is zero on the family of parallel lines

$$\boldsymbol{s}\cdot\boldsymbol{\omega} = 2\pi n, \quad n = \pm 1, \pm 2, \ldots \tag{3.18}$$

which are orthogonal to the vector $\boldsymbol{s}$. The FT of the noise-free image $g^{(0)}(\boldsymbol{x})$ is also zero on these lines, because we have $\hat{g}^{(0)}(\boldsymbol{\omega}) = \hat{K}(\boldsymbol{\omega})\hat{f}^{(0)}(\boldsymbol{\omega})$. However, this is not true for the FT of the noisy image $\hat{g}(\boldsymbol{\omega})$, as given by equation (3.8). The noise perturbs the lines of zeros but, if it is not too large, the perturbed lines are close to the unperturbed ones.

In figure 3.4 we provide an example of an image blurred by a horizontal linear motion and contaminated by white Gaussian noise. As we see, in this case the lines of zeros of $\hat{g}(\boldsymbol{\omega})$ are rather close to straight lines, so that, if the displacement vector $\boldsymbol{s}$ is not known, these lines of zeros can provide a reasonable way for estimating the components of $\boldsymbol{s}$. The problem of fitting straight lines to these lines of zeros is greatly simplified by the knowledge that the lines are parallel and uniformly separated. Once the fitted straight lines are drawn, the components s_1, s_2 of $\boldsymbol{s}$ can be easily found.

In figure 3.4 we consider a case of horizontal motion. This simplification is not restrictive because, by means of a rotation of the photograph, an arbitrary

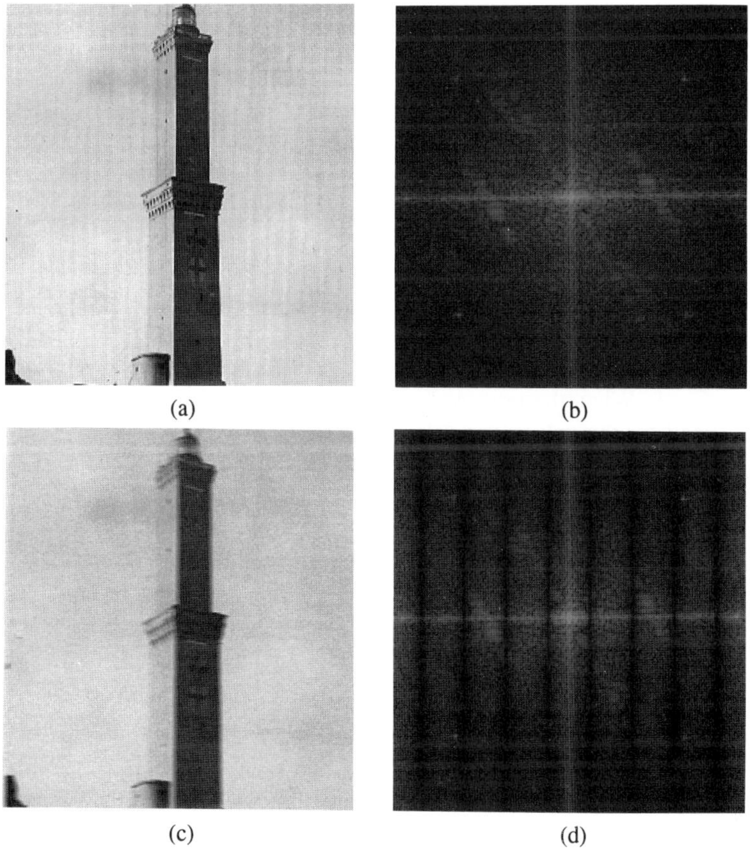

Figure 3.4. Example of linear motion blur: (a) the object: a picture of 'Lanterna', the lighthouse of the Genoa port; (b) the modulus of the FT of the object; (c) the blurred and noisy image; (d) the modulus of the FT of (c).

linear motion can always be reduced to this one. In such a case the blurring (3.12) can be written as a 1D convolution for each horizontal line

$$g^{(0)}(x_1, x_2) = \int_{-\infty}^{\infty} K(x_1 - x_1')f^{(0)}(x_1', x_2)dx_1', \tag{3.19}$$

the 1D PSF being given by ($s = cT$)

$$K(x) = \begin{cases} 1/s, & 0 \leq x \leq s \\ 0, & \text{elsewhere} . \end{cases} \tag{3.20}$$

The corresponding discrete PSF is given by

$$K_n = \begin{cases} 1/S, & 0 \leq n \leq S-1 \\ 0, & S < n \leq N-1 \end{cases} \tag{3.21}$$

where S is the number of pixels corresponding to the displacement distance s, i.e. $S = s/\delta$ if δ is the sampling distance.

The DFT of K_n can be easily computed and is given by

$$\hat{K}_m = \frac{1}{S}\exp\left[-i\frac{\pi}{N}(S-1)m\right]\frac{\sin\left(\frac{\pi S}{N}m\right)}{\sin\left(\frac{\pi}{N}m\right)}. \tag{3.22}$$

When the ratio N/S is integer, then $\hat{K}_m$ is zero for $m = Nk/S$ with $k = 1, 2, \ldots, S-1$. When the ratio N/S is not integer, then $\hat{K}_m$ is never zero but it can be small for the values of m close to Nk/S.

3.3 Out-of-focus blur

In the case of a lens an object point is in focus if its distance from the lens, d_0, satisfies the lens conjugation law

$$\frac{1}{d_0} + \frac{1}{d_i} = \frac{1}{d_f} \tag{3.23}$$

where d_i is the distance between the lens and the image plane and d_f is the focal length of the lens. If this condition is satisfied then, according to geometrical optics, the image of a point is again a point (the case where diffraction effects are important will be briefly discussed in the next section). If condition (3.23) is not satisfied, then the image of the point, always according to geometrical optics, is a disc, called the *circle of confusion* (COC), whose radius can be computed as a function of d_0, d_i, d_f and of the effective lens diameter [1]. The radius of the COC can also depend on the wavelength through the refractive index of the lens. Finally, as concerns the intensity distribution within the COC, it follows from geometrical optics that it is approximately uniform over the COC so that the image of an out-of-focus point, located on the optical axis, is given by

$$K(\boldsymbol{x}) = \frac{1}{\pi D^2}\chi_{\mathrm{coc}}(\boldsymbol{x}). \tag{3.24}$$

D is the radius of the COC and $\chi_{\mathrm{coc}}(\boldsymbol{x})$ is the characteristic function of the COC. A more accurate computation of $K(\boldsymbol{x})$ [2] includes the effects of diffraction (discussed in the next section) which are neglected in the present simplified treatment.

When a 3D scene is imaged by a camera some of the points are in focus and others are out-of-focus, so that the blurring of the image is, in general, space-variant. Only in the following case one has a space-invariant blurring: an out-of-focus flat object whose plane is parallel to the image plane. In such a case, the PSF is given by equation (3.24) and therefore the TF can be expressed in terms of the Bessel function of order 1, $J_1(\xi)$ (see also equation (2.7))

$$\hat{K}(\omega) = 2\frac{J_1(D|\omega|)}{D|\omega|}. \tag{3.25}$$

The function $J_1(\xi)/\xi$ is plotted in figure 3.5. This function has an infinity

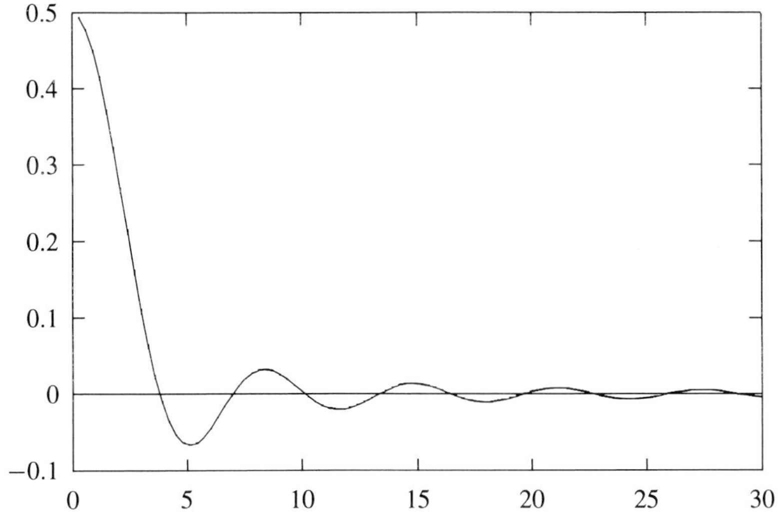

Figure 3.5. Plot of the function $J_1(\xi)/\xi$

of zeros, $\xi_n = \pi x_n$, $n = 1, 2, 3, \ldots$, which, for large n, are approximately equispaced, with spacing π. We also recall that the asymptotic behaviour of $J_1(\xi)$ for large ξ is given by

$$J_1(\xi) \simeq \sqrt{\frac{2}{\pi\xi}} \cos\left(\xi - \frac{3}{4}\pi\right), \quad \xi \to \infty \tag{3.26}$$

while its asymptotic behaviour for small ξ is given by

$$J_1(\xi) \simeq \frac{1}{2}\xi, \quad \xi \to 0. \tag{3.27}$$

The first three zeros of $J_1(\xi)$ correspond to $x_1 = 1.220, x_2 = 2.233$ and $x_3 = 3.238$. As we see $x_3 - x_2$ is already very close to 1.

From these properties of the Bessel function $J_1(\xi)$ it follows that the TF of the out-of-focus blur is zero on a family of circumferences with centre at the origin and radii

$$\omega_n = \frac{\pi}{D} x_n; \quad n = 1, 2, 3, \ldots \tag{3.28}$$

Even for moderate values of n these circumferences are approximately equispaced, in the sense that $\omega_{n+1} - \omega_n \simeq \pi / D$. Moreover, for large values of $|\omega|$, the TF tends to zero as given by the asymptotic behaviour (3.26)

$$\hat{K}(\omega) \simeq \frac{1}{\sqrt{\pi}} \left(\frac{2}{D|\omega|} \right)^{3/2} \cos\left(D|\omega| - \frac{3}{4}\pi \right). \tag{3.29}$$

As in the case of the linear motion blur the FT of the noise-free image $g^{(0)} = K * f^{(0)}$ is zero on the lines where the TF is zero while this property does not hold for the noisy image because the noise perturbs these lines. In figure 3.6 we give an example of an image degraded by out-of-focus blurring and by the addition of white Gaussian noise. In this example the lines of zeros of $\hat{g}(\omega)$ are rather close to concentric circumferences. This property of $\hat{g}(\omega)$ and the relation (3.28) can be used to estimate the radius of the COC when this radius is not known. Indeed, its theoretical determination would require a precise knowledge of camera misadjustment and object position.

3.4 Diffraction-limited imaging systems

According to geometrical optics the image of a point source provided by an optical instrument is a perfectly sharp point. However, because of diffraction effects, the image of a point, which is in focus, is not a point but a small light patch, called the *diffraction pattern*. This is precisely the PSF as defined in section 3.1. Optical instruments where diffraction effects are important are also called diffraction-limited imaging systems. Examples are provided by instruments with a very high resolving power (i.e. ability to separate the images of two neighbouring points) such as microscopes and telescopes.

Diffraction-limited imaging systems are usually treated by Fourier methods, and the corresponding theory is called Fourier optics [3]. In this theory an optical system is just viewed as a linear system, i.e. as a black box fully characterized by its PSF. This description is correct in the case of spatially coherent or spatially incoherent illumination. An optical system is not linear when the so-called partial-coherence effects are not negligible. The case of incoherent illumination is perhaps the more frequent one even if coherent illumination is often important in microscopy and has gained additional importance since the discovery of the laser.

We assume for simplicity that the scalar theory of light diffraction can be used. In this theory monochromatic light is represented by a scalar function which is usually written as a complex-valued function, called the *complex amplitude* (or phasor) whose modulus and phase are respectively the amplitude and phase of the light disturbance. In the case of *spatially coherent illumination*

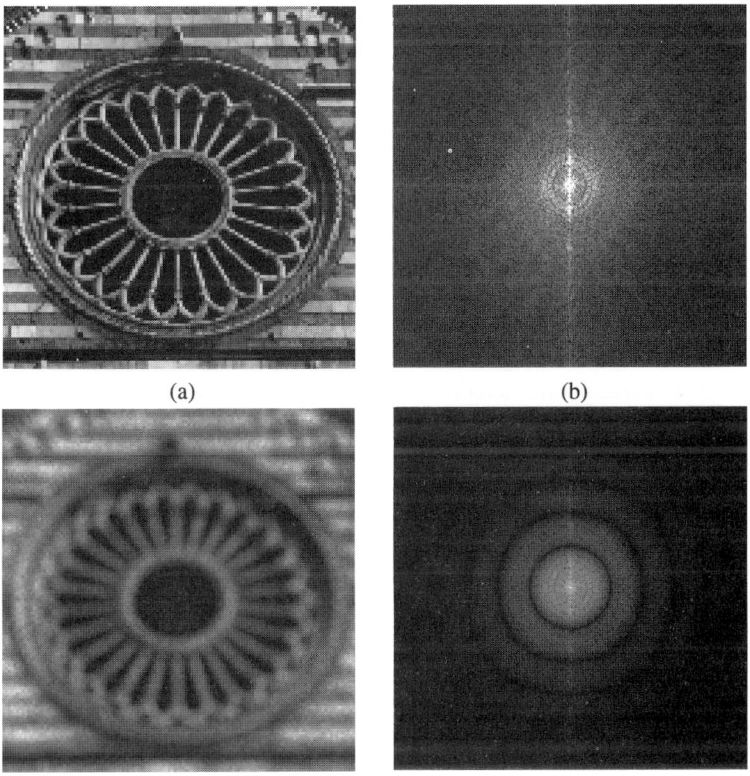

(a) (b) (c) (d)

Figure 3.6. Example of out-of-focus blur: (a) the object: a picture of the rose-window of S. Lorenzo cathedral in Genoa; (b) the modulus of the FT of the object; (c) the blurred and noisy image; (d) the modulus of the FT of (c).

(for short, coherent illumination) the relative phase of two object points is constant in time, i.e. even if the two phases can vary randomly in time, they vary in identical fashion. In the case of *spatially incoherent illumination* (for short, incoherent illumination) the phases of all points are varying in statistically independent fashion.

We consider systems producing real (non-virtual) images. Indeed, even in the case of a virtual image, the detection system must contain a lens converting the virtual image to a real one. Then this lens can be considered as the final element of the imaging system. We also assume that the system is *isoplanatic*, i.e. space-invariant. In practice, optical imaging systems are seldom isoplanatic over the whole object field, but it is usually possible to divide the object field into regions within which the system is approximately space-invariant (see section 8.1).

Finally we assume that the magnification factor of the optical system has been reduced to one by a rescaling of the space variables of the image plane.

(A) *Coherent illumination.* In this case the system is linear in the complex amplitude. Then the process of image formation is described by an equation like (3.1) if one takes as object $f^{(0)}(\boldsymbol{x})$ the complex amplitude of the ideal image predicted by geometrical optics, while $g^{(0)}(\boldsymbol{x})$ is the complex amplitude of the diffraction-limited image.

For monochromatic light with wavelength λ and in the absence of aberrations, i.e. in the case of a system consisting of perfectly corrected lenses, the PSF is, through a change of variable, the inverse FT of the *pupil function*, which is the characteristic function $\chi_{\mathcal{P}}(\boldsymbol{x})$ of the exit pupil $\mathcal{P}$ of the instrument. Therefore an ideal diffraction-limited system is a bandlimited system which behaves like a perfect low-pass Fourier filter.

The change of variable relating the band $\mathcal{B}$ of the system to the exit pupil $\mathcal{P}$ is the following

$$\boldsymbol{\omega} = \frac{2\pi}{\lambda d_i}\boldsymbol{x}. \tag{3.30}$$

Here $\boldsymbol{x}$ is a point of $\mathcal{P}$, $\boldsymbol{\omega}$ is the corresponding point of $\mathcal{B}$, λ is the wavelength of the light and d_i is the distance between the image plane and the exit pupil plane. Then the PSF is given by

$$K(\boldsymbol{x}) = \frac{1}{(2\pi)^2}\int \chi_{\mathcal{B}}(\boldsymbol{\omega})e^{i\boldsymbol{x}\cdot\boldsymbol{\omega}}d\boldsymbol{\omega}. \tag{3.31}$$

In the case of a square exit pupil of side $2a$ one has

$$K(\boldsymbol{x}) = \frac{\sin(\Omega x_1)}{\pi x_1}\,\frac{\sin(\Omega x_2)}{\pi x_2} \tag{3.32}$$

the cut-off frequency Ω being given by (see equation (3.30))

$$\Omega = \frac{2\pi}{\lambda}\,\frac{a}{d_i}, \tag{3.33}$$

while in the case of a circular pupil of radius a one has

$$K(\boldsymbol{x}) = \frac{\Omega}{2\pi}\,\frac{J_1(\Omega|\boldsymbol{x}|)}{|\boldsymbol{x}|}, \tag{3.34}$$

the cut-off frequency being given again by (3.33). The properties of the Bessel function $J_1(\xi)$ have been discussed in section 3.3. The first zero of $K(\boldsymbol{x})$,

$$\rho_1 = 1.22\frac{\pi}{\Omega} = 1.22\frac{d_i}{a}\frac{\lambda}{2}, \tag{3.35}$$

is the famous *Rayleigh resolution distance*, which is usually viewed as a measure of the resolving power of the instrument.

The effect of aberrations can be described by means of a phase distortion over the band of the instrument, so that the TF can be written in the following form

$$\hat{K}(\omega) = \chi_B(\omega)\exp[ik\phi(\omega)] \tag{3.36}$$

where $k = 2\pi/\lambda$ is the wave-number of the light radiation. The phase distortion depends on the type of aberration. In the case of that due to a focusing error, the phase is given by

$$\phi(\omega) = \frac{\varepsilon}{2}\left(\frac{d_i}{k}\right)^2 |\omega|^2 \tag{3.37}$$

where, in the case of a lens,

$$\varepsilon = \frac{1}{d_0} + \frac{1}{d_i} - \frac{1}{d_f}, \tag{3.38}$$

d_0 being the distance between the object plane and the plane of the lens (see section 3.3). Equations (3.36) and (3.37) are the starting point for the analysis of the out-of-focus blur in the case of diffraction-limited systems [2].

(B) *Incoherent illumination.* In this case the system is linear with respect to the light intensity, which is proportional to the square of the modulus of the complex amplitude. In equation (3.1) $f^{(0)}(\boldsymbol{x})$ represents now the intensity distribution of the ideal image as predicted by geometrical optics while $g^{(0)}(\boldsymbol{x})$ is the intensity distribution of the diffraction-limited image.

The PSF of the system in the incoherent case is proportional to the square of the modulus of the PSF of the system in the coherent case [3]. If we denote by $K_c(\boldsymbol{x})$ the coherent PSF of the system and by $K_{\text{inc}}(\boldsymbol{x})$ the incoherent PSF of the same system, we have

$$K_{\text{inc}}(\boldsymbol{x}) = |K_c(\boldsymbol{x})|^2. \tag{3.39}$$

Therefore the relationship between the two transfer functions is the following one

$$\hat{K}_{\text{inc}}(\omega) = \frac{1}{(2\pi)^2}\int \hat{K}_c(\omega + \omega')\hat{K}_c^*(\omega')d\omega'. \tag{3.40}$$

Very often, in Fourier optics, the following normalized transfer function is used

$$\hat{H}(\omega) = \frac{\hat{K}_{\text{inc}}(\omega)}{\hat{K}_{\text{inc}}(0)} \tag{3.41}$$

which is called the *optical transfer function* (OTF). It is easy to show that this function always satisfies the condition $|\hat{H}(\omega)| \leq 1$.

We point out that the band of an optical system in the case of incoherent illumination is broader than the band of the same optical system in the case of coherent illumination. For instance, in the case of a square pupil, equation (3.32), with a cut-off frequency Ω, the cut-off frequency for the incoherent illumination is 2Ω. A similar result holds true for the circular pupil, (3.34). However, while the TF is constant over the band in the coherent case, it tends to zero at the boundary of the band in the incoherent one. In figure 3.7 we compare the coherent and incoherent TF for a circular pupil in the absence of aberrations.

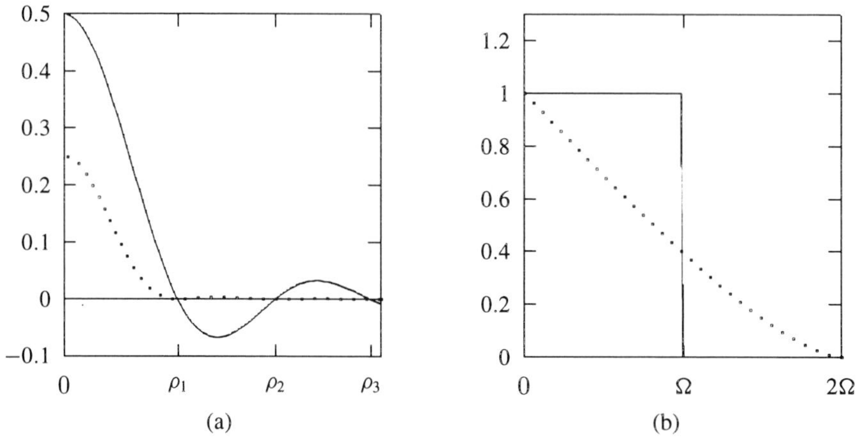

Figure 3.7. a) PSF of a circular pupil both in the coherent (full line) and in the incoherent case (dotted line); (b) the corresponding transfer functions. ρ_1 denotes the Rayleigh distance associated with the bandwidth Ω.

The effect of aberrations in the incoherent case is not simply a phase modulation as in the coherent one—see equation (3.36). Indeed, from equation (3.40), with $\hat{K}_c(\omega)$ given by equation (3.36), we see that aberrations can also produce variations of the modulus of the OTF.

In the case of a focusing error, equation (3.37), and of a circular pupil, one can find an expression of the OTF in terms of slowly convergent series [2]. The case of a square pupil is much simpler, because the OTF is given by [3]

$$\hat{H}(\omega) = \left(1 - \frac{|\omega_1|}{2\Omega}\right)\left(1 - \frac{|\omega_2|}{2\Omega}\right) \tag{3.42}$$
$$\times \operatorname{sinc}\left[\frac{8W}{\lambda}\left(\frac{\omega_1}{2\Omega}\right)\left(1 - \frac{|\omega_1|}{2\Omega}\right)\right] \operatorname{sinc}\left[\frac{8W}{\lambda}\left(\frac{\omega_2}{2\Omega}\right)\left(1 - \frac{|\omega_2|}{2\Omega}\right)\right]$$

where Ω is the cut-off frequency as given by equation (3.33) and $W = \varepsilon a^2/8$ (with ε defined in equation (3.38)) is a number which provides a convenient indication of the severity of the focusing error. In figure 3.8 we plot $\hat{H}(\omega_1, 0)$ for some values of W. When $W > \lambda/2$, the OTF can take negative values. In the limit $W \gg 1$, i.e. when the focusing error is very severe, one can approximate

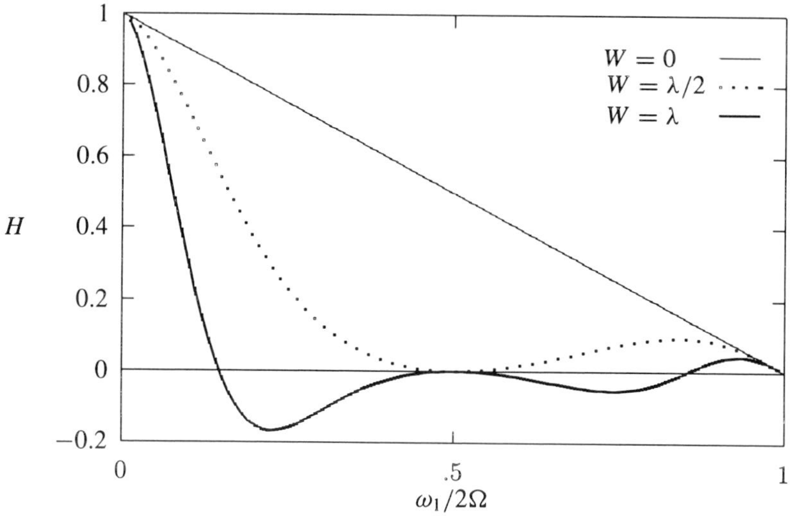

Figure 3.8. Cross-section of the OTF, as a function of the normalized frequency $\omega_1/2\Omega$, for various focusing errors (case of a square pupil).

the OTF (3.42) by means of its low frequency behaviour

$$\hat{H}(\omega) = \operatorname{sinc}\left(\frac{8W}{\lambda}\frac{\omega_1}{2\Omega}\right)\operatorname{sinc}\left(\frac{8W}{\lambda}\frac{\omega_2}{2\Omega}\right) \tag{3.43}$$

and therefore one obtains the PSF predicted by geometrical optics, i.e. the characteristic function of a square. Indeed, the predictions of geometrical optics provide good approximations whenever aberrations of any sort are very severe.

The correction of aberrations, by means of methods of image deconvolution, has been recently attempted in the case of the images of the *Hubble Space Telescope* (HST) [4]. HST is an optical observatory launched in a low-earth orbit outside the disturbing atmosphere (see the discussion of the next section). Soon after launch, in early 1990, a severe optical fault, producing a pronounced blurring of the images, was detected in the telescope. Spherical aberration, caused by a manufacturing error, was taking light away from the central core of the PSF. Indeed, according to the original design, about 70% of the light had to be concentrated into a disc with diameter of 0.1 arcseconds (diameter of the central lobe of the PSF). However, in the configuration corresponding to the so-called Faint Object Camera (FOC) only 16% of the light was concentrated into that disc, while the rest was diffused over a broader disc with diameter of about 2 arcseconds.

The originally designed capability of the telescope was restored by implementing the COSTAR corrective optics during the shuttle servicing mission

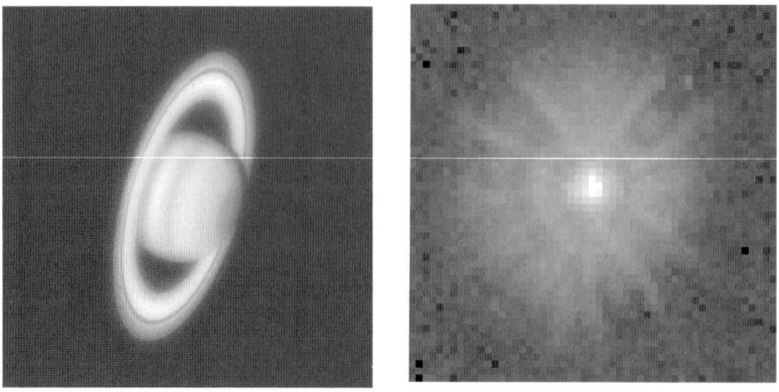

(a) (b)

Figure 3.9. (a) 512×512 image of Saturn provided by the wide field planetary camera (WF/PC) of HST; (b) 53×53 corresponding PSF (Images produced by AURA/STScI).

of December 1993. In the four years between the launch and the correction, several efforts were made to improve the quality of the images and to this purpose both computed and measured PSF were used. A few examples of HST images will be used in the following chapters of this book to illustrate some of the deconvolution methods.

In figure 3.9 we show an example of pre-COSTAR image produced by HST. It is a wide-field/planetary camera image of Saturn (512×512). In the same figure we also give the corresponding observed PSF (53×53).

The *confocal scanning laser microscope* (CSLM) [5] is another important example of a diffraction-limited imaging system where image deconvolution can be useful [6]. The most interesting property of a CSLM is the optical sectioning of a fluorescent object. As a result, three-dimensional imaging of biological objects becomes possible. Even if a CSLM has a resolution better than that of a conventional microscope, confocal images are still affected by a significant blurring which makes difficult subsequent quantitative analysis. An example of a 3D image provided by a confocal microscope is shown in figure 3.10

Image formation by CSLM in the case of a fluorescent object is described by equation (3.6) where $f^{(0)}(\boldsymbol{x})$ denotes the density of the fluorescent material in the microscopic volume while $g^{(0)}(\boldsymbol{x})$ is the intensity distribution in the image. The PSF $K(\boldsymbol{x})$ has a rather involved expression in terms of the properties of the optical instrument. Its main feature is that it is a bandlimited function with a band $\mathcal{B}$ having the following structure. If $\boldsymbol{x} = \{x_1, x_2, x_3\}$ are the space coordinates (with x_3 coordinate in the direction of the optical axis of the instrument) and if $\boldsymbol{\omega} = \{\omega_1, \omega_2, \omega_3\}$ are the corresponding space frequencies, then $\mathcal{B}$ is rotationally invariant around the ω_3-axis (a section in the plane $\omega_2 = 0$ is shown in

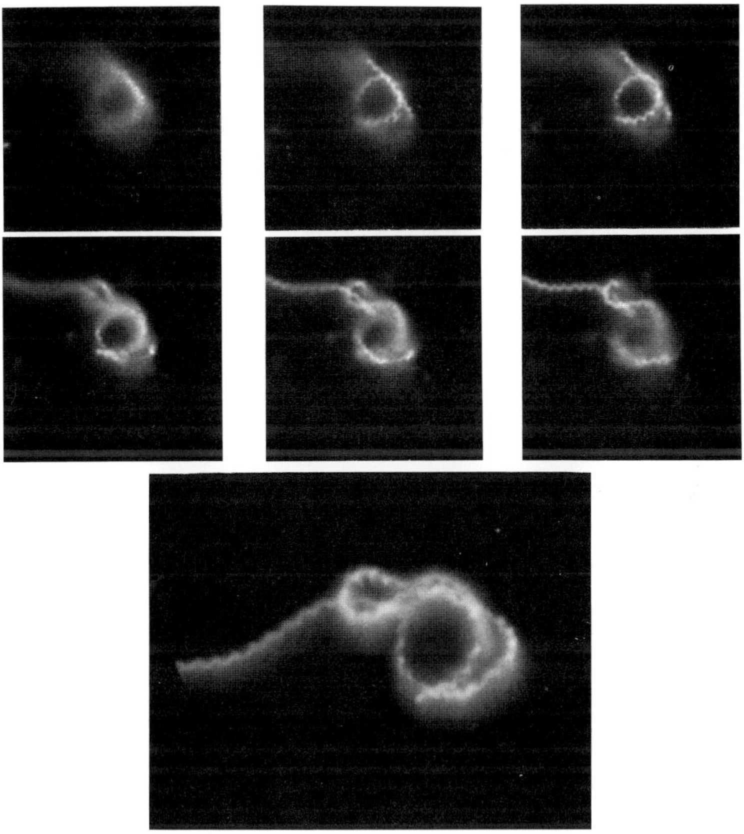

Figure 3.10. Confocal images of six slices of a helical sperm head from an octopus *Eledone Cirrhosa* and 3D representation of the same object obtained from 64 slices (images kindly provided by A. Diaspro).

figure 3.11) but its size in the directions orthogonal to ω_3 is larger than its size in the direction of ω_3. This property implies that the axial resolution of a CSLM is poorer than the lateral resolution and thus the microscope provides orientation-dependent images. Image restoration is especially important for improving the quality of the images in the axial direction.

3.5 Atmospheric turbulence blur

Turbulent velocity fluctuations in the atmosphere generate a statistical temperature field and thus give rise to random inhomogeneities in the index of refraction. As a consequence they cause distortions in the propagation of

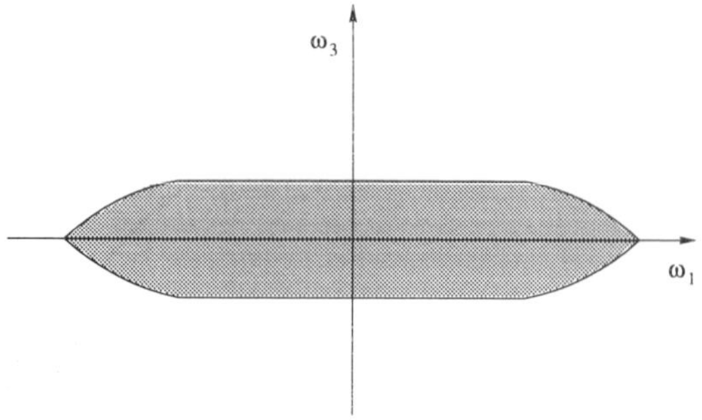

Figure 3.11. Schematic picture of the section, in the plane $\omega_2 = 0$, of the band $\mathcal{B}$ of a confocal microscope.

electromagnetic and acoustic fields at all wavelengths and for this reason their effect is important in imaging through an atmosphere, as occurs in optical and radio astronomy, in remote sensing, in target identification, etc [7].

We restrict our examples to the case of light propagation. Indeed, atmospheric turbulence is a major problem in optical astronomy because it drastically reduces the angular resolution of the ground-based telescopes. In the absence of turbulence (and also of aberrations in the optical components), the quality of the image is limited by the diffraction effects and the angular resolution is given by $\Delta\theta = 0.61(\lambda/a)$, where λ is the wavelength of the observed light and a is the radius of the objective aperture. When turbulence is present, it produces a large and irregular blur. For instance, under good weather conditions at the best observatories, the angular resolution allowed by this blur varies approximately from 0.6 to 1.2 arcseconds. But according to the previous formula with $\lambda \simeq 5.6 \times 10^{-5}$ cm, these resolutions correspond to telescope diameters ranging from about 30 to about 10 cm (the diameter of the Mount Palomar telescope is about 5 metres). Therefore it is clear that the resolution limits due to atmospheric turbulence are much more severe than those due to diffraction effects.

Another feature of the atmospheric turbulence blur is that it changes rapidly with time. The characteristic time of these variations is about 1 millisecond. Therefore it is usual to distinguish between *long-exposure images* (the exposure time is large with respect to the characteristic time) and *short-exposure images* (the exposure time is comparable with the characteristic time of the fluctuations).

In a conventional astronomical photograph, the exposure time can be of the order of few seconds and therefore the recorded image is not random. This is a typical example of a long-exposure image. The basic result in this case is that the OTF $\hat{K}(\omega)$ of the whole system, telescope and atmosphere, is the product of the OTF of the telescope, $\hat{H}(\omega)$, and of an atmospheric transfer function, $\hat{B}(\omega)$:

$$\hat{K}(\omega) = \hat{H}(\omega)\hat{B}(\omega) \tag{3.44}$$

(we notice that here the components of ω are frequencies associated with angular variables). Then, it can be shown that [7]

$$\hat{B}(\omega) = \exp\left[-3.44\left(\frac{\lambda|\omega|}{r_0}\right)^{5/3}\right] \tag{3.45}$$

where λ is the wavelength of the observed radiation and r_0 is a parameter, called *critical diameter*, which depends on λ and is roughly proportional to $\lambda^{6/5}$. If the diameter of the telescope is small with respect to the critical diameter, then the effect of turbulence is negligible while it is dominant and determines the resolving power of the telescope when the critical diameter is smaller than the diameter of the telescope. As follows from equation (3.45), the atmospheric turbulence blur can be roughly approximated by a Gaussian one

$$\hat{B}(\omega) = \exp(-s^2|\omega|^2). \tag{3.46}$$

For short exposure times, i.e. of the order of 1 ms, the images look random and therefore also the PSF and the OTF of atmospheric turbulence look random. It can be shown that the ensemble average of the OTF is given again by equation (3.44) where, under near-field conditions

$$\hat{B}(\omega) = \exp\left\{-3.44\left(\frac{\lambda|\omega|}{r_0}\right)^{5/3}\left[1-\left(\frac{|\omega|}{\Omega}\right)^{1/3}\right]\right\} \tag{3.47}$$

(Ω being the cut-off frequency of the telescope) while, under far-field conditions,

$$\hat{B}(\omega) = \exp\left\{-3.44\left(\frac{\lambda|\omega|}{r_0}\right)^{5/3}\left[1-\frac{1}{2}\left(\frac{|\omega|}{\Omega}\right)^{1/3}\right]\right\}. \tag{3.48}$$

It is important to note that, at space frequencies close to the cut-off frequency of the optical system, the long-exposure OTF is considerably smaller than the short-exposure average. Therefore, short-exposure pictures are preferred because they can provide a better resolution than long-exposure pictures. In figure 3.12 we compare the atmospheric turbulence TF for long-exposure images, equation (3.45), with that for short-exposure images in the case of far-field conditions, equation (3.48). It is assumed that $r_0/\lambda = \Omega/2$.

3.6 Near-field acoustic holography

Near-field acoustic holography (NAH) was proposed in 1980 by Williams and Maynard [8] as a technique which can produce high-resolution images of sound

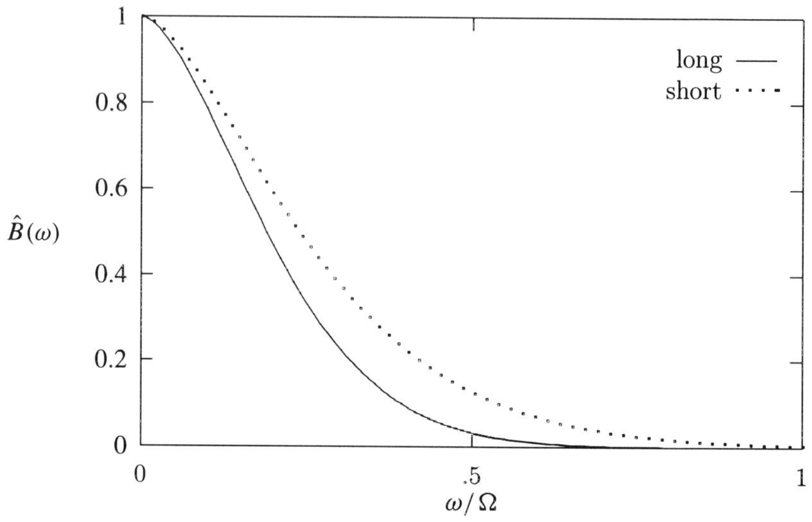

Figure 3.12. Comparison of the TF for long and short exposure time, as functions of ω/Ω, in the case $r_0/\lambda = \Omega/2$.

sources. Since the resolution which can be achieved is much better than the wavelength of the radiation used, this method can provide a satisfactory resolution also in the case of long wavelengths.

The basic principle underlying NAH is that the acoustic pressure is measured (for instance by means of an array of microphones) on a surface, containing the acoustic source inside, such that its distance from the surface of the source is smaller than the wavelength of the emitted radiation. Then the measured data are used for computing the acoustic pressure on the surface of the source. The good resolution achieved is due to the use of the information conveyed by the so-called *evanescent waves*, which are still present in the radiated field when the distance from the source is smaller than the wavelength. This is the same principle used in scanning near-field optical microscopy (SNOM) [9].

NAH has gained wide popularity in recent years as an experimental tool for studying acoustical radiation from finite sources, including also industrial applications. From the mathematical point of view, the underlying mathematical problem is that of *inverse diffraction* which will be discussed in chapter 8. Here we consider the simple case where the wave-field propagates from a plane, say $x_3 = 0$, to another parallel plane, say $x_3 = a$. The geometry of this problem is illustrated in figure 3.13.

More precisely, we assume that the sources are located in the half-space $x_3 < 0$, so that the plane $x_3 = 0$ is the boundary of the source region. We

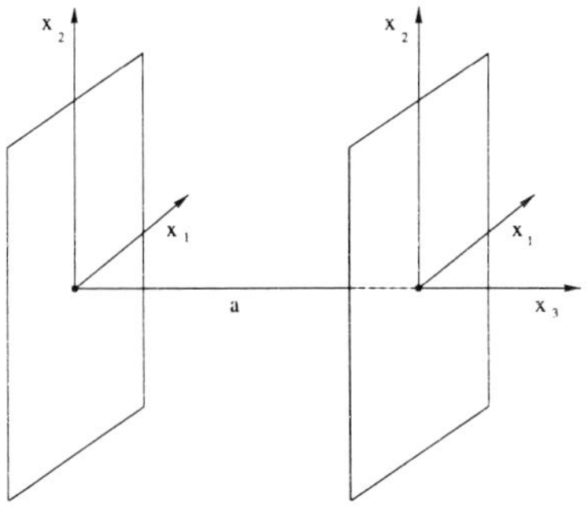

Figure 3.13. Schematic representation of the geometry of the problem of inverse diffraction

also assume that the sources emit a scalar monochromatic field, propagating in the half space $x_3 > 0$, where its complex amplitude $u(\boldsymbol{x}) = u(x_1, x_2, x_3)$ is a solution of the Helmholtz equation

$$\Delta u + k^2 u = 0. \tag{3.49}$$

As usual, we denote by k the wave number, which is associated to the frequency ν of the monochromatic field by $k = 2\pi\nu/c$, c being the sound velocity. The wave-length λ is defined by $\lambda = 2\pi/k$ and it is the characteristic length of the problem.

The solution of equation (3.49) is uniquely determined by the following boundary conditions:
(1) the values $f^{(0)}(x_1, x_2)$ of the complex amplitude on the boundary plane $x_3 = 0$

$$u(x_1, x_2, 0) = f^{(0)}(x_1, x_2); \tag{3.50}$$

(2) the Sommerfeld radiation condition at infinity

$$\lim_{r\to\infty}\left[r\left(\frac{\partial u}{\partial r} - iku\right)\right] = 0, \ |\theta| < \pi/2 \tag{3.51}$$

where r, θ are polar coordinates (the polar axis is the x_3-axis).

In other words there exists a unique solution of equation (3.49), taking given values on the source place $x_3 = 0$ and behaving at infinity as an outgoing spherical wave. This is the forward problem, or also the direct problem in the sense of section 1.1. It is natural to assume that the function $f^{(0)}$ is square-integrable because its L^2-norm is proportional to the energy associated with the field.

It is possible to give an integral representation of the solution of this problem in terms of the boundary values $f^{(0)}$, using the appropriate Green function

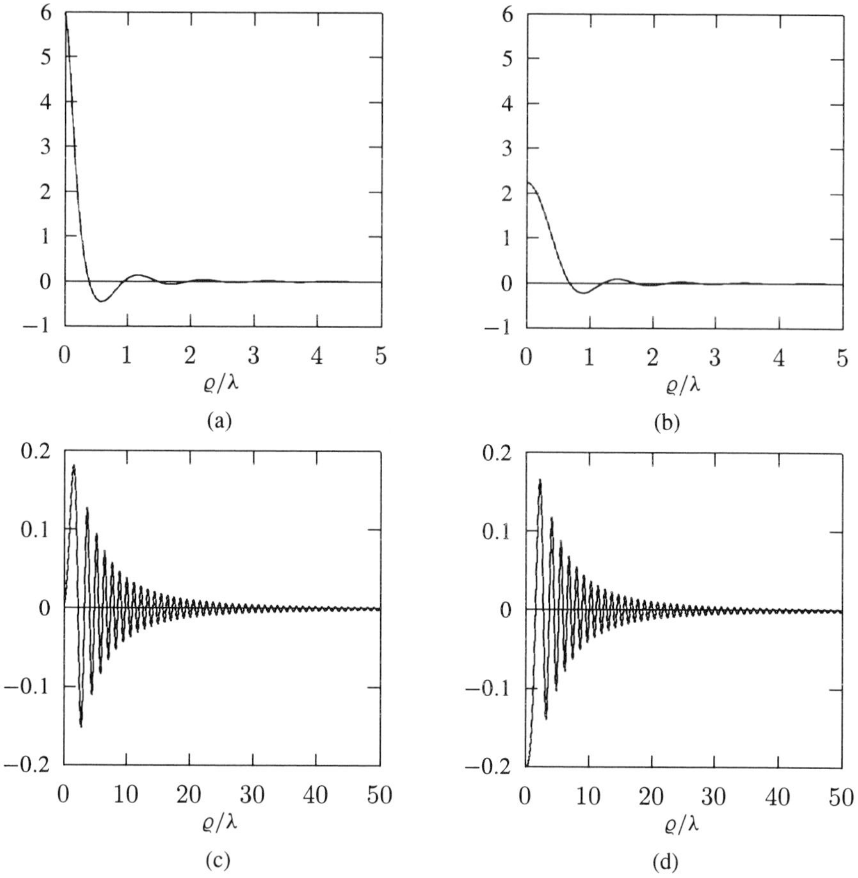

Figure 3.14. Plot of $\lambda K_a^{(+)}(\varrho)$ as a function of ϱ/λ. Top: real part (a) and imaginary part (b) in the case $a = \lambda/5$; bottom: real part (c) and imaginary part (d) in the case $a = 5\lambda$.

[10]. Here we write the solution in a form suitable for the problem of inverse diffraction. Let us denote by $g_a^{(0)}(x_1, x_2)$ the values of the scalar field on the plane $x_3 = a$:

$$g_a^{(0)}(x_1, x_2) = u(x_1, x_2, a). \tag{3.52}$$

Then, by solving the forward problem one gets [10] ($\boldsymbol{x} = \{x_1, x_2\}$)

$$g_a^{(0)}(\boldsymbol{x}) = \int K_a^{(+)}(\boldsymbol{x} - \boldsymbol{x}') f^{(0)}(\boldsymbol{x}') d\boldsymbol{x}' \tag{3.53}$$

where

$$K_a^{(+)}(\boldsymbol{x}) = -\frac{1}{2\pi} \frac{\partial}{\partial a} \frac{e^{ikr_a}}{r_a}, \quad r_a = \sqrt{|\boldsymbol{x}|^2 + a^2}. \tag{3.54}$$

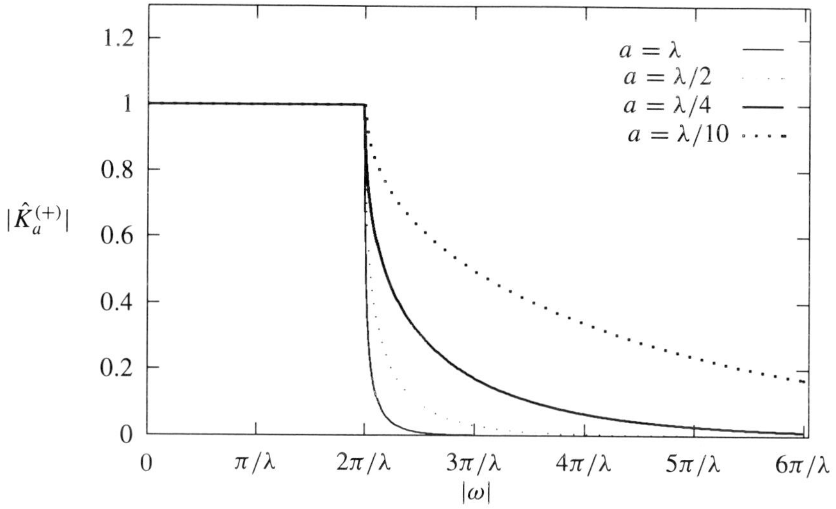

Figure 3.15. Behaviour of the modulus of the transfer function $\hat{K}_a^{(+)}(\omega)$ for various values of a.

The FT of this function has a very simple expression [11], [12]

$$K_a^{(+)}(\omega) = e^{iam(\omega)} \tag{3.55}$$

where

$$m(\omega) = \begin{cases} \sqrt{k^2 - |\omega|^2}, & |\omega| \leq k \\ i\sqrt{|\omega|^2 - k^2}, & |\omega > k. \end{cases} \tag{3.56}$$

The representation of $g_a^{(0)}(\boldsymbol{x})$ obtained from the equations (3.53),(3.56) and the convolution theorem is also called the *angular-spectrum representation* of the propagating field

$$g_a^{(0)}(\boldsymbol{x}) = \frac{1}{(2\pi)^2} \int e^{iam(\omega)} \hat{f}^{(0)}(\omega) e^{i\boldsymbol{x}\cdot\omega} d\omega. \tag{3.57}$$

In figure 3.14 we plot the real and imaginary parts of the PSF (3.54) for $a = \lambda/5$ and $a = 5\lambda$ while in figure 3.15 we plot the modulus of the TF (3.55) for various values of a.

The problem of *inverse diffraction from plane to plane* [13] is the problem of estimating $f^{(0)}$ from given and noisy values of $g_a^{(0)}$, i.e. $g_a = g_a^{(0)} + w$. Of course one must know the amplitude and phase of the scalar field (for instance the acoustic pressure) and therefore this problem is quite similar to the problem of coherent imaging, considered in section 3.4.

In the present case, however, we have to deal with a PSF which is not rigorously bandlimited although it becomes practically bandlimited when a is sufficiently large. Indeed, the modulus of the TF has an exponential decay when $|\omega| > k$ (see figure 3.15). In practice, when $a > \lambda$, the TF drops to zero at $|\omega| = k$ and therefore for such values of a the PSF $K_a^{(+)}(\boldsymbol{x})$ is essentially bandlimited, the band being the disc of radius k. The cut-off frequency is $\Omega = k$ and the corresponding Rayleigh distance is $\rho_1 = 1.22(\lambda/2)$. Moreover, on the band, the exponential $\exp[iam(\omega)]$ is a pure phase factor. If we compare with equation (3.36) we see that the effect of propagation is analogous to that of an aberrated optical system with a phase distortion given by

$$\phi(\omega) = a\sqrt{1 - |\omega|^2/k^2}. \tag{3.58}$$

In the case where the object $f^{(0)}$ is bandlimited with a band whose size is much smaller than k, then, for small spatial frequencies one can approximate equation (3.58) by $\phi(\omega) \simeq a(1 - |\omega|^2/2k^2)$ and therefore the aberration due to propagation turns out to be equivalent to an aberration due to a focusing error (equation (3.37) with $\varepsilon d_i^2 = -a$).

References

[1] Patmesil M and Chakravarty I 1982 *ACM Trans. Graph.* **1** 25
[2] Stokseth P A 1969 *J. Opt. Soc. Am.* **59** 1314
[3] Goodman J W 1968 *Introduction to Fourier Optics* (New York: McGraw-Hill)
[4] Adorf H M 1995 *Inverse Problems* **11** 639
[5] Wilson T ed. 1990 *Confocal Microscopy* (London: Academic Press)
[6] van der Voort H T M and Strasters K C 1995 *J. Microsc.* **178** 165
[7] Roddier F 1981 *Progress in Optics* vol XIX (ed E Wolf) 281
[8] Williams E G and Maynard J D 1980 *Phys. Rev. Lett.* **45** 554
[9] Betzig E and Trautman T D 1992 *Science* **257** 189
[10] Luneburg R K 1964 *Mathematical Theory of Optics* (Berkeley: University of California Press)
[11] Sherman G C 1967 *J. Opt. Soc. Am.* **57** 1490
[12] Shewell J R and Wolf E 1968 *J. Opt. Soc. Am.* **58** 1596
[13] Baltes H P (ed) 1978 *Inverse Source Problems in Optics* (Berlin: Springer)

Chapter 4

The ill-posedness of image deconvolution

Ill-posedness characterizes linear inverse problems and, in particular, image deconvolution. It implies that the solution of the problem may not exist or, when it exists as in the case of discrete problems, is completely deprived of physical meaning as a consequence of error propagation and amplification from the data to the solution. It is essentially a consequence of the fact that the system does not transmit complete information about the object. Therefore, even in the absence of noise one could find difficulties in solving the problem.

The first consequence of the situation described above is that one must reject the concept of an exact solution of the problem and look for approximate solutions, i.e. objects which reproduce approximately the noisy images. This is the first basic point in the treatment of ill-posed problems. However, another consequence of ill-posedness is that the set of approximate solutions is too broad and contains not only physically acceptable objects but also objects which are too large and wildly oscillating. Then the second basic point is that one must use the knowledge of additional properties of the unknown object (constraints) for selecting from the set of approximate solutions those which are physically meaningful. This is the role played by so-called *a priori information*.

4.1 Formulation of the problem

We consider a space-invariant imaging system, such that the model of image formation is given by equation (3.6). The linear convolution operator associated with the PSF $K(\boldsymbol{x})$ and given by

$$(Af)(\boldsymbol{x}) = \int K(\boldsymbol{x} - \boldsymbol{x}')f(x')dx' \tag{4.1}$$

defines the mapping from the object space into the image space, along the general lines discussed in section 1.2. In our treatment of image deconvolution we assume that these spaces coincide in the sense that both are spaces of square-integrable functions, equipped with the scalar product (2.11).

In the image space, the subset of the noise-free images is the set of all square-integrable functions given by $g^{(0)} = Af^{(0)}$, where $f^{(0)}$ is an arbitrary square-integrable object. Then the noisy images are given by equation (3.6) which will be rewritten as

$$g = Af^{(0)} + w. \tag{4.2}$$

In the case of 2D discrete objects and images, the mapping is given by the 2D cyclic matrix $\mathbf{A}$ associated with the cyclic PSF $\boldsymbol{K}$, as defined in section 3.1

$$(\mathbf{A}\boldsymbol{f})_{m,n} = \sum_{k,l=0}^{N-1} K_{m-k,n-l} f_{k,l}. \tag{4.3}$$

The object and the image spaces are now Euclidean spaces of 2D arrays, equipped with the scalar product (2.56). Again the subset of the noise-free images is the set of all 2D arrays given by $\boldsymbol{g}^{(0)} = \mathbf{A}\boldsymbol{f}^{(0)}$, while the noisy images are given by equation (3.10) which will be rewritten as

$$\boldsymbol{g} = \mathbf{A}\boldsymbol{f}^{(0)} + \boldsymbol{w}. \tag{4.4}$$

In rather broad terms, the problem of image deconvolution can now be formulated as the problem of estimating the unknown object $f^{(0)}(\boldsymbol{x})$, or its discrete version $\boldsymbol{f}^{(0)}$, given the noisy image $g(\boldsymbol{x})$, or its discrete version $\boldsymbol{g}$. Moreover it must be assumed that the PSF $K(\boldsymbol{x})$, or the cyclic PSF $\boldsymbol{K}$, is also known.

Remark 4.1. In some cases one does not know the PSF or one only knows that it belongs to a class of functions depending on a certain number of parameters. For instance, one knows that the blur is due to uniform motion but the parameters of the motion are not known. Section 3.2 contains a short discussion of this problem and a method for its solution, by means of processing the image g. This is only a particular case of a more general problem, which is known as image identification (equivalent names: blur identification, image-blur identification, a posteriori restoration, blind deconvolution) and consists of the attempt to estimate both $K(\boldsymbol{x})$ and $f^{(0)}(\boldsymbol{x})$ with the unique knowledge of $g(\boldsymbol{x})$. It should be obvious that to solve such a problem one needs a lot of additional information about the two unknown functions $K(\boldsymbol{x})$ and $f^{(0)}(\boldsymbol{x})$. In any case this problem is not considered in this book where we always assume that the PSF or a sufficiently good approximation of the PSF itself is known.

In equation (4.2), as well as in equation (4.4), both $f^{(0)}$ and w are unknown, but if the SNR, as defined in equation (3.5), is sufficiently large, it may seem reasonable at first sight to neglect the noise term in these equations. If we make this approximation, then we can formulate the problem of image restoration as that of solving the linear equation

$$Af = g \tag{4.5}$$

or, in the discrete case, as that of solving the linear algebraic system

$$\mathbf{A}f = g. \tag{4.6}$$

As concerns equation (4.5), since we assume that $g(\boldsymbol{x})$ is a square-integrable function we must search for a solution $f(\boldsymbol{x})$ which is also square-integrable. If we can find such a function, we expect that it provides an approximation of the unknown object $f^{(0)}(\boldsymbol{x})$. As concerns the discrete equation (4.6), we also expect that the solution $\boldsymbol{f}$ provides an approximation of the discrete object $\boldsymbol{f}^{(0)}$.

If we use the FT, equation (4.5) becomes rather trivial, since we get

$$\hat{K}(\boldsymbol{\omega})\hat{f}(\boldsymbol{\omega}) = \hat{g}(\boldsymbol{\omega}). \tag{4.7}$$

Analogously, by means of the DFT, from equation (4.6) we have

$$\hat{K}_{k,l}\hat{f}_{k,l} = \hat{g}_{k,l}. \tag{4.8}$$

These equations are the starting points of the analysis of the next sections.

4.2 Well-posed and ill-posed problems

According to a definition which is now classical, a mathematical problem is said to be *well-posed* in the sense of Hadamard if it satisfies the following conditions:

- the solution of the problem is unique
- the solution exists for any data
- the solution depends with continuity on the data.

The meaning of the first requirement is obvious. In the case of problem (4.5) the second requirement implies that the solution should exist for any square-integrable image $g(\boldsymbol{x})$. Finally, the third requirement has the following precise formulation: let g_n be a sequence of square-integrable images such that $\|g_n\| \to 0$ when $n \to \infty$, the norm being defined in equation (2.12); then the corresponding solutions f_n of the equation must have the same property, i.e. $\|f_n\| \to 0$ when $n \to \infty$.

The meaning of this property is clearly illustrated by the following statement of Courant [1]: 'the third requirement, particularly incisive, is necessary if the mathematical formulation is to describe observable natural phenomena. Data in nature cannot possibly be conceived as rigidly fixed; the mere process of measuring them involves small errors. Therefore a mathematical problem cannot be considered as realistically corresponding to physical phenomena unless a variation of the given data in a sufficiently small range leads to an arbitrary small change in the solution. This requirement of *stability* is not only essential for meaningful problems in mathematical physics, but also for approximation methods'.

If one of the three conditions above is not satisfied, the problem is said to be *ill-posed*. Therefore an ill-posed problem is a problem whose solution is not unique and/or does not exist for any data and/or does not depend continuously on the data. In this and next section we will show that the problem (4.5) is always ill-posed. The peculiar features of its discrete version, problem (4.6), which can be well-posed in the sense specified above, will be discussed in section 4.4.

We first discuss the question of uniqueness. If the solution of equation (4.5) is not unique, then there exist at least two distinct objects, say f_1 and f_2, such that $Af_1 = g$ and $Af_2 = g$. Since the operator A is linear, we obtain $A(f_1 - f_2) = 0$ and therefore $f = f_1 - f_2$ is a non-trivial solution of the homogeneous equation $Af = 0$. Conversely, if this equation has a non-trivial solution f and if f_1 is a solution of equation (4.5), i.e. $Af_1 = g$, then $f_2 = f_1 + f$ is also a solution of the same equation because: $Af_2 = Af_1 + Af = Af_1 = g$. We see that *the solution of the equation $Af = g$ is unique if and only if the equation $Af = 0$ has only the solution $f = 0$.*

The previous remark implies that the uniqueness or non-uniqueness of the solution can be established by investigating the solutions of the homogeneous equation. When there exist non-trivial solutions, these constitute a linear subspace called the *null space* of the operator A or also the space of the *invisible objects*, i.e. objects whose noise-free image is exactly zero—see section 2.3.

In order to establish the existence of invisible objects it is convenient to write the equation $Af = 0$ in terms of Fourier transforms. We obtain

$$\hat{K}(\omega)\hat{f}(\omega) = 0. \tag{4.9}$$

The existence of non-trivial solutions of this equation is related to the support of $\hat{K}(\omega)$, as defined in section 2.2. If the support of $\hat{K}(\omega)$ coincides with the whole frequency space, then $\hat{f}(\omega) = 0$. Among the examples of chapter 3, this is the case, for instance, of linear motion blur, of out-of-focus blur, of atmospheric turbulence blur and of near-field acoustic holography. In all these cases the solution of the restoration problem is unique. On the other hand, in the case of diffraction-limited imaging systems and in the case of far-field acoustic holography, the support $\mathcal{B}$ of $\hat{K}(\omega)$ is a bounded subset of the frequency space and therefore invisible objects exist. An example in one dimension is given in figure 2.5.

We can conclude that, in the case of a bandlimited system, the solution of the restoration problem is not unique and the first condition required for well-posedness is not satisfied. The obvious reason is that the imaging system does not transmit information about the object at the frequencies ω outside the band of the instrument.

As concerns the discretized problem (4.8), the equation $\mathbf{A}f = 0$ implies

$$\hat{K}_{k,l}\hat{f}_{k,l} = 0 \tag{4.10}$$

and therefore the solution of the problem (4.8) is unique if and only if $\hat{K}_{k,l} \neq 0$ for any value of k and l.

It is important to notice that, even if the problem (4.6) is a discrete version of the problem (4.5), the uniqueness of the solution of (4.6) is not directly related to the uniqueness of the solution of (4.5). Indeed, when the PSF $K(\boldsymbol{x})$ is discretized and the DFT of the discrete PSF is computed along the lines described in section 2.6, as a consequence of the discretization errors all values of the discrete TF $\hat{K}_{k,l}$ can be different from zero even if $K(\boldsymbol{x})$ is bandlimited. On the other hand, in the case of linear motion blur where uniqueness holds true, it is possible that some values of $\hat{K}_m$ are zero for the discrete problem (see equation (3.22)) and therefore in such a case uniqueness does not hold.

4.3 Existence of the solution and inverse filtering

In this section we investigate the existence of the solution of the problem of image restoration in the case where uniqueness holds true. Indeed, as was shown in the previous section, the uniqueness of the solution can be investigated independently of its existence. We consider first the case of equation (4.5) and then the case of its discrete version, equation (4.6).

If the support of $\hat{K}(\boldsymbol{\omega})$ is the whole $\boldsymbol{\omega}$-space, equation (4.7) implies that the FT of the solution of the problem is given by

$$\hat{f}(\boldsymbol{\omega}) = \frac{\hat{g}(\boldsymbol{\omega})}{\hat{K}(\boldsymbol{\omega})}. \tag{4.11}$$

However the solution of the problem exists if and only if the r.h.s. of equation (4.11) defines the FT of a function. In order to investigate this point, let us recall the model of image formation discussed in section 3.1. According to this model, the FT of the image $\hat{g}(\boldsymbol{\omega})$ is given by

$$\hat{g}(\boldsymbol{\omega}) = \hat{K}(\boldsymbol{\omega})\hat{f}^{(0)}(\boldsymbol{\omega}) + \hat{w}(\boldsymbol{\omega}) \tag{4.12}$$

where $\hat{f}^{(0)}(\boldsymbol{\omega})$ is the FT of the true object and $\hat{w}(\boldsymbol{\omega})$ is the FT of the noise contribution. If we substitute equation (4.12) into equation (4.11) we get

$$\hat{f}(\boldsymbol{\omega}) = \hat{f}^{(0)}(\boldsymbol{\omega}) + \frac{\hat{w}(\boldsymbol{\omega})}{\hat{K}(\boldsymbol{\omega})}. \tag{4.13}$$

We conclude that the function $\hat{f}(\boldsymbol{\omega})$ is the sum of the FT of the true object $\hat{f}^{(0)}(\boldsymbol{\omega})$ plus a term which comes from the inversion of the noise contribution. This second term may be responsible for the non-existence of the solution, i.e. of the inverse FT of $\hat{f}(\boldsymbol{\omega})$.

We first consider the case where $\hat{K}(\boldsymbol{\omega})$ is zero for some values of $\boldsymbol{\omega}$ (this situation applies, for instance, to the linear motion blur and to the out-of-focus blur). In this case the FT of the noise term in general is not zero at the frequencies where the TF is zero because the noise is a process which is independent of the imaging process. As a consequence in equation (4.13) we have division by zero and $\hat{f}(\boldsymbol{\omega})$ has singularities at the zeros of $\hat{K}(\boldsymbol{\omega})$. This fact implies that the inverse FT of $\hat{f}(\boldsymbol{\omega})$ may not exist and therefore equation (4.5) may not have a solution.

Moreover, even if $\hat{K}(\omega)$ is not zero for some values of ω, it tends to zero when $|\omega| \to \infty$. This property holds true for all TF considered in Chapter 3. Since the behaviour of $\hat{w}(\omega)$ for large values of $|\omega|$ is not related to the behaviour of $\hat{K}(\omega)$, the ratio $\hat{w}(\omega)/\hat{K}(\omega)$ may not tend to zero. Depending on the relationship between the high frequency behaviour of the noise and that of the TF, this ratio may tend to infinity, or to a constant, or to zero. Possibly it may have no limit. The typical situation is that the inverse FT of this ratio does not exist and therefore no solution of the problem exists.

The previous discussion can be synthesized in a precise mathematical form as follows: the solution of equation (4.5) exists and is square-integrable if and only if the image $g(\boldsymbol{x})$ satisfies the following condition

$$\int \left| \frac{\hat{g}(\omega)}{\hat{K}(\omega)} \right|^2 d\omega < \infty; \tag{4.14}$$

in this case the solution (which is denoted by $f^\dagger$ in order to have the same notation as the one used in section 4.5 for the generalized solution) is given by

$$f^\dagger(\boldsymbol{x}) = \frac{1}{(2\pi)^q} \int \frac{\hat{g}(\omega)}{\hat{K}(\omega)} e^{i\boldsymbol{x}\cdot\omega} d\omega. \tag{4.15}$$

Since condition (4.14) is not satisfied by arbitrary square-integrable functions $\hat{g}(\omega)$ and since, in general, noisy images do not satisfy this condition, the solution of equation (4.5) may not exist. If we recall the general definition of noise-free images, given at the beginning of section 4.1, we can conclude that the solution $f^\dagger$ exists if and only if g is a noise-free image.

As concerns the continuous dependence of the solution on the data, it is sufficient to observe that, if we consider, for instance the object $f^{(0)}(\boldsymbol{x}) = A_0 \exp(i\boldsymbol{x} \cdot \omega_0)$, with a sufficiently large frequency ω_0, then its image is given by $g^{(0)}(\boldsymbol{x}) = A_0 \hat{K}(\omega_0) \exp(i\boldsymbol{x} \cdot \omega_0)$ and therefore its amplitude can be very small even if the amplitude A_0 of the object can be very large (for instance, one can take $A_0 = |\hat{K}(\omega_0)|^{-1/2}$).

The previous discussion makes clear that both the non-existence of the solution and the lack of uniqueness discussed in the previous section are due to the fact that the imaging system does not transmit complete information about the Fourier transform of the object at certain frequencies. This lack of information is a point which is very important and which must never be forgotten. To this purpose one must always remember a very incisive statement of Lanczos [2]: 'a lack of information cannot be remedied by any mathematical trickery'. Indeed, the methods developed for solving ill-posed problems are not based on mathematical trickeries but rather on a reformulation of the problem based on the use of additional information on the object. This additional information is compensating, in some way, the lack of information due to the imaging system. This point will be further discussed in section 4.6.

We conclude that, when uniqueness holds true, the problem of image restoration is ill-posed because the second and third conditions for well-posedness are not satisfied.

Consider now the discrete equation (4.6). In this case the situation is quite different. We know that the solution is unique if and only if $\hat{K}_{k,l} \neq 0$ for all values of k and l. If this condition is satisfied, from equation (4.8) we get

$$\hat{f}_{k,l} = \frac{\hat{g}_{k,l}}{\hat{K}_{k,l}} \tag{4.16}$$

and, by taking the inverse DFT of this equation, we obtain

$$(f^\dagger)_{m,n} = \frac{1}{N^2} \sum_{k,l=0}^{N-1} \frac{\hat{g}_{k,l}}{\hat{K}_{k,l}} \exp\left[i\frac{2\pi}{N}(km + ln)\right]. \tag{4.17}$$

This solution exists for any noisy image g and also depends continuously on the data. Therefore we conclude that, when uniqueness holds true, the discrete problem (4.6) is well posed.

The procedure outlined above is usually called *inverse filtering* in the literature on image restoration. We also observe that equation (4.17) can be written in terms of the inverse of the cyclic matrix $\mathbf{A}$ associated with the cyclic PSF K

$$f^\dagger = \mathbf{A}^{-1} g. \tag{4.18}$$

The matrix $\mathbf{A}^{-1}$ is the 2D cyclic matrix generated by the 2D array

$$(K^{-1})_{m,n} = \frac{1}{N^2} \sum_{k,l=0}^{N-1} \frac{1}{\hat{K}_{k,l}} \exp\left[i\frac{2\pi}{N}(km + ln)\right]. \tag{4.19}$$

It follows that equation (4.18) can be written as a cyclic convolution

$$(f^\dagger)_{m,n} = \sum_{k,l=0}^{N-1} (K^{-1})_{m-k,n-l}\, g_{k,l}. \tag{4.20}$$

The array K^{-1} is also called the *inverse filter* or the *inverse PSF*.

We have given the equations for the 2D case. Their modification in order to cover the 1D and 3D case is easy, since one has simply to use, in these cases, the 1D and 3D DFT respectively. An alternative expression of $\mathbf{A}^{-1}$ in the 1D case is given in equation (2.71). A similar representation can be obtained in the 2D case by using the orthonormal arrays (2.57).

4.4 Discretization: from ill-posedness to ill-conditioning

A puzzling implication of the analysis performed in the previous section is that, even if equation (4.5) is ill-posed because, in general, its solution does not exist, its discrete version (4.6) is well-posed because it has a unique solution which depends continuously on the data. In this section we show that this solution is, in general, unacceptable from the physical point of view because it is completely corrupted by noise. To this purpose we must investigate error propagation from the data to the solution. The usual analysis proceeds as follows.

Let us consider a (small) variation $\delta \boldsymbol{g}$ of the discrete image $\boldsymbol{g}$. The corresponding variation of the solution $\delta \boldsymbol{f}^{\dagger}$ is given by

$$\delta \boldsymbol{f}^{\dagger} = \mathbf{A}^{-1} \delta \boldsymbol{g} \tag{4.21}$$

thanks to the linearity of equation (4.18). If we use the bound (2.73) we have

$$\|\delta \boldsymbol{f}^{\dagger}\| \leq \frac{1}{\hat{K}_{min}} \|\delta \boldsymbol{g}\| \tag{4.22}$$

where $\hat{K}_{min}$ is now the minimum value of $|\hat{K}_{k,l}|$. On the other hand, from equation (4.6) and the bound (2.67) we get

$$\|\boldsymbol{g}\| \leq \hat{K}_{max} \|\boldsymbol{f}^{\dagger}\| \tag{4.23}$$

where $\hat{K}_{max}$ is now the maximum value of $|\hat{K}_{k,l}|$. By combining the two bounds we obtain

$$\frac{\|\delta \boldsymbol{f}^{\dagger}\|}{\|\boldsymbol{f}^{\dagger}\|} \leq \frac{\hat{K}_{max}}{\hat{K}_{min}} \frac{\|\delta \boldsymbol{g}\|}{\|\boldsymbol{g}\|}. \tag{4.24}$$

The quantity

$$\alpha = \frac{\hat{K}_{max}}{\hat{K}_{min}} \tag{4.25}$$

is called the *condition number* of the problem and is the quantity which controls error propagation from the data to the solution.

Let us first remark that the inequality (4.24) cannot be improved because, in some cases, equality holds true. To this purpose let us consider, for simplicity, the 1D case and let us use the expansions of $\boldsymbol{g}$ and $\boldsymbol{f}$ in terms of the orthonormal vectors (2.46) (see equations (2.48) and (2.49)). Moreover let us also assume that $\hat{K}_{max} = |\hat{K}_0|$ (i.e. the maximum value of the TF corresponds to the zero frequency component) while $\hat{K}_{min} = |\hat{K}_{N/2}|$ (i.e. the minimum value of the TF corresponds to the highest frequency component). This assumption is not very restrictive because very often in the practical problems the maximum corresponds to a low-frequency component while the minimum corresponds to a high-frequency component.

We consider now the following particular case: a data vector $\boldsymbol{g}$ such that only its component associated with $\boldsymbol{v}_0$ is different from zero and an error vector $\delta \boldsymbol{g}$ such that only its component associated with $\boldsymbol{v}_{N/2}$ is different from zero:

$$\boldsymbol{g} = \frac{1}{\sqrt{N}} \hat{g}_0 \boldsymbol{v}_0, \quad \delta \boldsymbol{g} = \frac{1}{\sqrt{N}} \delta \hat{g}_{N/2} \boldsymbol{v}_{N/2}. \tag{4.26}$$

The corresponding solution $\boldsymbol{f}^\dagger$ and solution error $\delta \boldsymbol{f}^\dagger$ are given by

$$\boldsymbol{f}^\dagger = \frac{1}{\sqrt{N}} \frac{\hat{g}_0}{\hat{K}_0} \boldsymbol{v}_0, \quad \delta \boldsymbol{f}^\dagger = \frac{1}{\sqrt{N}} \frac{\delta \hat{g}_{N/2}}{\hat{K}_{N/2}} \boldsymbol{v}_{N/2} \tag{4.27}$$

so that

$$\|\delta \boldsymbol{f}^\dagger\| = \frac{1}{\sqrt{N}} \|\delta \boldsymbol{g}\| / \hat{K}_{min}, \ \|\boldsymbol{f}^\dagger\| = \frac{1}{\sqrt{N}} \|\boldsymbol{g}\| / \hat{K}_{max}. \tag{4.28}$$

These equations imply that, for this particular example, equality holds true in equation (4.24). This example, however, makes clear that even if it is not possible to improve the inequality (4.24), this inequality can be pessimistic in a practical situation whenever this does not correspond exactly to the limiting case considered above.

The second remark is that, *when discretizing an ill-posed problem, the condition number of the corresponding discrete problem can be very large and also extremely large.* Indeed, as follows from the discussion of section 2.6, the values of the discrete TF are approximations of sampled values of $\hat{K}(\omega)$, which can be zero at certain frequencies and, in any case, tend to zero when $|\omega| \to \infty$. Moreover, we expect that, if we discretize more accurately the problem, we will obtain even smaller values of $\hat{K}_{min}$. Consider, for example, a PSF with a Gaussian shape, i.e. $K(x) = (2\pi)^{-1/2} \exp(-x^2/2)$. If we sample this function in 32 points in the interval $[-3, 3]$ and compute the DFT of these samples, we find that the value of the condition number (4.25) is $\alpha = 4.35 \times 10^3$. Then, if we sample the same function in 128 points in the interval $[-6, 6]$ and compute again the DFT of these samples, we find that the condition number is now $\alpha = 3.42 \times 10^7$. In conclusion, *the finer the discretization of the ill-posed problem, the larger the condition number of the corresponding discrete problem.*

A problem with a large condition number is called *ill-conditioned* while a problem whose condition number is close to one is called *well-conditioned.* A well-conditioned problem is, for instance, a problem of image restoration where the TF is a phase factor, i.e. $\hat{K}_m = \exp(i\phi_m)$ with ϕ_m real. Examples of such problems will be discussed in the next section.

The previous analysis implies that when discretizing an ill-posed problem we usually get an ill-conditioned problem and, in fact, a very ill-conditioned one if the discretization is very accurate. If the discretization is rough, the discrete problem can be moderately ill-conditioned and, possibly, nearly well-conditioned if the discretization is very rough. It should be obvious that a large condition number implies *numerical instability*: for instance, if $\alpha = 10^6$, a relative error on the data of the order of 10^{-6} may imply an error of 100% on the solution. Therefore we see that *continuous dependence of the solution on the data is necessary but not sufficient to guarantee numerical stability.*

An example of the numerical instability of the inverse filter is provided in figure 4.1. In this figure we give two restorations of the object of figure 3.4. In (a) we reproduce the object while in (b) we give its image corrupted by linear

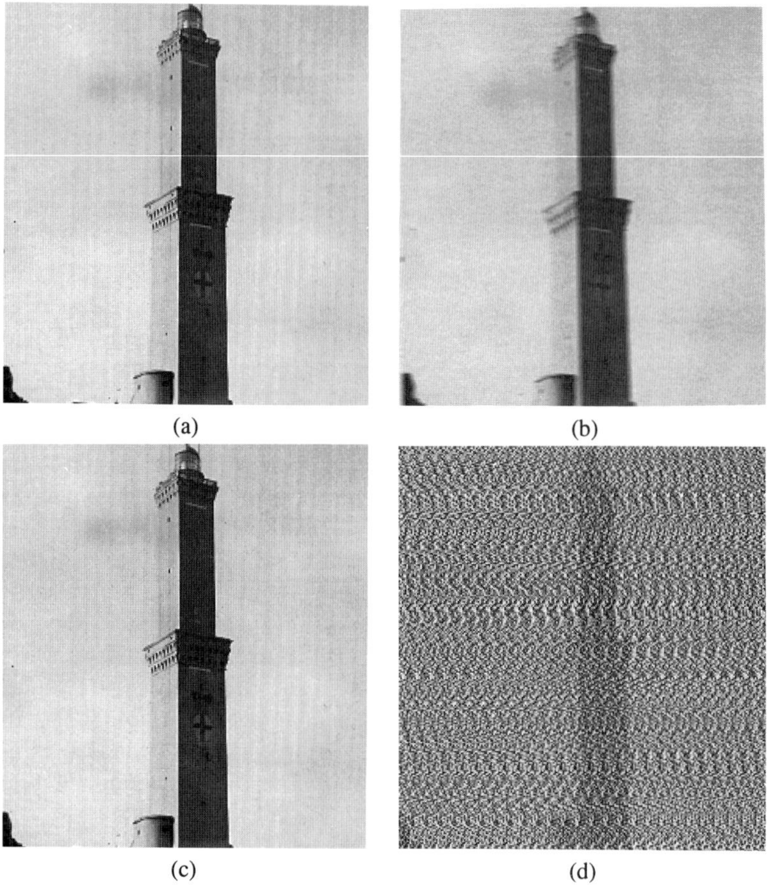

Figure 4.1. (a) The object of figure 3.4; (b) noise-free image corrupted by linear motion blur; (c) solution obtained by inverse filtering in the noise-free case; (d) solution obtained by inverse filtering in the noisy case (white Gaussian noise added to the image in (b))

motion blur. If no noise is added to this image, the result obtained by applying the inverse filtering is shown in (c). It is evident that the restoration is quite good. If white Gaussian noise is added to the image in (b) and inverse filtering is applied again, the result is shown in (d). It is obvious that the restoration is quite bad, even if it is difficult to represent graphically its main feature: large positive and negative values in adjacent pixels. In fact, instability appears, in general, in the form of wild oscillations superimposed on the image since it is generated mainly by the noise corrupting the high-frequency components of the image.

In the case of synthetic data, where the undistorted object $f^{(0)}$ is available (see the discussion in chapter 12) a measure of the improvement introduced by the process of restoration is given by the so-called *Mean Square Error Improvement Factor* (MSEIF) defined by

$$\text{MSEIF} = 20\ \log_{10} \frac{\|g - f^{(0)}\|}{\|f - f^{(0)}\|} (\text{dB}) \tag{4.29}$$

where $f^{(0)}$ is the object, g is the noisy image and f is the restored image. In the case of figure 4.1 we have MSEIF = 88 dB for the restoration obtained from the noise-free image and MSEIF = −19 dB for the restoration from the noisy image. In such a case the inverse filtering has degraded the quality of the image.

Remark 4.2. The condition number (4.25) quantifies error propagation not only in the solution of the inverse problem but also in the solution of the direct one, which consists of computing the image $g^{(0)} = \mathbf{A}f^{(0)}$. Indeed inequality (4.22) holds true when $\delta f^{\dagger}$ and δg are replaced respectively by $f^{(0)}$ and $g^{(0)}$, while inequality (4.23) holds true when g and $f^{\dagger}$ are replaced by $\delta g^{(0)}$ and $\delta f^{(0)}$. As a consequence we obtain

$$\frac{\|\delta g^{(0)}\|}{\|g^{(0)}\|} \leq \frac{\hat{K}_{max}}{\hat{K}_{min}} \frac{\|\delta f^{(0)}\|}{\|f^{(0)}\|}. \tag{4.30}$$

We should conclude that, in the case of a large condition number, the direct problem is affected by uncontrolled error propagation from the data, $f^{(0)}$, to the solution, $g^{(0)}$, even if this result seems to be in conflict with the everyday experience that the image $g^{(0)}$ can be accurately computed also when the object $f^{(0)}$ is affected by errors.

The inequality (4.30) can be understood by looking at the case where equality holds true. Under the same assumptions used for equation (4.24), we have equality in equation (4.30) when $f^{(0)}$ is proportional to $v_{N/2}$ (and therefore is a rapidly oscillating vector) while $\delta f^{(0)}$ is proportional to v_0 (and therefore is a constant vector representing, for instance, a systematic error); indeed, we have $\|g^{(0)}\| = \hat{K}_{min}\|f^{(0)}\|$ and $\|\delta g^{(0)}\| = \hat{K}_{max}\|\delta f^{(0)}\|$.

In conclusion, uncontrolled error propagation appears in the solution of the inverse problem when the noise-free image is smooth and the error is rapidly varying (and this is precisely the usual situation in practice) while it appears in the solution of the direct problem when the object is rapidly varying and the error is smooth (and this is a very unusual situation). When we compute the solution of the direct problem in the case of a reasonably smooth object, the small round-off errors (which can be rapidly oscillating) not only do not produce numerical instability, but are in fact reduced by the small values of $\hat{\mathbf{K}}$ at high frequencies.

This remark further emphasizes the fact that the condition number can be a pessimistic estimate of error propagation even if, in the case of the inverse problem, the practical situation is close to the situation corresponding to equality in equation (4.24). A better estimate of error propagation is the so-called Average Relative Error Magnification Factor (AREMF) introduced by Twomey [3]. For image restoration it is given by

$$\beta = \frac{1}{N}\left(\sum_{m=0}^{N-1}|\hat{K}_m|^2\right)^{1/2}\left(\sum_{m=0}^{N-1}|\hat{K}_m|^{-2}\right)^{1/2}. \tag{4.31}$$

This formula applies to the 1D case. A more general expression is the following one

$$\beta = \frac{1}{N}\|\boldsymbol{K}\|\;\|\boldsymbol{K}^{-1}\| \tag{4.32}$$

where the Parseval equality (2.41) has been used as well as the definition (4.19) of the inverse PSF.

For example in the case of the Gaussian PSF considered above when $\alpha = 4.35 \times 10^3$ we have $\beta = 6.25 \times 10^2$ and when $\alpha = 3.42 \times 10^7$ we have $\beta = 1.83 \times 10^6$.

4.5 Bandlimited systems: least-squares solutions and generalized solution

In section 4.2 we proved that, if the imaging system is bandlimited, then the solution of the image deconvolution problem is not unique as a consequence of the existence of invisible objects. We discuss now the existence of the solution and we introduce a modification of the concept of solution which is very useful in many circumstances.

Let us consider first equation (4.5) and its version in terms of Fourier transforms, equation (4.7). When ω does not belong to the band $\mathcal{B}$ of the instrument, the first member of this equation is zero because $\hat{K}(\omega)$ is zero. On the other hand the second member $\hat{g}(\omega)$ is given by equation (4.12) and therefore, in general, is not zero because the noise term $\hat{w}(\omega)$, due to the recording process, may be different from zero. Equation (4.7) then becomes $0 = \hat{w}(\omega)$ at these frequencies and therefore it is inconsistent, i.e. the solution of equation (4.5) does not exist. The function $\hat{w}_{out}(\omega)$, which is zero when ω is in $\mathcal{B}$ and coincides with $\hat{w}(\omega)$ when ω is not in $\mathcal{B}$, is called the *out-of-band noise*. We notice that a realization of the out-of-band noise is also an invisible object, in the sense that, if it is imaged by the bandlimited imaging system, its image should be zero. Therefore, the out-of-band noise is always orthogonal to the noise-free image, i.e. the image which is in the range of the operator A (see the remarks at the end of section 2.3) and, more generally, to all bandlimited functions with band $\mathcal{B}$.

If the non-existence of the solution is only due to out-of-band noise, then it is easy to remove this obstacle. We introduce the projection operator over the subspace of the bandlimited functions whose band coincides with or is contained in the band $\mathcal{B}$ of the imaging system (see appendix B for a short discussion of projection operators)

$$(P^{(\mathcal{B})}g)(\boldsymbol{x}) = \frac{1}{(2\pi)^q}\int \chi_{\mathcal{B}}(\boldsymbol{\omega})\hat{g}(\boldsymbol{\omega})e^{i\boldsymbol{x}\cdot\boldsymbol{\omega}}d\boldsymbol{\omega} \tag{4.33}$$

where $\chi_{\mathcal{B}}(\boldsymbol{\omega})$ is the characteristic function of the set $\mathcal{B}$. The operator $P^{(\mathcal{B})}$ can be called the *bandlimiting operator*. If the function $g(\boldsymbol{x})$ is not bandlimited, it is obvious that $P^{(\mathcal{B})}g$ is bandlimited and orthogonal to the out-of-band noise. Then, a quite natural idea is to modify equation (4.5) by suppressing the out-of-band noise by the use of the operator $P^{(\mathcal{B})}$, i.e. to replace equation (4.5) by the following one

$$Af = P^{(\mathcal{B})}g. \tag{4.34}$$

The FT of both members of this equation are zero outside the band while on the band we re-obtain equation (4.7). At this point one can repeat the discussion of section 4.3 and conclude that a solution of equation (4.34) exists if and only if the following condition is satisfied

$$\int_{\mathcal{B}} |\frac{\hat{g}(\boldsymbol{\omega}}{\hat{K}(\boldsymbol{\omega})}|^2 d\omega < \infty \tag{4.35}$$

which is analogous to condition (4.14), the integration now being restricted to the band $\mathcal{B}$ of the imaging system.

If condition (4.35) is satisfied for any g, then equation (4.34) has a solution for any g. If we look at the examples of chapter 3, we find that condition (4.35) is satisfied for any g in the case of a diffraction-limited system with coherent illumination (with or without aberrations) as well as in the case of far-field acoustic holography. In all these cases we have $|\hat{K}(\boldsymbol{\omega})| = 1$. On the other hand, if the bandlimited PSF is integrable, then thanks to the Riemann–Lebesgue theorem (see section 2.1), the TF $\hat{K}(\boldsymbol{\omega})$ is continuous and therefore, since it is zero outside the band it must be zero also on the boundary of the band. Examples are provided by diffraction-limited imaging systems in the case of incoherent illumination. In these cases, equation (4.35) is not satisfied by a noisy image because the noise, in general, is not zero on the boundary of the band. As a consequence equation (4.34) has no solution.

It is easy to prove that equation (4.34) is equivalent to the following one

$$A^*Af = A^*g \tag{4.36}$$

in the sense that any solution of this equation is also a solution of equation (4.34) and conversely. Indeed, by taking the Fourier transform of both sides of equation (4.36) we obtain

$$\left|\hat{K}(\boldsymbol{\omega})\right|^2 \hat{f}(\boldsymbol{\omega}) = \hat{K}^*(\boldsymbol{\omega})\hat{g}(\boldsymbol{\omega}); \tag{4.37}$$

when $\hat{K}(\omega) \neq 0$, $\hat{f}(\omega)$ is given by equation (4.11) while, when $\hat{K}(\omega) = 0$, the equation (4.37) is identically satisfied and $\hat{f}(\omega)$ is arbitrary. These are precisely the properties of the solutions of equation (4.34)

In appendix E we prove that equation (4.36) is the Euler equation of the variational problem

$$\|Af - g\| = minimum \tag{4.38}$$

and therefore its solutions are the objects which provide the best approximation of the given image g. Any function f which solves this minimization problem is also called a *least-squares solution.* The name comes from the fact that, in the discrete case, such a solution minimizes the sum of the squares of the differences between the components of $\mathbf{A}\boldsymbol{f}$ and the components of $\boldsymbol{g}$.

Since any least-squares solution is also a solution of equation (4.34), from the previous analysis it follows that *least-squares solutions exist for a given g if and only if g satisfies condition (4.35).* The solution of the least-squares problem (4.38), however, is not unique because to any given least-squares solution we can add an arbitrary invisible object and obtain in this way another least-squares solution. If we do not have additional information on the object, it can be reasonable to consider the unique least-squares solution which is zero outside the band of the imaging system. When condition (4.35) is satisfied, this particular least-squares solution, which is called the *generalized solution* and denoted by $f^{\dagger}(\boldsymbol{x})$, is given by

$$f^{\dagger}(\boldsymbol{x}) = \frac{1}{(2\pi)^q} \int_{\mathcal{B}} \frac{\hat{g}(\boldsymbol{\omega})}{\hat{K}(\boldsymbol{\omega})} e^{i\boldsymbol{x}\cdot\boldsymbol{\omega}} d\boldsymbol{\omega}. \tag{4.39}$$

In appendix E we show that *the generalized solution is the unique least-squares solution of minimal norm. It is also the unique least-squares solution orthogonal to the subspace of all invisible objects.*

If the function $1/\hat{K}(\omega)$ is not bounded as a consequence of zeros of $\hat{K}(\omega)$ inside or at the boundary of the band $\mathcal{B}$, the generalized solution (4.39) does not exist for arbitrary $\hat{g}(\omega)$—and, in general, does not exist for noisy $\hat{g}(\omega)$—so that *the problem of determining the generalized solution is ill-posed.* However, equation (4.39) defines a linear mapping from the set of all noise-free images into the set of all bandlimited objects. This linear mapping is called the *generalized inverse operator* of the operator A and denoted by $A^{\dagger}$, so that $f^{\dagger} = A^{\dagger}g$. If we consider a noise-free image $g = Af$, then, from equation (4.39) with $\hat{g}(\omega) = \hat{K}(\omega)\hat{f}(\omega)$ we find that

$$A^{\dagger}g = P^{(\mathcal{B})}f \tag{4.40}$$

where $P^{(\mathcal{B})}$ is the bandlimiting operator defined in equation (4.33).

The problem of determining the generalized solution is well posed in the important cases, discussed in chapter 3, where $|\hat{K}(\omega)| = 1$ on the band $\mathcal{B}$. In these cases, if we introduce the *generalized inverse PSF* defined by

$$K^{\dagger}(\boldsymbol{x}) = \frac{1}{(2\pi)^q} \int_B \frac{1}{\hat{K}(\boldsymbol{\omega})} e^{i\boldsymbol{x}\cdot\boldsymbol{\omega}} d\boldsymbol{\omega} \tag{4.41}$$

we can write equation (4.39) as follows

$$f^{\dagger} = K^{\dagger} * g. \tag{4.42}$$

We find that the generalized inverse is a linear and continuous operator in the space of the square-integrable functions which is given by

$$A^{\dagger} g = K^{\dagger} * g. \tag{4.43}$$

The extension of this analysis to the discrete case is quite easy. The band is now the set B of the pairs of indices $\{k, l\}$ such that $\hat{K}_{k,l} \neq 0$. Then by an easy repetition of the various steps discussed above one concludes that there exists a unique generalized solution given by

$$(\boldsymbol{f}^{\dagger})_{m,n} = \frac{1}{N^2} \sum_B \frac{\hat{g}_{k,l}}{\hat{K}_{k,l}} \exp\left[i\frac{2\pi}{N}(km + ln)\right] \tag{4.44}$$

the difference with respect to equation (4.17) being that the summation is only extended to the pairs of indices in the set B. This generalized solution can also be written in terms of a *generalized inverse matrix* $\mathbf{A}^{\dagger}$ (also called *Moore–Penrose inverse matrix*)

$$\boldsymbol{f}^{\dagger} = \mathbf{A}^{\dagger}\boldsymbol{g}. \tag{4.45}$$

The matrix $\mathbf{A}^{\dagger}$ is the 2D cyclic matrix associated with the 2D array defined as in equation (4.19) but with the summation restricted to the set B. This array will be denoted by $\mathbf{K}^{\dagger}$ and will be called the *generalized inverse filter* or the *generalized inverse PSF*.

Error propagation from $\boldsymbol{g}$ to $\boldsymbol{f}^{\dagger}$ can be analysed along the lines indicated in section 4.4 and the result is that numerical stability is still controlled by the condition number (4.25), $\hat{K}_{min}$ being now the minimum of $|\hat{K}_{k,l}|$ over the set B.

Let us consider now a few examples where the generalized solution can be used. The first one is that of an ideal diffraction-limited system in the case of coherent illumination. The TF of the imaging system is simply the characteristic function of the band $\mathcal{B}$ of the instrument and therefore from equation (4.41) with $\hat{K}(\boldsymbol{\omega}) = \chi_{\mathcal{B}}(\boldsymbol{\omega})$ we see that $K^{\dagger}(\boldsymbol{x})$ coincides with the PSF of the imaging system. In such a case the generalized solution is simply the noisy image without the out-of-band noise.

More interesting results can be obtained in the case of aberrations or in the case of far-field acoustic holography. In the latter case, from equation (3.57) and the far-field approximation, we deduce that the generalized inverse PSF is given by

$$K^{\dagger}(\boldsymbol{x}) = \frac{1}{(2\pi)^2} \int_{|\boldsymbol{\omega}|\leq k} \exp(-ia(k^2 - |\boldsymbol{\omega}|^2)^{1/2}) e^{i\boldsymbol{x}\cdot\boldsymbol{\omega}} d\boldsymbol{\omega}. \tag{4.46}$$

The operator defined by the convolution of the image g with the generalized inverse PSF is usually called the *back-propagation operator* because it transforms the field amplitude on the plane $x_3 = a$ into the field amplitude on the plane $x_3 = 0$. In figure 4.2 we present the result of a numerical simulation. The original object is a binary object consisting of a square grid, with overall dimensions $10\lambda \times 10\lambda$. The size of the horizontal and vertical bars is λ and therefore is greater than the Rayleigh resolution distance (see section 3.4). The receiving plane is located at a distance 5λ from the source plane. At this distance the effect of evanescent waves is negligible. The image $g^{(0)}(\boldsymbol{x})$, which is the complex amplitude of the acoustic field in the receiving plane, has been corrupted by white Gaussian noise (SNR $\simeq$ 50 (dB)). The picture of $|g(\boldsymbol{x})|$, given in figure 4.2(c), is a picture of the shadow of the object. The generalized solution corresponding to the noisy image $g(\boldsymbol{x})$ is given in figure 4.2(d). The result is quite satisfactory and the reason is that the object considered does not contain details smaller than the Rayleigh resolution distance which is of the order of $\lambda/2$.

4.6 Approximate solutions and the use of *a priori* information

The analysis of the previous sections has shown that the solution of an ill-posed problem may not exist while the solution (or generalized solution) of the corresponding discrete problem always exists but may be completely corrupted by noise and therefore completely deprived of physical meaning. It should be clear, however, that the image g must contain some information about the object $f^{(0)}$ and therefore that it must be possible to find some estimate of $f^{(0)}$ better than g itself. It should also be clear that, to make some progress in this direction, one must modify the concept of *solution* of the problem. A first step in this direction has been already performed in the case of bandlimited imaging systems where the concept of a solution has been replaced by that of a least-squares solution and of a generalized solution. However this step is not sufficient, because the generalized solution can also be numerically unstable.

Since the solution is the object which reproduces exactly the noisy data, i.e. the noisy image g (the generalized solution is the object which reproduces the data in the best possible way), and since this solution is unacceptable, the only possible way is to look for solutions which reproduce only approximately the noisy image, i.e. for *approximate solutions*. We point out that the true object $f^{(0)}$ is such an approximate solution because, as a consequence of the noise, it does reproduce the noisy image g not exactly but only within the noise term w.

If f is an approximate solution, then the image Af associated with f does not coincide in general with g; the difference between the computed image Af and the measured image g is called *residual* and denoted by r

$$r = g - Af. \tag{4.47}$$

The square norm of the residual will be called the *discrepancy* of the object f

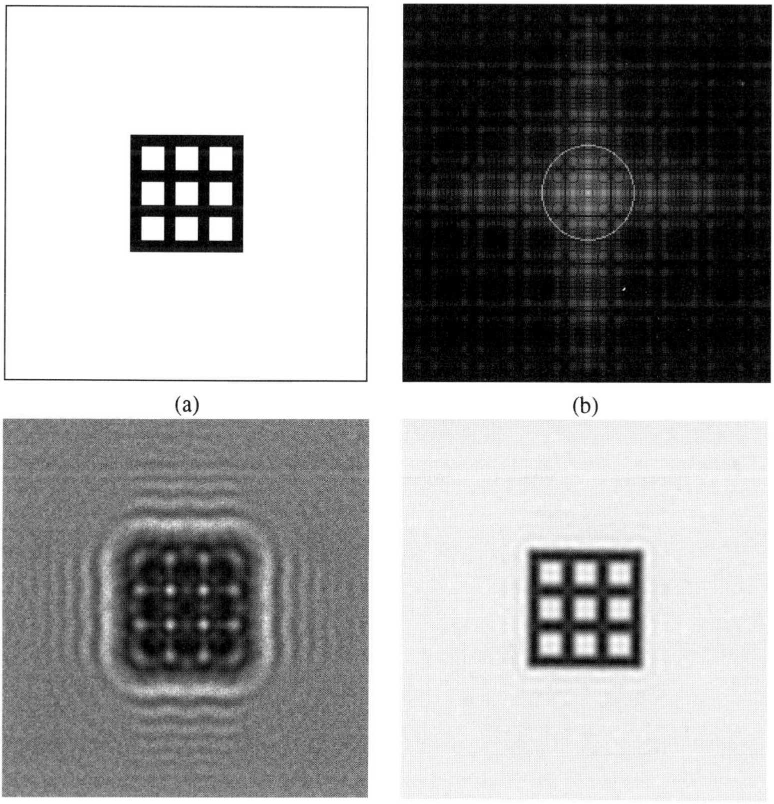

(a) (b) (c) (d)

Figure 4.2. Example of the use of the generalized solution in far-field acoustic holography: (a) the object, $10\lambda \times 10\lambda$ wide; (b) modulus of the FT of the object (the white circle indicates the band of the far-field data); (c) modulus of the noisy image at a distance 5λ from the source plane; (d) the reconstruction obtained by means of the generalized inverse (4.42).

and will be denoted by $\varepsilon^2(f; g)$. In the discrete case the norm of the residual $\boldsymbol{r}$ is the absolute root-mean-square (rms) error we commit when estimating $\boldsymbol{g}$ by means of $\mathbf{A}\boldsymbol{f}$.

The point now is to understand the structure of the set of all approximate solution compatible with the noisy image within a given noise level. In order to simplify the analysis we only consider the discrete case. Then, let us assume that we know an estimate ε of the absolute rms-error affecting the noisy image $\boldsymbol{g}$. Under these conditions, the set of all approximate solutions compatible with the noisy image is the set of all objects $\boldsymbol{f}$ whose discrepancy does not exceed

ε^2, i.e.

$$\|\mathbf{A}f - g\|^2 \leq \varepsilon^2. \tag{4.48}$$

If ε^2 is a reliable estimate then it is obvious that both the true object $f^{(0)}$ and the unphysical solution, discussed in the previous sections, belong to this set, which therefore contains completely different elements. As a consequence, we expect that this set is extremely large and this result can be quantified by a more detailed analysis.

For the sake of simplicity we consider the 1D case assuming that all components of the TF, $\hat{K}_m$, are different from zero. In such a case the solution of the discrete problem is given by

$$(f^\dagger)_n = \frac{1}{N}\sum_{m=0}^{N-1} \frac{\hat{g}_m}{\hat{K}_m} \exp\left(i\frac{2\pi}{N}nm\right). \tag{4.49}$$

Then, using the Parseval equality for the DFT, we can write condition (4.48) as follows

$$\frac{1}{N}\sum_{m=0}^{N-1} |\hat{K}_m \hat{f}_m - \hat{g}_m|^2 \leq \varepsilon^2 \tag{4.50}$$

or also

$$\sum_{m=0}^{N-1} \left(\frac{|\hat{K}_m|^2}{N\varepsilon^2}\right) |\hat{f}_m - \frac{\hat{g}_m}{\hat{K}_m}|^2 \leq 1. \tag{4.51}$$

This condition defines, in the space of the DFT of the objects, the set of the interior points of an N-dimensional ellipsoid whose centre is the DFT of the solution (4.49) and whose half-axes have lengths

$$a_m = \frac{\sqrt{N}\varepsilon}{|\hat{K}_m|}. \tag{4.52}$$

Since the DFT defines a unitary transformation, in the space of the objects we still have an N-dimensional ellipsoid with centre $f^\dagger$ and half-axes in the directions of the vectors v_m, equation (2.46), whose lengths are given again by equation (4.52). This ellipsoid is the inverse image of a small ball of radius ε, centred at the noisy image g (see figure 4.3). The ratio between the maximum half-axis and the minimum half-axis of the ellipsoid is precisely the condition number introduced in section 4.4.

As we see, this set is too broad and it is hopeless to extract a useful estimate of $f^{(0)}$ from it without using additional information. Indeed when we say that the solution or generalized solution $f^\dagger$ is unacceptable from the physical point of view, some additional information on the object $f^{(0)}$ is implicit in the statement. In particular we know that the components of $f^\dagger$ are too large and too rapidly oscillating to correspond to a physical object.

This remark suggests a first remedy to the pathology of the problem. If we know an upper bound on the norm of the unknown object $f^{(0)}$, then we can

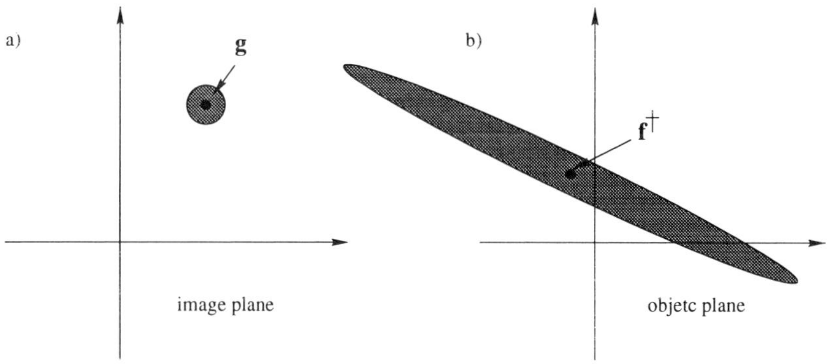

Figure 4.3. Two-dimensional representation of the situation described by equation (4.48). The set in (b) is the set of all objects which correspond to images in the small circle around g in (a).

look only for the objects f which are compatible both with the data and with this bound (see figure 4.4). More generally, any kind of property of the solution which can be expressed in a suitable mathematical form should be used. In the

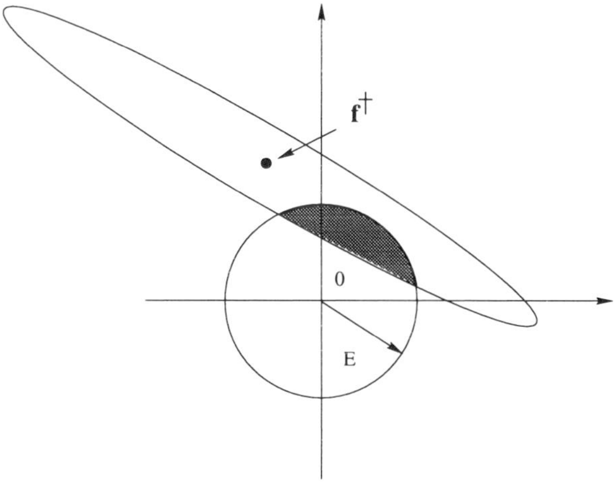

Figure 4.4. Two-dimensional representation of the effect of a bound on the norm of the unknown object. The estimates of the object compatible with the data and the bound E on their norm are the points in the intersection of the ellipse with the disc of radius E.

next section we will give a list of such possible properties. Here we merely discuss the basic idea.

The picture of figure 4.3 visualizes the fact that the image does not contain sufficient information on the object. A very large set of objects (side (b) of the figure) is compressed by the imaging system in a very small set of images (side (a) of the same figure). This loss of information must be compensated in some way by additional information, coming from other sources or from previous information we have on the object. Mathematically such information will be expressed in the form of constraints on the solution. This is usually called *a priori* information (or prior information) and, in a more or less explicit form, this is the basic principle on which all inversion methods rely. In figure 4.4 we see that this additional information can considerably reduce the uncertainty about the unknown object. The objects compatible with the noisy image are represented by the points interior to the ellipse while the objects compatible with the constraint are represented by the points interior to the circle of radius E. Any point of the shaded region may be considered as a satisfactory estimate of the unknown object because it is compatible both with the data and with the additional information.

Let us remark that information about the noise, in particular about its statistical properties, is also useful. This is the starting point of the maximum-likelihood methods which will be discussed in Chapter 7. However, it must be clear that information about noise can modify the structure of the set of objects compatible with the data but does not reduce in a significant way the lack of information about the object. This is an intrinsic property of the imaging system, i.e. of the PSF, and it cannot be compensated by information about the noise. *A priori* information about the object is also required and this is the main difference between the Wiener filter method and the maximum-likelihood methods, as discussed in chapter 7.

4.7 Constrained least-squares solutions

The most convenient way for expressing *a priori* information about the solution of the image restoration problem is to state that the object must belong to some given subset of the space of all possible objects. We give here some examples of sets which are used very frequently in image restoration and, more generally, in the solution of inverse problems. We consider mainly the case of functions but the analogous sets for discrete images can be derived in a straightforward way.

- The set of all functions whose 'energy', defined as the square of their L^2-norm, does not exceed a prescribed value

$$\|f\|^2 = \int |f(\boldsymbol{x})|^2 d\boldsymbol{x} \leq E^2. \tag{4.53}$$

  This is the ball, with radius E and the zero function as centre, in the space of all square-integrable functions.

- The set of all functions whose derivatives up to a certain order satisfy some prescribed bound. In general, if D is a differential operator (for instance, the Laplace operator, $D = \Delta$) then the set is defined by

$$\|Df\|^2 = \int |(Df)(x)|^2 dx \leq E^2. \tag{4.54}$$

  Examples will be considered in the next chapter.
- The set of all functions which are zero outside a bounded region $\mathcal{D}$. This constraint can be used in image restoration when we know that the object is localized in one or more small portions of its domain of definition.
- The set of all functions whose Fourier transform is zero outside a bounded domain $\mathcal{B}$. In signal processing this constraint is used for the extrapolation of bandlimited signals.
- The set of all functions which are non-negative. This is a constraint which is, in general, very important in image restoration.
- The set of all functions which coincide with a prescribed function on some bounded region $\mathcal{D}$ or which take some prescribed value in some points of the domain where they are defined.
- A set defined as in the previous case but with the Fourier transform at the place of the function itself.
- Any intersection of the previous sets.

A common feature of the above sets is that they are *closed* and *convex*. If we denote by $\mathcal{C}$ such a set, then closure means that, if we consider a sequence of functions f_n which belong to $\mathcal{C}$ and if this sequence is convergent to a function f, then also f belongs to $\mathcal{C}$. On the other hand convexity is a geometrical property which can be formulated as follows: if $f^{(1)}$ and $f^{(2)}$ are two functions in $\mathcal{C}$, then all functions f given by

$$f = tf^{(1)} + (1-t)f^{(2)}, \quad 0 \leq t \leq 1 \tag{4.55}$$

also belong to $\mathcal{C}$. This means that the segment of straight line joining $f^{(1)}$ to $f^{(2)}$ belongs also to $\mathcal{C}$ (see figure 4.5).

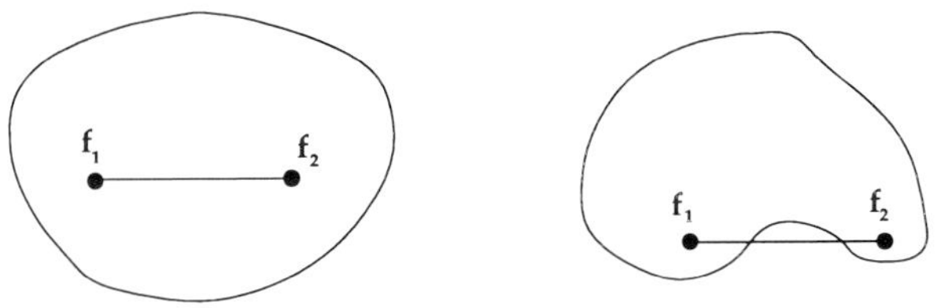

Figure 4.5. Two-dimensional example of a convex set (a) and of a non-convex set (b).

A basic property of a closed and convex set C in a space of functions where a scalar product has been defined is that, given an arbitrary element f of the space there exists a unique element f_C of C which has minimal distance from f [4] (see appendix F). This element is called the *convex projection* (or metric projection) of f onto C. The operator which associates to any object f its convex projection f_C is called projection operator onto the set C and denoted by P_C

$$P_C f = f_C. \tag{4.56}$$

This operator, which will be used in section 6.2, is, in general, nonlinear. For example, in the case where the set C is defined by the condition (4.53), the operator P_C is given by

$$P_C f = \frac{E}{\|f\|} f \tag{4.57}$$

while, in the case of the set of non-negative functions, the operator P_C is given by

$$(P_C f)(\boldsymbol{x}) = \begin{cases} f(\boldsymbol{x}) & \text{if } f(\boldsymbol{x}) > 0 \\ 0 & \text{if } f(\boldsymbol{x}) \le 0. \end{cases} \tag{4.58}$$

The effect of this projection operator is illustrated in figure 4.6.

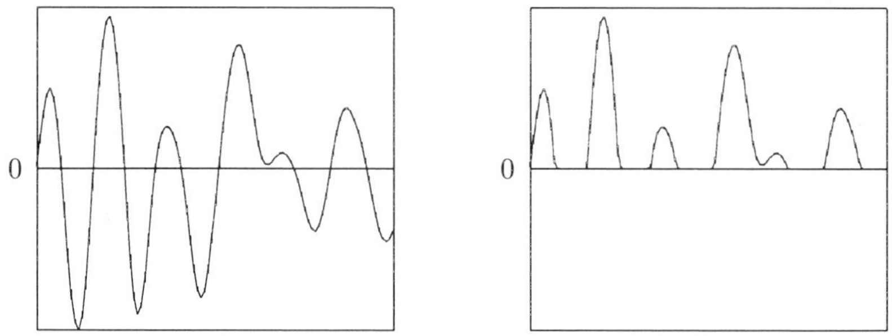

(a) (b)

Figure 4.6. Illustrating the effect of the projection onto the convex set of non-negative functions: (a) the original function; (b) the projected function.

If we know that the solution of the image restoration problem must belong to some closed and convex set C, it is quite natural to look, among all possible objects satisfying this condition, for that or those which minimize the discrepancy, i.e. for the solutions of the following constrained problem

$$\|Af - g\| = minimum, \quad f \in C. \tag{4.59}$$

Any solution of this problem is called a *constrained least-squares solution*.

As we will discuss in the following, this problem can be ill-posed or well-posed, according to properties of the set C and of the operator A. If it is well-posed, it provides a way for obtaining a reasonable solution to the problem of image deconvolution.

References

[1] Courant R and Hilbert D 1962 *Methods of Mathematical Physics* vol II (New York: Interscience) 227
[2] Lanczos C 1961 *Linear Differential Operators* (London: Van Nostrand) p 132
[3] Twomey S 1974 *Applied Optics* **13** 942
[4] Balakrishnan A V 1976 *Applied Functional Analysis* (New York: Springer)

Chapter 5

Regularization methods

The first approach to linear ill-posed problems based on constrained least-squares solutions was proposed by V.K. Ivanov in 1962 [1]. In the same year D.L. Phillips [2] introduced a method based on the determination of the smoothest approximate solution compatible with the data within a given noise level. One year later the Russian mathematician A.N. Tikhonov proposed, independently, a general approach termed *regularization* or the *regularization method* of ill-posed problems. This approach also provided a unification of the methods of Ivanov and Phillips.

The basic idea of regularization consists of considering a family of approximate solutions depending on a positive parameter called the *regularization parameter*. The main property is that, in the case of noise-free data, the functions of the family converge to the exact solution of the problem when the regularization parameter tends to zero. In the case of noisy data one can obtain an optimal approximation of the exact solution for a non-zero value of regularization parameter. Moreover for suitable values of the regularization parameter one recovers both the solutions of Ivanov and of Phillips.

In this chapter we present these regularization methods in the case of image deconvolution. For this particular problem it is easy to show that the most simple form of Tikhonov regularization is equivalent to a filtering of the solution or generalized solution discussed in the previous chapter. This remark suggests the introduction of a wider class of regularization methods called *spectral filtering*.

The description of regularization in terms of a global PSF and the problem of the choice of the regularization parameter are also discussed.

5.1 Least-squares solutions with prescribed energy

We call the *energy* of a given object $f(x)$ the square of its L^2-norm

$$E^2(f) = \|f\|^2 = \int |f(x)|^2 dx. \tag{5.1}$$

In some cases (for instance, in the case of coherent imaging) this quantity is just proportional to the physical energy of the signal $f(\boldsymbol{x})$.

If we recall that a *functional* is a mapping which associates a real (or complex) number to each object, then the *energy functional* is the mapping which associates to each object the value of its energy. Analogously the *discrepancy functional* is the mapping which associates to each object the value of its discrepancy, which was defined in section 4.6 as the squared norm of the residual $r = g - Af$. This functional will be denoted by $\varepsilon^2(f; g)$.

Now a *least-squares solution with prescribed energy* is an object $\tilde{f}_1(\boldsymbol{x})$ which solves the following constrained least-squares problem: *find the minimum (or the minima) of the discrepancy functional*

$$\varepsilon^2(f; g) = \|Af - g\|^2 \tag{5.2}$$

in the set of all objects whose energy does not exceed a prescribed bound E^2

$$E^2(f) = \|f\|^2 \leq E^2. \tag{5.3}$$

This problem is a particular example of the problem (4.59) and is precisely the problem considered by Ivanov in the paper mentioned above, although in a more general setting.

The bound E^2 must be acceptable from the point of view of physics, i.e. it must be a reasonable value of the energy of the unknown object $f^{(0)}$. As discussed in chapter 4, the solution or generalized solution of equation (4.5) does not satisfy this constraint because of noise propagation, i.e.

$$E^2(f^\dagger) = \frac{1}{(2\pi)^q} \int_B \left| \frac{\hat{g}(\omega)}{\hat{K}(\omega)} \right|^2 d\omega > E^2 \tag{5.4}$$

(possibly $E^2(f^\dagger) = +\infty$). This fact implies that the unconstrained minimum of the discrepancy functional (5.2) is not interior to the ball defined by the inequality (5.3). If $E^2(f^\dagger) = +\infty$ the unconstrained minimum does not exist because, in such a case, no solution or least-squares solution exists, as explained in section 4.3 and section 4.5. It follows that the constrained minimum must stay on the surface of this ball (a pictorial justification of this point is given in figure 5.1) and therefore the inequality (5.3) can be replaced by the following equality

$$E^2(f) = E^2. \tag{5.5}$$

The problem of minimizing the functional (5.2) over the set of functions satisfying condition (5.5) can be solved by means of the method of *Lagrange multipliers*. According to this method we introduce a functional which is a linear combination of the functional to be minimized and of the functional expressing the constraint,

$$\Phi_\mu(f; g) = \varepsilon^2(f; g) + \mu E^2(f) = \|Af - g\|^2 + \mu\|f\|^2 \tag{5.6}$$

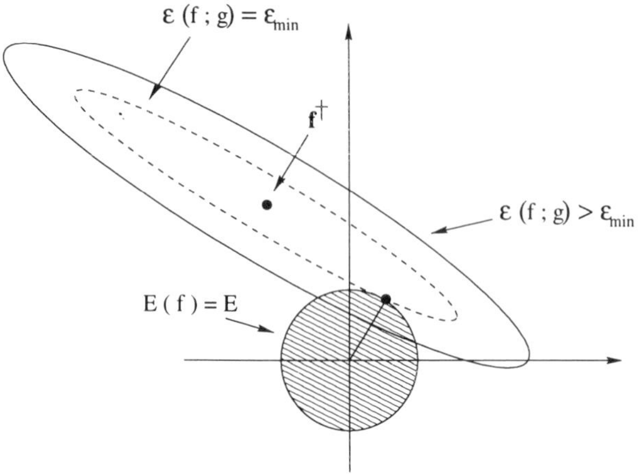

Figure 5.1. Two-dimensional picture showing that the minimum of the discrepancy is reached on the surface of the ball of radius E. The ellipses are the level curves of the discrepancy functional. The centre of these ellipses is the solution provided by the inverse filtering method.

where μ is an arbitrary real and positive number which is called the *Lagrange multiplier*. Then the proof of existence and uniqueness of the solution of the problem formulated above follows from the proof of the following points:

- for any $\mu > 0$, there exists a unique function $f_\mu(\boldsymbol{x})$ which minimizes the functional (5.6);
- there exists a unique value of μ, say $\mu_1 = \mu_1(E)$, such that

$$E(f_{\mu_1}) = E. \tag{5.7}$$

If these results are true, then the object $\tilde{f}_1 = f_{\mu_1}$ is the unique object which minimizes the discrepancy functional (5.2) under the condition (5.5). Once the two points above are proved, we can conclude that: *for any image g there exists a unique object $\tilde{f}_1$ which is the unique least-squares solution with prescribed energy E^2.*

For the problem of image deconvolution the previous points can be proved in a simple way. We assume that the imaging system is bandlimited and that $\mathcal{B}$ is its band while $\bar{\mathcal{B}}$ is the set of the out-of-band frequencies. This is not restrictive since a system which is not bandlimited can be viewed as the particular case where $\mathcal{B}$ coincides with the whole frequency space and $\bar{\mathcal{B}}$ is empty.

By the use of the Parseval equality we write the functional (5.6) in the following form (q-dimensional images):

$$\Phi_\mu(f;g) = \frac{1}{(2\pi)^q}\int_{\mathcal{B}} |\hat{K}(\omega)\hat{f}(\omega) - \hat{g}(\omega)|^2 d\omega + \frac{1}{(2\pi)^q}\int_{\bar{\mathcal{B}}} |\hat{g}(\omega)|^2 d\omega + \frac{\mu}{(2\pi)^q}\int_{\mathcal{B}} |\hat{f}(\omega)|^2 d\omega + \frac{\mu}{(2\pi)^q}\int_{\bar{\mathcal{B}}} |\hat{f}(\omega)|^2 d\omega. \tag{5.8}$$

The first two terms correspond to the discrepancy functional (the second is obtained by observing that $\hat{K}(\omega) = 0$ when ω is not in the band) while the two other terms correspond to the decomposition of the total energy of $f(\boldsymbol{x})$ into the sum of the in-band and out-of-band energy.

The integrands of the first and of the third term of equation (5.8) can be rearranged as follows

$$\begin{aligned} &|\hat{K}(\omega)\hat{f}(\omega) - \hat{g}(\omega)|^2 + \mu|\hat{f}(\omega)|^2 \\ &= (|\hat{K}(\omega)|^2 + \mu)|\hat{f}(\omega)|^2 - 2Re\{\hat{K}(\omega)\hat{f}(\omega)\hat{g}^*(\omega)\} + |\hat{g}(\omega)|^2 \\ &= (|\hat{K}(\omega)|^2 + \mu)\left|\hat{f}(\omega) - \frac{\hat{K}^*(\omega)\hat{g}(\omega)}{|\hat{K}(\omega)|^2 + \mu}\right|^2 + \frac{\mu}{|\hat{K}(\omega)|^2 + \mu}|\hat{g}(\omega)|^2 \end{aligned} \tag{5.9}$$

so that the functional (5.8) takes the following form

$$\begin{aligned} \Phi_\mu(f;g) &= \frac{1}{(2\pi)^q}\int_{\mathcal{B}} (|\hat{K}(\omega)|^2 + \mu)\left|\hat{f}(\omega) - \frac{\hat{K}^*(\omega)\hat{g}(\omega)}{|\hat{K}(\omega)|^2 + \mu}\right|^2 d\omega \\ &+ \frac{1}{(2\pi)^q}\int_{\bar{\mathcal{B}}} |\hat{f}(\omega)|^2 d\omega + \frac{\mu}{(2\pi)^q}\int_{\mathcal{B}} \frac{|\hat{g}(\omega)|^2}{|\hat{K}(\omega)|^2 + \mu} d\omega \\ &+ \frac{1}{(2\pi)^q}\int_{\bar{\mathcal{B}}} |\hat{g}(\omega)|^2 d\omega. \end{aligned} \tag{5.10}$$

From this expression it follows that, for any $\mu > 0$, there exists a unique function $f_\mu(\boldsymbol{x})$ which minimizes the functional. Its FT can be obtained by making the first two terms equal to zero and therefore $\hat{f}_\mu(\omega) = 0$ when ω is in $\bar{\mathcal{B}}$ and

$$\hat{f}_\mu(\omega) = \frac{\hat{K}^*(\omega)}{|\hat{K}(\omega)|^2 + \mu}\hat{g}(\omega) \tag{5.11}$$

when ω is in $\mathcal{B}$. Since $\hat{K}(\omega)$ is zero in $\bar{\mathcal{B}}$, the last expression can be used for any ω and therefore

$$f_\mu(\boldsymbol{x}) = \frac{1}{(2\pi)^q}\int \frac{\hat{K}^*(\omega)}{|\hat{K}(\omega)|^2 + \mu}\hat{g}(\omega)e^{i\boldsymbol{x}\cdot\omega} d\omega. \tag{5.12}$$

We point out that *if the imaging system is bandlimited then $f_\mu(\boldsymbol{x})$ is also bandlimited and has the same band of the system.* We also observe that the minimum value of the functional $\Phi_\mu(f; g)$, which is the sum of the last two terms in equation (5.10) can be written in the following form

$$\Phi_\mu(f_\mu; g) = \frac{\mu}{(2\pi)^q} \int \frac{|\hat{g}(\omega)|^2}{|\hat{K}(\omega)|^2 + \mu} d\omega \tag{5.13}$$

where we have used again the fact that $\hat{K}(\omega) = 0$ when ω is in $\bar{\mathcal{B}}$.

The function $f_\mu(\boldsymbol{x})$ has the following properties:

- it is square integrable for any g and any $\mu > 0$;
- it depends continuously on g;
- it is orthogonal to the subspace of all invisible objects.

The first two properties are a simple consequence of the following elementary inequality

$$\frac{\xi^2}{(\xi^2 + \mu)^2} = \frac{\xi^2}{(\xi^2 + \mu)} \frac{1}{(\xi^2 + \mu)} \leq \frac{1}{\mu}. \tag{5.14}$$

From this inequality with $\xi = |\hat{K}(\omega)|$ and from the Parseval equality we get

$$\begin{aligned} E^2(f_\mu) = \int |f_\mu(\boldsymbol{x})|^2 d\boldsymbol{x} &= \frac{1}{(2\pi)^q} \int_{\mathcal{B}} \frac{|\hat{K}(\omega)|^2}{(|\hat{K}(\omega)|^2 + \mu)^2} |\hat{g}(\omega)|^2 d\omega \\ &\leq \frac{1}{\mu} \frac{1}{(2\pi)^q} \int_{\mathcal{B}} |\hat{g}(\omega)|^2 d\omega \leq \frac{1}{\mu} \|g\|^2. \end{aligned} \tag{5.15}$$

This inequality proves the first two statements above. As concerns the third one, it is obvious because, as we already remarked, $f_\mu(\boldsymbol{x})$ is band-limited with the same band of the imaging system and therefore it is orthogonal to all invisible objects.

Having proved that the functional (5.6) has a unique minimum for any $\mu > 0$, in order to complete the program of the method of Lagrange multipliers we still have to show that there exists a unique value of μ satisfying equation (5.7). Now, from the expression of $E(f_\mu)$ provided by the Parseval equality—see equation (5.15)—it follows that $E(f_\mu)$ is a decreasing function of μ. Moreover it tends to $E(f^\dagger) > E$ (possibly, we have $E(f^\dagger) = \infty$) when $\mu \to 0$ and tends to zero when $\mu \to \infty$. It follows that there exists a unique value of μ, $\mu_1 = \mu_1(E)$, such that $E(f_{\mu_1}) = E$. In figure 5.2 we give the plot of the function $E(f_\mu)$ in the case of the numerical example discussed in section 5.6. We conclude that, for any given g, there exists a unique least-squares solution with prescribed energy.

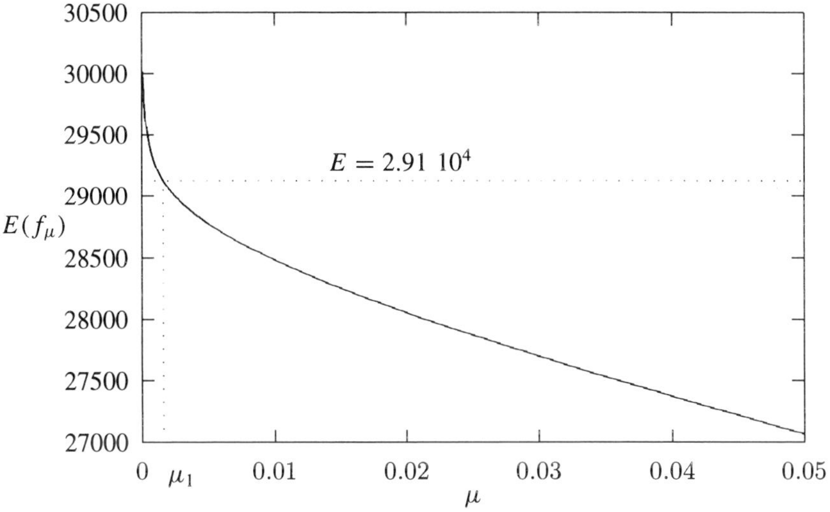

Figure 5.2. Behaviour of $E(f_\mu)$ as a function of μ in the case of the example of section 5.6. The point μ_1 where $E(f_{\mu_1}) = E$ is also indicated. In such a case we have $\mu_1 = 1.6\ 10^{-3}$.

Remark 5.1. The continuous dependence of $\tilde{f}_1 = f_{\mu_1}$ on the data as well as its convergence to the exact generalized solution $(f^{(0)})^\dagger = P^{(\mathcal{B})} f^{(0)}$, as given in equation (4.40) ($P^{(\mathcal{B})}$ is the band-limiting operator defined in equation (4.33)), when the error on the data tends to zero are rather subtle questions and go beyond the scope of this book (see, for instance, [1] and [4]). It is possible to prove that $\|\tilde{f}_1 - P^{(\mathcal{B})} f^{(0)}\|$ tends to zero when $g \to g^{(0)}$, if and only if E is a precise estimate of the energy of $P^{(\mathcal{B})} f^{(0)}$, i.e. $\|P^{(\mathcal{B})} f^{(0)}\| = E$ [4]. If this energy has been overestimated, i.e. $\|P^{(\mathcal{B})} f^{(0)}\| < E$, then it is only possible to prove convergence in a weaker sense, namely that the scalar products $(\tilde{f}_1 - P^{(\mathcal{B})} f^{(0)}, h)$ tend to zero, when $g \to g^{(0)}$, for any arbitrary square integrable function $h(\boldsymbol{x})$. These results about convergence are important because they ensure that, when the noise is small, the least-squares solution of minimal energy is close to the in-band component of the true object $f^{(0)}$.

The method considered above can be easily applied to discrete images if the L^2-norm is replaced by the canonical Euclidean norm and the FT is replaced by the DFT. The function (5.12) is now replaced by the following array (2D image)

$$(\boldsymbol{f}_\mu)_{m,n} = \frac{1}{N^2} \sum_{k,l=1}^{N} \frac{\hat{K}^*_{k,l}}{|\hat{K}_{k,l}|^2 + \mu} \hat{g}_{k,l} \exp\left[i \frac{2\pi}{N}(km + ln) \right]. \tag{5.16}$$

Since the Euclidean norm of $\boldsymbol{f}_\mu$ is a decreasing function of μ, one can easily prove again that there exists a unique least-squares solution with prescribed energy. If E is an upper bound of the Euclidean norm of $\boldsymbol{f}^{(0)}$ this least-squares solution converges to $(\boldsymbol{f}^{(0)})^\dagger$ when the error on the data tends to zero.

5.2 Approximate solutions with minimal energy

An alternative to the least-squares solution with prescribed energy is provided by a method investigated by Ivanov [5] which is a simplified version of the method proposed by Phillips [2]. The starting point is the assumption that an estimate ε^2 of the energy of the noise is known. Then it is quite natural to consider the set of the approximate solutions such that their discrepancy does not exceed ε^2. This set has been already investigated in the case of discrete images—see section 4.6—and we know that it is very broad and that it contains both physical and unphysical approximate solutions as well as the true object $f^{(0)}(\boldsymbol{x})$. Since the unphysical solutions are those which are too large and oscillatory it is quite natural to look for the approximate solution which minimizes the energy functional (5.1).

An *approximate solution of minimal energy* is an object $\tilde{f}_2(\boldsymbol{x})$ which solves the following problem: *find the minimum (or the minima) of the energy functional*

$$E^2(f) = \|f\|^2 \tag{5.17}$$

in the set of all objects whose discrepancy does not exceed a prescribed bound ε^2

$$\varepsilon^2(f;g) = \|Af - g\|^2 \leq \varepsilon^2. \tag{5.18}$$

It is obvious that, if the set (5.18) contains $f = 0$, then the solution of minimal energy is just $f = 0$. But this is a trivial case. If $f = 0$ belongs to the set (5.18), this means that $\|g\| \leq \varepsilon$, i.e. the image g consists only of noise and does not contain a detectable signal. If we exclude this situation, then it is rather intuitive, as shown in figure 5.3, that the solution of minimal energy stays on the boundary of the set defined by equation (5.18) so that the inequality (5.18) can be replaced by the following equality

$$\varepsilon^2(f;g) = \|Af - g\|^2 = \varepsilon^2. \tag{5.19}$$

This result can be proved rigorously (see, for instance, [5]) and is related to the fact that a closed and convex set of a complete Euclidean space has always a point with minimum distance from the origin and that this point lies on the boundary of the set, if the set does not contain $f = 0$.

The problem we are considering is, in a sense, the dual of the problem considered in section 5.1 because the role of the two functionals is exchanged: *in the previous one the discrepancy is minimized, given the value of the energy, while now the energy is minimized, given the value of the discrepancy.* This

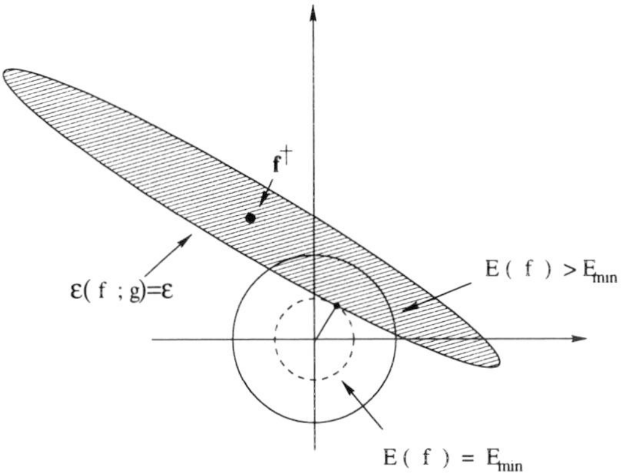

Figure 5.3. Two-dimensional picture showing that the point of minimal energy (= minimal distance from the origin) is on the boundary of the ellipse representing the set of approximate solutions whose discrepancy does not exceed ε^2. The centre of the ellipse is the solution provided by the inverse filtering method. The circles are the level lines of the energy functional.

remark makes obvious that we can use again the method of Lagrange multipliers for solving the new problem.

We consider the functional (5.6), with μ real and positive, and the function $f_\mu(\boldsymbol{x})$, given by equation (5.12), which is the unique minimizer of this functional. Then the second point of the method of Lagrange multipliers, as formulated in section 5.1, must be modified as follows:

- there exists a unique value of μ, let us say $\mu_2 = \mu_2(\varepsilon)$, such that

$$\varepsilon(f_{\mu_2}; g) = \varepsilon. \tag{5.20}$$

If such a value exists, then $\tilde{f}_2 = f_{\mu_2}$ is the unique object of minimal energy which is compatible with the image g within the error ε. Once this result is proved, we can conclude that: *for any image g there exists a unique object $\tilde{f}_2$ which is the approximate solution of minimal energy for a prescribed value of the discrepancy.*

In order to prove the existence of a unique value μ_2 such that equation (5.20) holds true, we compute $\varepsilon(f_\mu; g)$ using again the Parseval equality; if we take into account that $\hat{K}(\omega) = 0$ when ω is in $\bar{\mathcal{B}}$, we get

$$\varepsilon^2(f_\mu; g) = \frac{1}{(2\pi)^2} \int \left| \frac{|\hat{K}(\omega)|^2}{|\hat{K}(\omega)|^2 + \mu} \hat{g}(\omega) - \hat{g}(\omega) \right|^2 d\omega \tag{5.21}$$

$$= \frac{1}{(2\pi)^2} \int_{\mathcal{B}} \frac{\mu^2 |\hat{g}(\omega)|^2}{(|\hat{K}(\omega)|^2 + \mu)^2} d\omega + \frac{1}{(2\pi)^2} \int_{\bar{\mathcal{B}}} |\hat{g}(\omega)|^2 d\omega.$$

It follows that $\varepsilon(f_\mu; g)$ is an increasing function of μ which takes the value $\|g_{out}\|$ (norm of the out-of-band noise; if the system is not band-limited, then $\|g_{out}\| = 0$) for $\mu = 0$ and tends to $\|g\|$ when $\mu \to \infty$. In figure 5.4 we plot the behaviour of the function $\varepsilon(f_\mu; g)$, again in the case of the numerical example of section 5.6.

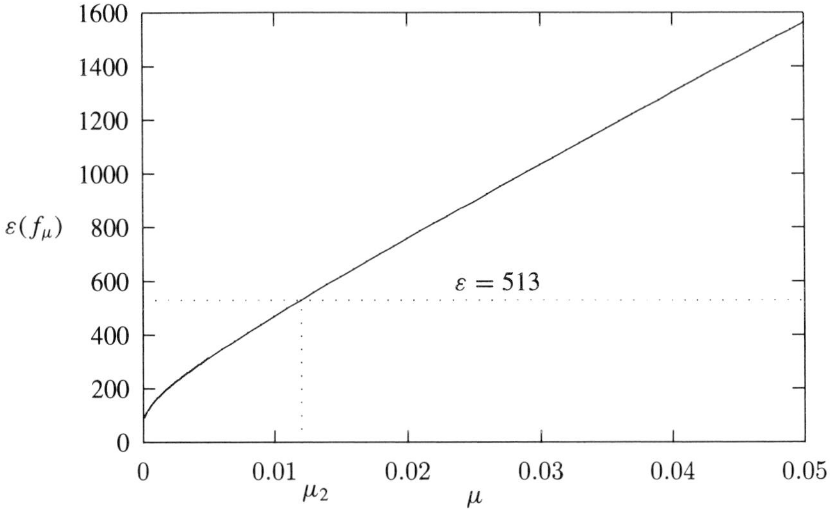

Figure 5.4. Behaviour of $\varepsilon(f_\mu; g)$ as a function of μ in the case of the numerical example of section 5.6. The point μ_2 where $\varepsilon(f_{\mu_2}; g) = \varepsilon$ is also indicated. In such a case we have $\mu_2 = 1.2 \times 10^{-2}$.

Therefore, if ε satisfies the conditions

$$\|g_{out}\| < \varepsilon < \|g\| \tag{5.22}$$

there exists a unique value of μ, $\mu_2 = \mu_2(\varepsilon)$, such that $\varepsilon(f_{\mu_2}; g) = \varepsilon$. This value is also indicated in figure 5.4.

We point out that the conditions (5.22) make sense from the physical point of view. Indeed the condition $\varepsilon < \|g\|$ must be satisfied if the image contains not only noise but also a detectable signal (this point has been already discussed). Moreover the condition $\varepsilon > \|g_{out}\|$ means that the energy of the total noise must be greater than the energy of the out-of-band noise.

Remark 5.2. As concerns the continuous dependence of $\tilde{f}_2$ on the data and its convergence to the exact generalized solution $(f^{(0)})^\dagger = P^{(\mathcal{B})} f^{(0)}$, it is possible *to prove that* $\|\tilde{f}_2 - P^{(\mathcal{B})} f^{(0)}\|$ *tends to zero when the error on the data tends to zero (see for instance [4] or [5]). Therefore the convergence properties of* $\tilde{f}_2$ *are stronger than those of* $\tilde{f}_1$.

All results can be easily extended to the case of discrete images using the expression (5.16) for $\boldsymbol{f}_\mu$.

5.3 Regularization algorithms in the sense of Tikhonov

The method of Lagrange multipliers used for the solution of the constrained minimization problems of the previous sections introduces, as an intermediate step, the family of functions defined in equation (5.12) (in equation (5.16) for discrete images). This family of approximate solutions has very interesting properties which have been elucidated and stressed by Tikhonov [3], who based on these properties the general definition of a *regularization algorithm* [6]. In order to introduce the basic idea of this method it is convenient to write the functions $f_\mu(\boldsymbol{x})$ in a slightly different form.

In appendix E we prove that the function $f_\mu(\boldsymbol{x})$ is the solution of the Euler equation associated with the quadratic functional (5.6)

$$(A^*A + \mu I) f_\mu = A^* g \tag{5.23}$$

where I denotes the identity operator. This result can be easily verified in the case of image deconvolution because, by taking the FT of both members of this equation, we get

$$(|\hat{K}(\boldsymbol{\omega})|^2 + \mu)\hat{f}_\mu(\boldsymbol{\omega}) = \hat{K}^*(\boldsymbol{\omega})\hat{g}(\boldsymbol{\omega}) \tag{5.24}$$

and therefore the expressions (5.11) and (5.12) are re-obtained. It is also evident that, by putting $\mu = 0$ in equation (5.23), we re-obtain the equation of the least-squares solutions, equation (4.36).

Equation (5.23) implies that f_μ can be written in the following form

$$f_\mu = R_\mu g \tag{5.25}$$

where

$$R_\mu = (A^*A + \mu I)^{-1} A^*. \tag{5.26}$$

This is a different way of writing equation (5.12). It is a useful way because the operator R_μ is an approximation, in a sense which will be specified in a moment, of the generalized inverse operator $A^\dagger$ defined in equation (4.43). In order to investigate this point we observe that the operator R_μ is a convolution operator

$$R_\mu g = K_\mu^\dagger * g \tag{5.27}$$

where

$$K_\mu^\dagger(\boldsymbol{x}) = \frac{1}{(2\pi)^q} \int \frac{\hat{K}^*(\boldsymbol{\omega})}{|\hat{K}(\boldsymbol{\omega})|^2 + \mu} e^{i\boldsymbol{x}\cdot\boldsymbol{\omega}} d\boldsymbol{\omega}. \tag{5.28}$$

Thanks to the factor $\hat{K}^*(\boldsymbol{\omega})$ the integral is extended only to the band $\mathcal{B}$ of the imaging system. Then, in the limit $\mu = 0$, we re-obtain equation (4.41) if the integral in this equation is convergent. Otherwise the limit does not exist. For $\mu > 0$, $K_\mu^\dagger(\boldsymbol{x})$ is a well-defined function and this implies that R_μ is a bounded operator; more precisely, from equations (5.25) and (5.15) we get

$$\|R_\mu g\| \leq \frac{1}{\sqrt{\mu}} \|g\|. \tag{5.29}$$

From the previous remarks it follows that the family of linear operators R_μ has the following properties:

- for any $\mu > 0$, R_μ is a linear and continuous operator
- for any noise-free image $g^{(0)} = Af^{(0)}$

$$\lim_{\mu \to 0} R_\mu g^{(0)} = P^{(\mathcal{B})} f^{(0)} = A^\dagger g^{(0)} \tag{5.30}$$

  where the limit is in the L^2-norm and $P^{(\mathcal{B})}$ is the band-limiting operator defined in equation (4.33).

The second property clarifies in what sense R_μ is an approximation of $A^\dagger$: when R_μ is applied to a noise-free image $g^{(0)}$ then one gets an approximation of the generalized solution $f^\dagger = A^\dagger g^{(0)}$ and the approximation error tends to zero when μ tends to zero.

According to Tikhonov [6], any family $\{R_\mu\}$ of linear operators, depending on a parameter $\mu > 0$, is called a *linear regularization algorithm for the solution of the ill-posed problem* $Af = g$ if it satisfies the two conditions above. The parameter μ is called the *regularization parameter* and the functions $f_\mu = R_\mu g$ are called *regularized solutions*. In the particular example we have used to introduce this definition, the regularization parameter is just the Lagrange multiplier introduced in section 5.1. Other examples of regularization algorithms with regularization parameters not related to Lagrange multipliers will be presented and investigated in the next sections and chapters. We also observe that the condition of linearity can be dropped in the definition of regularization algorithm. Hence one can also consider nonlinear regularization algorithms. In the next chapter, we will provide examples of nonlinear regularization methods for the solution of linear inverse problems.

We now come back to the particular example of regularization algorithm provided by equations (5.26)–(5.28); this is usually called the *Tikhonov regularization algorithm*. We already know that, in the case of noise-free images $g^{(0)}$, the regularized solutions $f_\mu = R_\mu g^{(0)}$ provide better and better approximations of the true solution (or generalized solution) when the regularization parameter μ tends to zero. The important question is: what happens in the case of a noisy image g?

If we use the model (4.2) for image formation we have

$$R_\mu g = R_\mu A f^{(0)} + R_\mu w. \tag{5.31}$$

From equation (5.30) we know that we can restore at most $P^{(\mathcal{B})} f^{(0)}$ when the noise is zero and μ tends to zero. When the image is noisy and μ is not zero, the error we commit in the restoration of $P^{(\mathcal{B})} f^{(0)}$ is given by

$$R_\mu g - P^{(\mathcal{B})} f^{(0)} = (R_\mu A f^{(0)} - P^{(\mathcal{B})} f^{(0)}) + R_\mu w. \tag{5.32}$$

The first term in the second member of this equation can be called the *approximation error* because this is the error due to the fact that we approximate the operator $A^\dagger$ by means of R_μ. This error depends only on the value of μ and does not depend on the noise. The second term $R_\mu w$ is the error due to the noise corrupting the image and it can be called the *noise-propagation error* since it quantifies the propagation of the noise from the image g to the regularized solution f_μ. It depends both on μ and on w.

If we use the triangular inequality for the norm (see appendix A) we have

$$\| R_\mu g - P^{(\mathcal{B})} f^{(0)} \| \leq \| R_\mu A f^{(0)} - P^{(\mathcal{B})} f^{(0)} \| + \| R_\mu w \| \tag{5.33}$$

i.e. *the norm of the restoration error is bounded by the sum of the norm of the approximation error and of the norm of the noise-propagation error.* In order to understand the behaviour of the restoration error as a function of μ, we investigate the behaviour of the two terms in the r.h.s. of equation (5.33).

If we use again the Parseval equality, we find that the norm of the approximation error is given by

$$\begin{aligned} \| R_\mu A f^{(0)} - P^{(\mathcal{B})} f^{(0)} \|^2 &= \frac{1}{(2\pi)^q} \int_B \left| \frac{|\hat{K}(\omega)|^2}{|\hat{K}(\omega)|^2 + \mu} \hat{f}^{(0)}(\omega) - \hat{f}^{(0)}(\omega) \right|^2 d\omega \\ &= \frac{1}{(2\pi)^q} \int_B \frac{\mu^2}{(|\hat{K}(\omega)|^2 + \mu)^2} |\hat{f}^{(0)}(\omega)|^2 d\omega. \end{aligned} \tag{5.34}$$

Therefore the norm of the approximation error is a continuous and increasing function of μ, equal to zero for $\mu = 0$ and tending to $\| P^{(\mathcal{B})} f^{(0)} \|$ when $\mu \to \infty$. On the other hand, the norm of the noise-propagation error is given by

$$\| R_\mu w \|^2 = \frac{1}{(2\pi)^q} \int_B \frac{|\hat{K}(\omega)|^2}{(|\hat{K}(\omega)|^2 + \mu)^2} |\hat{w}(\omega)|^2 d\omega \tag{5.35}$$

and therefore it is a continuous and decreasing function of μ, which tends to a very large value (possibly $+\infty$) when $\mu \to 0$ and tends to zero when $\mu \to +\infty$.

We find that the norm of the approximation error and the norm of the noise-propagation error have opposite behaviour as functions of μ. If we wish to reduce the approximation error, we must choose a small value of the regularization parameter but then the noise-propagation error is very large. On the other hand, if we wish to reduce the noise-propagation error, we must choose a large value of μ but then the approximation error is too large. In figure 5.5 we plot the behaviour of these functions in the case of the numerical example of section 5.6.

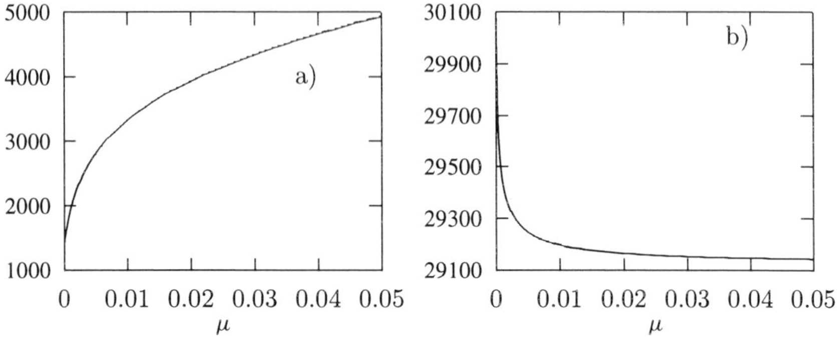

Figure 5.5. Behaviour of the norm of the approximation error, (a), and of the norm of the noise-propagation error, (b), as functions of μ, in the case of the example of section 5.6.

The r.h.s. of the inequality (5.33) has a minimum since it is the sum of an increasing function of μ and of a decreasing one. The same conclusion holds true for the norm of the restoration error because this norm is approximately equal to the norm of the noise-propagation error (decreasing) for small μ and to the norm of the approximation error (increasing) for large μ. The behaviour of the restoration error in the case of the numerical example of section 5.6 is given in figure 5.6. Here we plot the relative norm of the restoration error, i.e. the quantity $\|f_\mu - f^{(0)}\|/\|f^{(0)}\|$.

The value of μ minimizing the restoration error corresponds to the regularized solution which provides the best approximation of $\mathcal{P}^{(\mathcal{B})} f^{(0)}$ and hence the best compromise between approximation error and noise-propagation error. This value of μ can be called *optimal value* of the regularization parameter and it will be denoted by μ_{opt}. The corresponding *optimal regularized solution* will be denoted by $\tilde{f}_{opt}$.

The existence of a minimum of the restoration error and the related existence of an optimal value of the regularization parameter for any noisy image is a basic property of the Tikhonov regularization algorithm. This result implies that for decreasing values of μ the regularized solutions first approach the exact object $f^{(0)}(\boldsymbol{x})$ and then go away. One may call this property *semiconvergence* so that one has convergence in the case of noise-free images and semiconvergence in the case of noisy images. Semiconvergence is an additional property one should require for any regularization algorithm to be used in practice.

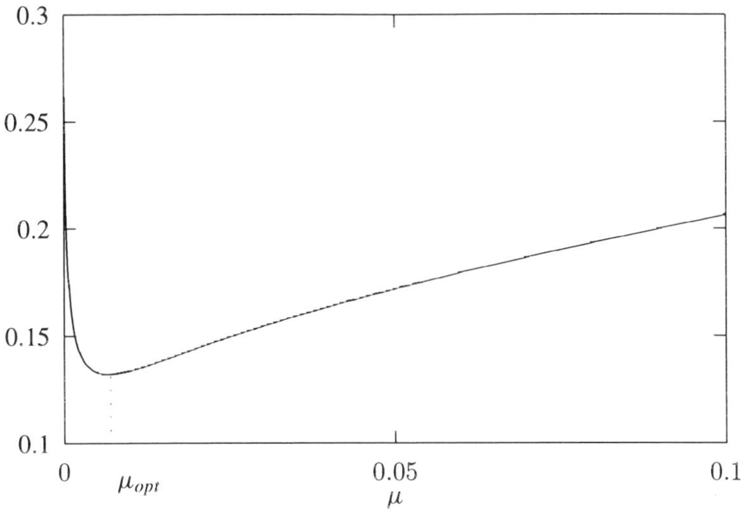

Figure 5.6. Plot of the relative norm of the restoration error, as a function of μ, in the case of the example of section 5.6. The optimal value of the regularization parameter, corresponding to the minimum of the function is also indicated. In such a case we have $\mu_{opt} = 7 \times 10^{-3}$.

It is obvious that the optimal value of the regularization parameter cannot be determined in the case of a real image because one does not know $f^{(0)}$. It can only be determined in the case of numerical simulations which are very useful for the understanding of any specific inverse problems. In the case of real images we need methods for estimating sensible values of the regularization parameter and this point will be discussed in section 5.6.

It is interesting to describe the behaviour of the regularized solutions $f_\mu(\boldsymbol{x})$ in terms of trajectories in the space of all possible square-integrable functions or also the behaviour of the regularized 2D arrays, $\boldsymbol{f}_\mu$, given by equation (5.16), in terms of trajectories in the space of all possible $N \times N$ arrays. Indeed, to each $f_\mu(\boldsymbol{x})$ or $\boldsymbol{f}_\mu$ there corresponds a point in the appropriate object space and, as μ varies, this point moves describing a trajectory in this space.

Consider for simplicity the discrete case and let $\boldsymbol{f}^\dagger$ be the solution of equation (4.6), as given by equation (4.17). As we know, this solution provides, in general, an unacceptable restoration of the true object $\boldsymbol{f}^{(0)}$. However this solution is just the starting point of the trajectory at $\mu = 0$. As μ grows, the distance of the point from the origin, i.e. $\|\boldsymbol{f}_\mu\|$, decreases (this is precisely the behaviour which can be derived from equation (5.15)); first the point approaches $\boldsymbol{f}^{(0)}$, the true object, and then it goes away, arriving at the origin only in the limit $\mu \to \infty$. This is the geometric description of the semiconvergence property. A two-dimensional picture of this behaviour is given in figure 5.7 where we also indicate the ellipse, which is the boundary of the set of approximate solutions defined by equation (5.18), as well as the circle which is the boundary of the set of images with prescribed energy defined by equation (5.3). We assume that the true object $f^{(0)}$ belongs to both sets so that their intersection is not empty (and is indicated by the shaded region in figure 5.7).

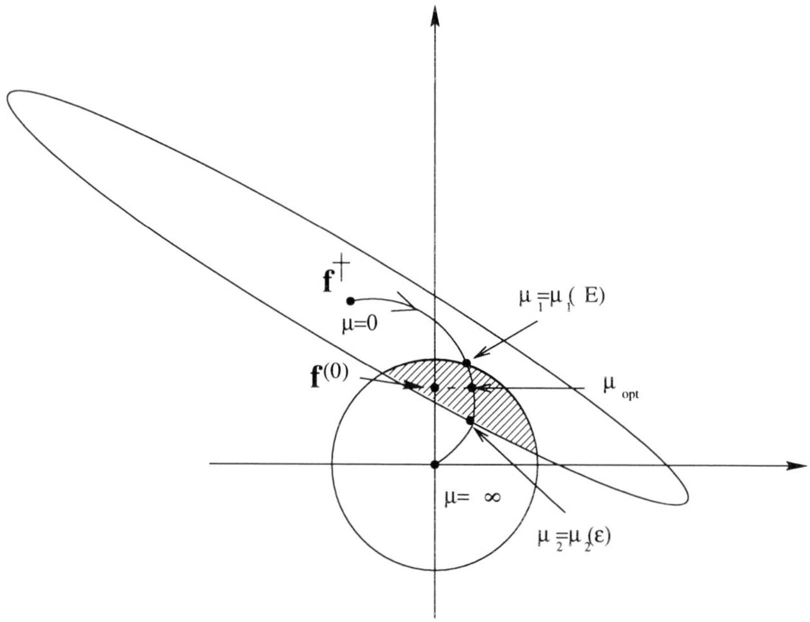

Figure 5.7. Two-dimensional picture of the trajectory described by the regularized solutions. The ellipse is the boundary of the set (5.18) while the circle is the boundary of the set (5.3). We indicate on the trajectory the three points corresponding to μ_1, μ_{opt} and μ_2.

The picture of figure 5.7 suggests that, if the intersection of the two sets (5.3) and (5.18) is not too broad, then the approximate solutions $\tilde{f}_1$ and $\tilde{f}_2$ considered in the previous sections can provide satisfactory approximations of the optimal regularized solution and that, in general, $\|\tilde{f}_1\|$ will be greater than $\|\tilde{f}_2\|$

$$\|\tilde{f}_2\| \leq \|\tilde{f}_1\| \tag{5.36}$$

with the corresponding inequality between the values of the regularization parameter

$$\mu_1(E) \leq \mu_2(\varepsilon). \tag{5.37}$$

It can be proved [7] that these inequalities hold true whenever the conditions indicated above are satisfied—see also section 5.6. Moreover figure 5.7 suggests that the optimal value of the regularization parameter should be intermediate between $\mu_1(E)$ and $\mu_2(\varepsilon)$. Such a relationship can be investigated by means of numerical simulations and it turns out that it is not always true (even if it is verified in the case of the numerical example of section 5.6).

The sequence of pictures of figure 5.8 is another way to visualize the variations of the regularized solutions f_μ when μ moves from 0 to ∞. Also in this case the existence of an optimum value of the regularization parameter is evident. The restorations corresponding to small values of the regularization parameter are very noisy. The noise propagation decreases when μ increases and for a certain value of μ a satisfactory restoration of the image appears. If we further increase the value of μ then the restored image becomes more and more blurred as an effect of the increase of the approximation error.

5.4 Regularization and filtering

The methods of Phillips [2] and Tikhonov [3] were formulated for linear integral equations of the first kind in one variable and were based on the use of differential operators for defining the smoothness condition. In this section we apply these methods to the problem of image deconvolution.

For 1D deconvolution problems, the method of Phillips consists of looking for the approximate solution (in the sense of equation (5.19)) which minimizes the L^2-norm of the second derivative

$$\|f''\|^2 = \int_{-\infty}^{+\infty} |f''(x)|^2 dx. \tag{5.38}$$

This requirement replaces that of the minimization of the energy considered in section 5.2. Roughly speaking, by minimizing the functional (5.38) with the condition (5.19), one looks for the approximate solution having minimal curvature in the sense of the quadratic mean.

In order to solve the problem we can use again the method of Lagrange multipliers and, therefore, minimize the following functional

$$\begin{aligned}\Phi_\mu(f; g) &= \|Af - g\|^2 + \mu\|f''\|^2 \qquad (5.39)\\ &= \frac{1}{2\pi}\int_{-\infty}^{+\infty} |\hat{K}(\omega)\hat{f}(\omega) - \hat{g}(\omega)|^2 d\omega + \frac{\mu}{2\pi}\int_{-\infty}^{+\infty} \omega^4 |\hat{f}(\omega)|^2 d\omega.\end{aligned}$$

The Parseval equality has been used again and also the fact that the FT of $f''(x)$ is $-\omega^2\hat{f}(\omega)$.

By repeating computations analogous to those of section 5.1, we easily find that, if $\hat{K}(0) \neq 0$, then, for any $\mu > 0$, there exists a unique minimum of the

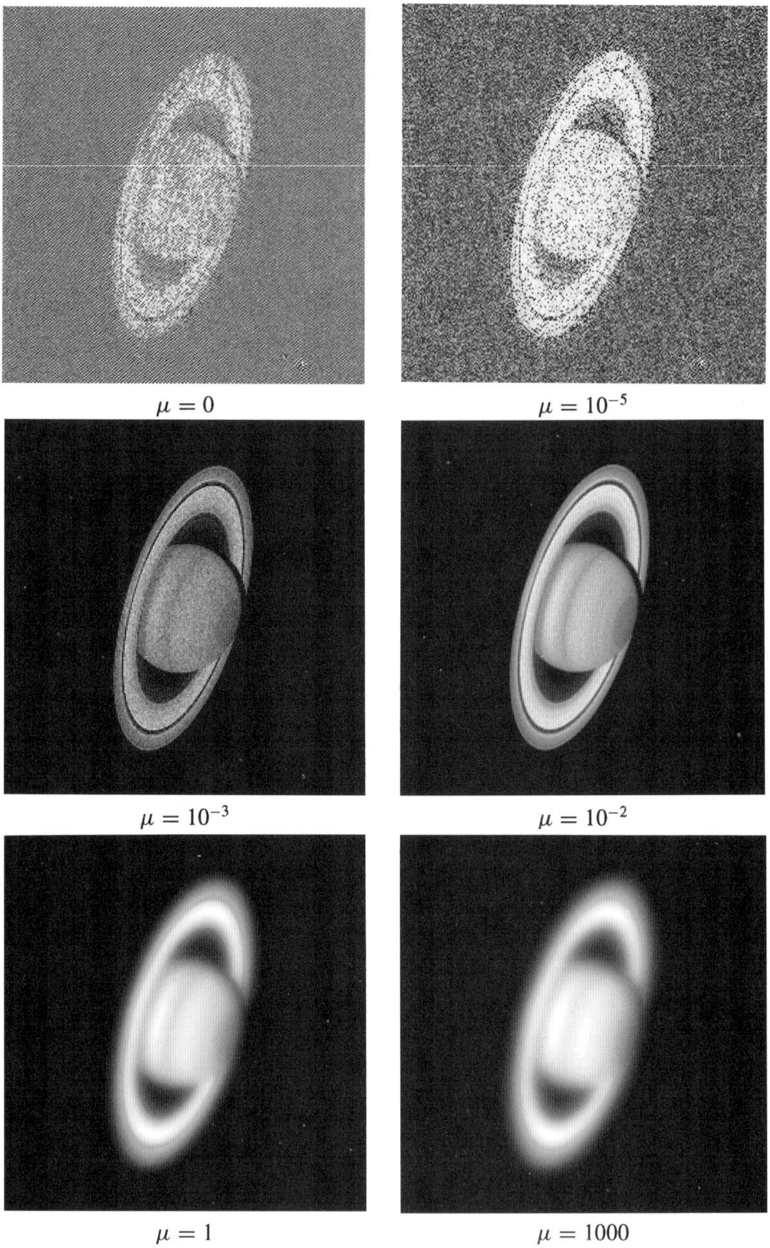

Figure 5.8. Sequence of regularized solutions obtained from the image of Saturn given in figure 3.10

functional (5.39) given by

$$f_\mu(x) = \frac{1}{2\pi}\int_{-\infty}^{+\infty} \frac{\hat{K}^*(\omega)}{|\hat{K}(\omega)|^2 + \mu\omega^4}\hat{g}(\omega)e^{ix\omega}d\omega. \tag{5.40}$$

If the imaging system is bandlimited, this function is also bandlimited and has the same band of the system. Moreover we find again that the discrepancy function $\varepsilon^2(f_\mu; g)$ is an increasing function of μ and therefore that there exists a unique value of μ, say $\mu_3 = \mu_3(\varepsilon)$, such that $\varepsilon(f_{\mu_3}; g) = \varepsilon$.

The family of approximate solutions defined by equation (5.40) provides another example of a regularization algorithm. However, the first example given by Tikhonov [3] was based on the minimization of a functional containing the L^2-norm of the function and of its first derivative. In the case of 1D deconvolution problems the corresponding family of regularized solutions is the family of the minimizers of the following functionals

$$\begin{aligned}\Phi_\mu(f; g) &= \|Af - g\|^2 + \mu(\|f\|^2 + a\|f'\|^2) \qquad (5.41)\\ &= \frac{1}{2\pi}\int_{-\infty}^{+\infty} |\hat{K}(\omega)\hat{f}(\omega) - \hat{g}(\omega)|^2 d\omega \\ &+ \frac{\mu}{2\pi}\int_{-\infty}^{+\infty} (1 + a\omega^2)|\hat{f}(\omega)|^2 d\omega\end{aligned}$$

where a is a given positive constant defining the relative weight of $\|f\|^2$ and $\|f'\|^2$.

Also in this case it is easy to show that the minimizer is given by

$$f_\mu(x) = \int_{-\infty}^{+\infty} \frac{\hat{K}^*(\omega)}{|\hat{K}(\omega)|^2 + \mu(1 + a\omega^2)}\hat{g}(\omega)e^{ix\omega}d\omega \tag{5.42}$$

and that this family of functions defines a regularization algorithm in the sense of section 5.3. Again, if the imaging system is bandlimited, the functions (5.42) have the same band as the system.

The extension to 2D and 3D problems of the previous regularization algorithms is rather obvious. If we consider derivatives up to the second order, then we can introduce the following functional

$$\Sigma^2(f) = a_0\|f\|^2 + \sum_{i=1}^{q} a_i \left\|\frac{\partial f}{\partial x_i}\right\|^2 + \sum_{i,j=1}^{q} a_{i,j} \left\|\frac{\partial f}{\partial x_i \partial x_j}\right\|^2 \tag{5.43}$$

where a_0, a_i and a_{ij} are given non-negative numbers and the a_{ij} define a positive semi-definite matrix. In the case $a_0 = 1, a_i = 0, a_{ij} = 0$ we re-obtain the energy functional considered in the previous sections. It is also evident that, in the 1D case, the functional (5.43) contains, as particular cases, the functionals of Phillips and Tikhonov.

The regularized solutions are defined as the minimizers of the functionals

$$\Phi_\mu(f;g) = \varepsilon^2(f;g) + \mu\Sigma^2(f) \tag{5.44}$$

and are given by

$$f_\mu(\boldsymbol{x}) = \frac{1}{(2\pi)^q}\int \frac{\hat{K}^*(\boldsymbol{\omega})}{|\hat{K}(\boldsymbol{\omega})|^2 + \mu\hat{P}(\boldsymbol{\omega})}\hat{g}(\boldsymbol{\omega})e^{i\boldsymbol{x}\cdot\boldsymbol{\omega}}d\boldsymbol{\omega} \tag{5.45}$$

where $\hat{P}(\boldsymbol{\omega})$ is the polynomial

$$\hat{P}(\boldsymbol{\omega}) = a_0 + \sum_{i=1}^{q} a_i\omega_i^2 + \sum_{i,j=1}^{q} a_{ij}(\omega_i\omega_j)^2. \tag{5.46}$$

The derivation of equation (5.45) is left to the reader.

The expression (5.45) contains as particular cases all regularized solutions considered previously. This equation can also be written in the same form as equation (5.25)

$$f_\mu = R_\mu g \tag{5.47}$$

where R_μ is a convolution operator

$$R_\mu g = K_\mu^\dagger * g, \tag{5.48}$$

the convolution kernel $K_\mu^\dagger$ now being given by

$$K_\mu^\dagger(\boldsymbol{x}) = \frac{1}{(2\pi)^q}\int \frac{\hat{K}^*(\boldsymbol{\omega})}{|\hat{K}(\boldsymbol{\omega})|^2 + \mu\hat{P}(\boldsymbol{\omega})}e^{i\boldsymbol{x}\cdot\boldsymbol{\omega}}d\boldsymbol{\omega}. \tag{5.49}$$

If we compute the norm of the approximation error and the norm of the noise-propagation error, we find that they have behaviours, as a function of μ, similar to those plotted in figure 5.5. As a consequence, the norm of the restoration error has a minimum, as shown in figure 5.6, so that an optimal value of the regularization parameter also exists for these regularization algorithms. In other words all regularization methods considered so far have the semiconvergence property.

An alternative way of writing equation (5.45), which suggests further generalizations, is the following one

$$f_\mu(\boldsymbol{x}) = \frac{1}{(2\pi)^q}\int_B \hat{W}_\mu(\boldsymbol{\omega})\frac{\hat{g}(\boldsymbol{\omega})}{\hat{K}(\boldsymbol{\omega})}e^{i\boldsymbol{x}\cdot\boldsymbol{\omega}}d\boldsymbol{\omega} \tag{5.50}$$

where

$$\hat{W}_\mu(\boldsymbol{\omega}) = \frac{|\hat{K}(\boldsymbol{\omega})|^2}{|\hat{K}(\boldsymbol{\omega})|^2 + \mu\hat{P}(\boldsymbol{\omega})}. \tag{5.51}$$

This form makes clear that a family of regularized solutions is a family of filtered versions of the solution or generalized solution of equation (4.5) (see equation (4.15) or (4.39)), which can be obtained by a family of *window functions* satisfying the following conditions

- $|\hat{W}_\mu(\omega)| \leq 1$, for any $\mu > 0$
- $lim_{\mu \to 0} \hat{W}_\mu(\omega) = 1$, for any ω such that $\hat{K}(\omega) \neq 0$
- $\hat{W}_\mu(\omega)/\hat{K}(\omega)$ bounded for any $\mu > 0$.

For the family of window functions defined by equation (5.51) the first two conditions are obvious. The third one is satisfied if the denominator in equation (5.51) is neither zero.

Conversely, one can prove [4] that any family of window functions satisfying the three conditions above defines a linear regularization algorithm since the family of linear operators

$$(R_\mu g)(\boldsymbol{x}) = \frac{1}{(2\pi)^q} \int_B \hat{W}_\mu(\omega) \frac{\hat{g}(\omega)}{\hat{K}(\omega)} e^{i\boldsymbol{x}\cdot\omega} d\omega \tag{5.52}$$

satisfies the conditions stated in section 5.3.

This remark suggests many other regularization methods. A very simple one is obtained by chopping off the spatial frequencies where the TF $\hat{K}(\omega)$ is smaller than some threshold value. This threshold plays the role of the regularization parameter. The corresponding regularization algorithm, investigated by Miller in a more general setting [8], is defined by the following family of window functions which will be called the *truncated window functions*

$$\hat{W}_\mu(\omega) = \begin{cases} 1 & \text{if } |\hat{K}(\omega)| > \sqrt{\mu} \\ 0 & \text{if } |\hat{K}(\omega)| \leq \sqrt{\mu}. \end{cases} \tag{5.53}$$

These window functions, as well as those defined by equation (5.51), depend on the TF $\hat{K}(\omega)$. If $\hat{K}(\omega)$ does not vanish, as in the case, for instance, of the Gaussian blur or of the near-field acoustic holography, then it is also possible to use window functions which are independent of $\hat{K}(\omega)$. In particular, it is possible to use window functions which have a bounded support so that they provide bandlimited regularized solutions in a case where the imaging system is not bandlimited.

For simplicity we give a list of these windows in the 1D case. They are different from zero on a bounded interval $[-\Omega, \Omega]$ and Ω can take any value from 0 to $+\infty$. In this way we introduce families of window functions depending on the parameter Ω. For any value of Ω the window functions satisfy the first and third conditions stated above. The second one is satisfied when $\Omega \to \infty$

so that the relationship between Ω and the regularization parameter μ, as it is usually defined, is given by

$$\mu = \frac{1}{\Omega}. \tag{5.54}$$

In the following we will indicate the window functions as depending on the parameter Ω, band-width of the corresponding regularized solutions.

- Rectangular window

$$\hat{W}_\Omega(\omega) = \begin{cases} 1 & \text{if} \quad |\omega| < \Omega \\ 0 & \text{if} \quad |\omega| \geq \Omega \end{cases} \tag{5.55}$$

- Triangular window

$$\hat{W}_\Omega(\omega) = \begin{cases} 1 - \frac{|\omega|}{\Omega} & \text{if} \quad |\omega| < \Omega \\ 0 & \text{if} \quad |\omega| \geq \Omega \end{cases} \tag{5.56}$$

- Generalized Hamming window

$$\hat{W}_\Omega(\omega) = \begin{cases} \alpha + (1-\alpha)\cos(\frac{\pi\omega}{\Omega}) & \text{if} \quad |\omega| < \Omega \\ 0 & \text{if} \quad |\omega| \geq \Omega \end{cases} \tag{5.57}$$

  For $\alpha = 0.5$ we get the Hanning window, while for $\alpha = 0.54$ we get the Hamming window [9].

The window functions which have a bounded support can be used for any TF (without zeros) independently of its behaviour for $|\omega| \to \infty$. It is also possible to use window functions which do not have a bounded support but depend on a parameter Ω defining an interval such that the corresponding window function is negligible outside this set. Their behaviour at infinity must be such that, for any $\Omega > 0$, they tend to zero more rapidly than $\hat{K}(\omega)$ when $|\omega| \to \infty$. We give a few examples.

- Gaussian window

$$\hat{W}_\Omega(\omega) = \exp\left[-\frac{1}{2}\left(\frac{\omega}{\Omega}\right)^2\right] \tag{5.58}$$

  This window can be used whenever $|\hat{K}(\omega)|$ tends to zero less rapidly than any Gaussian at infinity. An example is provided by near-field acoustic holography.

- Butterworth window

$$\hat{W}_\Omega(\omega) = \left[1 + \left(\frac{\omega}{\Omega}\right)^n\right]^{-1/2}. \tag{5.59}$$

This window function, which is frequently used in tomography (see chapter 11), can be used whenever $|\hat{K}(\omega)|$ tends to zero less rapidly than $|\omega|^{-n/2+1}$.

5.5 The global point spread function: resolution and Gibbs oscillations

The families of regularized solutions introduced in the previous sections are defined in terms of linear and bounded operators R_μ

$$f_\mu = R_\mu g \tag{5.60}$$

and these operators have the following properties: they are convolution operators

$$R_\mu g = K_\mu^\dagger * g \tag{5.61}$$

whose kernels $K_\mu^\dagger(\boldsymbol{x})$ are given by

$$K_\mu^\dagger(\boldsymbol{x}) = \frac{1}{(2\pi)^q}\int_B \frac{\hat{W}_\mu(\boldsymbol{\omega})}{\hat{K}(\boldsymbol{\omega})} e^{i\boldsymbol{x}\cdot\boldsymbol{\omega}} d\boldsymbol{\omega}, \tag{5.62}$$

$\hat{W}_\mu(\boldsymbol{\omega})$ being a family of window functions satisfying the conditions stated in the previous section. Analogously, in the discrete case we have

$$\boldsymbol{f}_\mu = \boldsymbol{R}_\mu \boldsymbol{g} = \boldsymbol{K}_\mu^\dagger * \boldsymbol{g} \tag{5.63}$$

where

$$(\boldsymbol{K}_\mu^\dagger)_{m,n} = \frac{1}{N^2}\sum_B \frac{(\hat{\boldsymbol{W}}_\mu)_{k,l}}{\hat{K}_{k,l}} \exp\left[i\frac{2\pi}{N}(mk+nl)\right]. \tag{5.64}$$

If we use the model discussed in chapter 3 for the process of image formation

$$g = Af^{(0)} + w \tag{5.65}$$

and if we substitute this equation in equation (5.60), we obtain

$$f_\mu = R_\mu A f^{(0)} + R_\mu w. \tag{5.66}$$

A similar equation holds true in the discrete case.

Equation (5.66) has the same structure as equation (5.65): the first term corresponds to the blurring while the second one corresponds to the noise. Therefore the operator $R_\mu A$ is a new blurring operator which describes the effect of two successive operations: the blurring due to the imaging operator and the partial deblurring due to the use of a regularized inversion algorithm. As we will see shortly this operator is again a convolution operator whose kernel is the PSF of a linear system consisting of two linear systems in cascade: the first one is the imaging system (without the recording system, whose effect is the addition of noise) while the second one is the computer where the linear inversion algorithm R_μ has been implemented. It follows that it is quite natural to say that the PSF associated with the operator $R_\mu A$ is the *global PSF* since it is the PSF of the global system described above. The scheme of this global system is represented in figure 5.9.

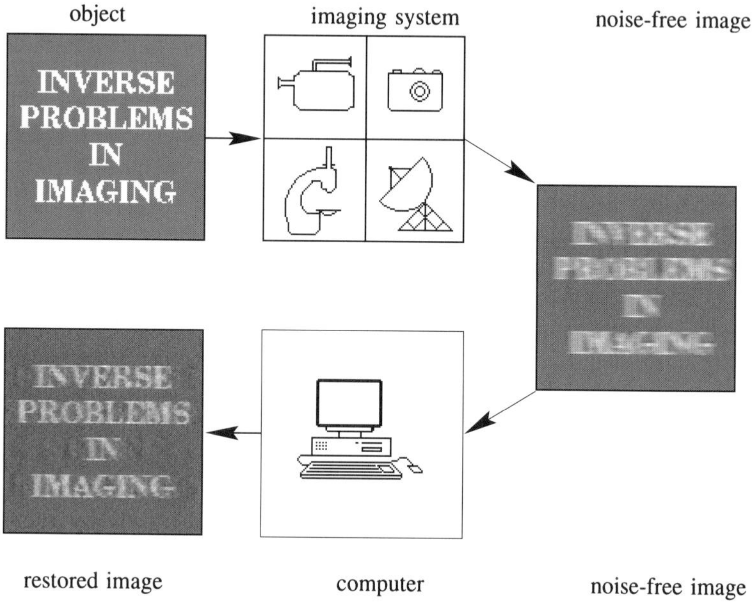

Figure 5.9. Scheme of the global system consisting of the imaging system and of the computer (where the inversion algorithm is implemented) in cascade.

In order to compute the global PSF we take the FT of both sides of equation (5.66). Using equations (5.61) and (5.62) we obtain

$$\hat{f}_\mu(\omega) = \hat{W}_\mu(\omega)\hat{f}^{(0)}(\omega) + \hat{W}_\mu(\omega)\frac{\hat{w}(\omega)}{\hat{K}(\omega)} \tag{5.67}$$

and this equation implies that the global TF is just the window function $\hat{W}_\mu(\omega)$ so that *the global PSF is the inverse FT of the window function:*

$$W_\mu(\boldsymbol{x}) = \frac{1}{(2\pi)^q}\int_B \hat{W}_\mu(\omega)e^{i\boldsymbol{x}\cdot\omega}d\omega. \tag{5.68}$$

In conclusion we obtain that

$$R_\mu A f^{(0)} = W_\mu * f^{(0)}. \tag{5.69}$$

Similar equations hold true in the discrete case where the discrete global PSF is the inverse DFT of the discrete window function

$$(\mathbf{W}_\mu)_{m,n} = \frac{1}{N^2} \sum_B (\hat{\mathbf{W}}_\mu)_{k,l} \, \exp\left[i \frac{2\pi}{N}(mk + nl) \right]. \tag{5.70}$$

The knowledge of the global PSF provides information about the approximation errors so that one can gain insight into their effect on image restoration. To this purpose we analyse two examples which can be viewed as paradigms of the PSF most frequently encountered in practice: Gaussian blur and linear motion blur.

(A) *Gaussian blur*

For simplicity we consider 1D Gaussian blur with unit variance so that the PSF is given by: $K(x) = (2\pi)^{-1/2} \exp(-x^2/2)$, and the corresponding TF is given by $\hat{K}(\omega) = \exp(-\omega^2/2)$. In figure 5.10 we plot the *Tikhonov window function*, as given by equation (5.51) with $\hat{P}(\omega) = 1$, in the case $\mu = 10^{-2}$; the *Phillips window function*, as given by equation (5.51) with $\hat{P}(\omega) = \omega^2$, in the case $\mu = 10^{-2}/2\ln(10)$; the *truncated window function*, defined in equation (5.53), also in the case $\mu = 10^{-2}$. In the same figure we give also the inverse FT of these functions, i.e. the corresponding global PSF. The values of the regularization parameters are chosen in such a way that the values of both the Tikhonov and the Phillips window functions are equal to 1/2 at the cut-off frequency of the truncated window function, which is given by $\Omega = [\ln(1/\mu)]^{1/2} = [2\ln(10)]^{1/2}$.

These window functions define low-pass filters. While the truncated window function defines a perfect low-pass filter, the Tikhonov window has a smoother behaviour and the Phillips window is intermediate between the two others. The corresponding global PSF have a central peak which is narrower than that of the Gaussian PSF. This means that in the restored image details are sharper than in the original blurred image. However all global PSF have rather important side lobes which can produce important effects.

A glance at figure 5.10(a) makes clear that, once the value of the regularization parameter has been chosen, the regularized solution is an essentially bandlimited function with an effective bandwidth Ω defined, for instance, by the condition $\hat{W}_\mu(\Omega) = 1/2$. Since $\hat{W}_\mu(\omega) \cong 1$ for $|\omega| < \Omega$, it follows that $\hat{f}(\omega)$ approximately coincides with $\hat{f}^{(0)}(\omega)$ for these values of ω (if we neglect noise contribution), i.e. $f_\mu(x)$ provides a Ω-bandlimited approximation of $f^{(0)}(x)$.

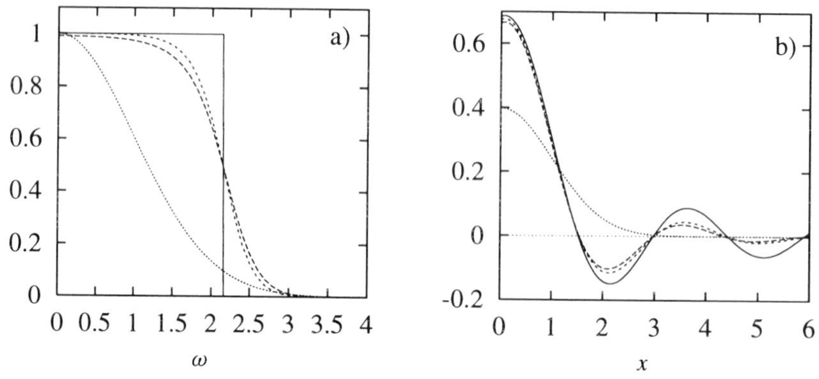

Figure 5.10. (a) Plot of the Gaussian TF (dotted line) compared with the Tikhonov window (dashed line), the Phillips window (thin dashed line) and the truncated window (solid line); (b) Plot of the Gaussian blur compared with the global PSF corresponding to the window functions in (a). The values of the regularization parameters are indicated in the text

The bandlimiting intrinsic to the regularization methods has two main effects:

- the object $f^{(0)}(x)$ can only be estimated with a limited resolution;
- the restored object may be affected by Gibbs oscillations.

Both effects are related to the behaviour of the global PSF as indicated above.

The problem of resolution will be mainly considered in section 11.4. Here we only observe that details of the object $f^{(0)}(x)$ smaller than the width of the central peak of the global PSF are usually lost in the restored image. We also observe that this width is approximately related to the effective bandwidth Ω by the sampling theorem, i.e. it is of the order of π/Ω.

As concerns the second effect, it is well-known that Gibbs oscillations are produced by the truncation of Fourier series or Fourier integrals in the case of discontinuous functions. Therefore they are also produced by the bandlimiting implied by regularization and are related to the side lobes of the global PSF. To this purpose it is important to note that a reduction of the value of μ (which is possible if the noise on the image is reduced) does not imply a reduction of the side lobes. A smaller value of μ, indeed, implies a narrower and higher central peak but also higher side lobes. In the particular case of the truncated window function, the global PSF is a sinc-function, i.e. $W_\mu(x) = \sin(\Omega x)/\pi x$ with $\Omega = [\ln(1/\mu)]^{1/2}$. Hence the modulus of the ratio between the value at the first side-lobe and the central value is $|\cos \xi_1|$, where ξ_1 is the root, between $\pi/2$ and $3\pi/2$, of the equation $tg(\xi) = \xi$. This ratio is independent of Ω and it does not change when Ω is increased as an effect of noise reduction. Similar considerations apply to Tikhonov and Phillips window functions.

Gibbs oscillations appear, for instance, in the restoration of a point-like object (a bright star on a black background): the side lobes of the global PSF originate a series of rings around the central spot, so that this feature of the restored image is also known as *ringing*.

Ringing also appears in the restoration of sharp intensity variations in the object. In figure 5.11 we give the restoration of a discontinuous function whose image is provided by the Gaussian blur considered above. The Gibbs oscillations

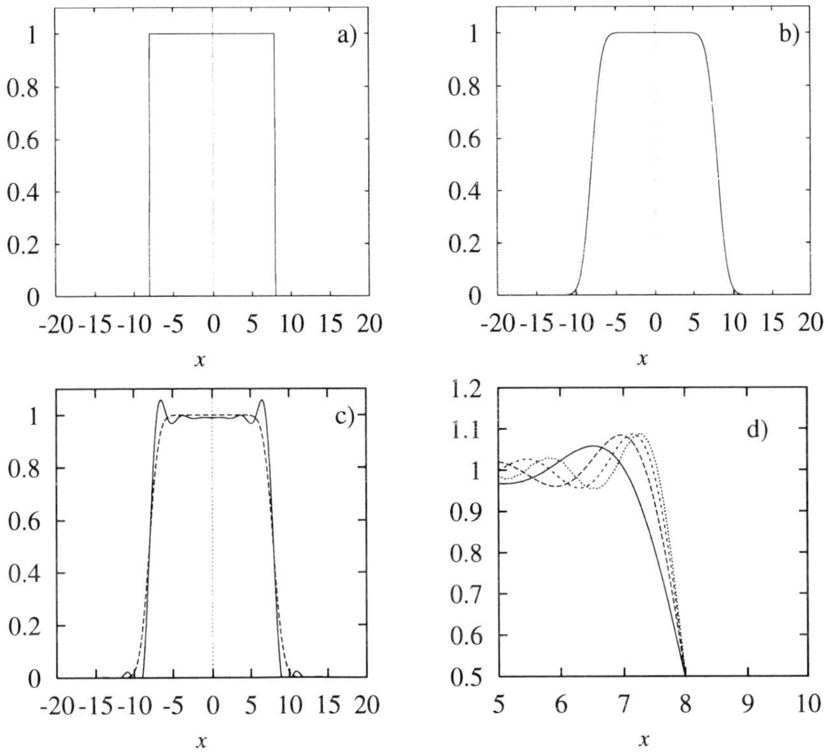

Figure 5.11. (a) The object; (b) the image provided by the Gaussian blur considered in the text; (c) Tikhonov restoration, with $\mu = 10^{-2}$, of the object in (a) (full line) compared with the image in (b) (dashed line); (d) magnification of the restorations obtained with $\mu = 10^{-2}, 10^{-4}, 10^{-6}, 10^{-8}$ in a neighbourhood of the discontinuity.

clearly appear at the edges of the squared object. The definition of the position of the discontinuity is more precise than in the blurred image but the restoration is distorted by this effect. If the value of μ is decreased, the width of the maxima and minima at the discontinuity decreases but their height does not decrease. In order to make evident this effect, which is due to the approximation error, we also provide in figure 5.11 some restorations obtained from noise-free data and corresponding to various values of the regularization parameter.

A reduction of the side lobes and therefore a reduction of the ringing can be obtained by the use of a triangular or of a Hanning window. In figure 5.12 these are compared with the rectangular window which, in the case of a Gaussian blur, is equivalent to the truncated window. It is evident that the side lobes generated by the triangular or by the Hanning window are smaller than the side lobes of the rectangular one, if we use the same band-width. However the width of their central peaks is about twice the width of the central peak of the rectangular window.

This remark suggests that, in the case of linear methods, there exists a trade-off between resolution and ringing: if we wish to reduce ringing we must lose resolution; if we wish to enhance resolution we must accept an increased ringing.

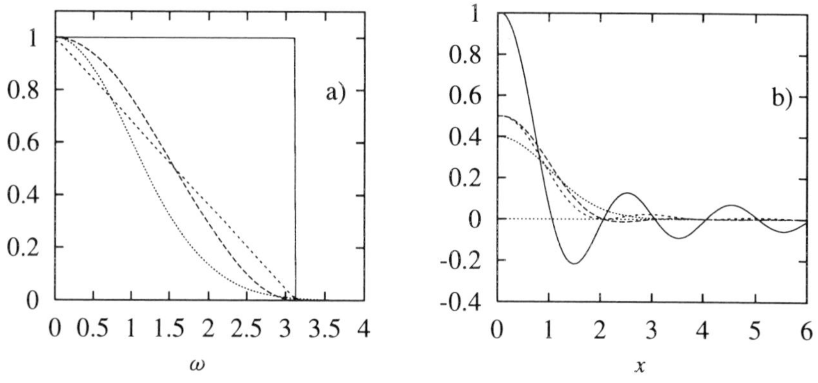

Figure 5.12. (a) Plot of the Gaussian TF (dotted line) compared with the rectangular window (solid line), triangular window (thin dashed line) and Hanning window (dashed line); (b) Plot of the Gaussian blur (dotted line) compared with the global PSF corresponding to the window functions in (a).

The previous analysis applies to all cases where the PSF is essentially bandlimited with a TF which does not vanish at low frequencies. In these cases the global PSF does not strongly depend on the PSF and, for some filtering methods, it is even independent of it.

(B) *Linear motion blur*

A quite different situation is that corresponding to the second example we are discussing, i.e. linear motion blur. If the PSF is given by equation (3.20), then from equation (3.17) we get that the modulus of the TF is given by

$$|\hat{K}(\omega)| = \left|\operatorname{sinc}\left(\frac{s\omega}{2\pi}\right)\right|. \tag{5.71}$$

In figure 5.13 we plot the Tikhonov window function for this problem (with $s = 16$, arbitrary units) and the truncated window function, both with $\mu = 10^{-2}$. We also give the corresponding global PSF. We point out that the window functions vanish at the zeros of the TF.

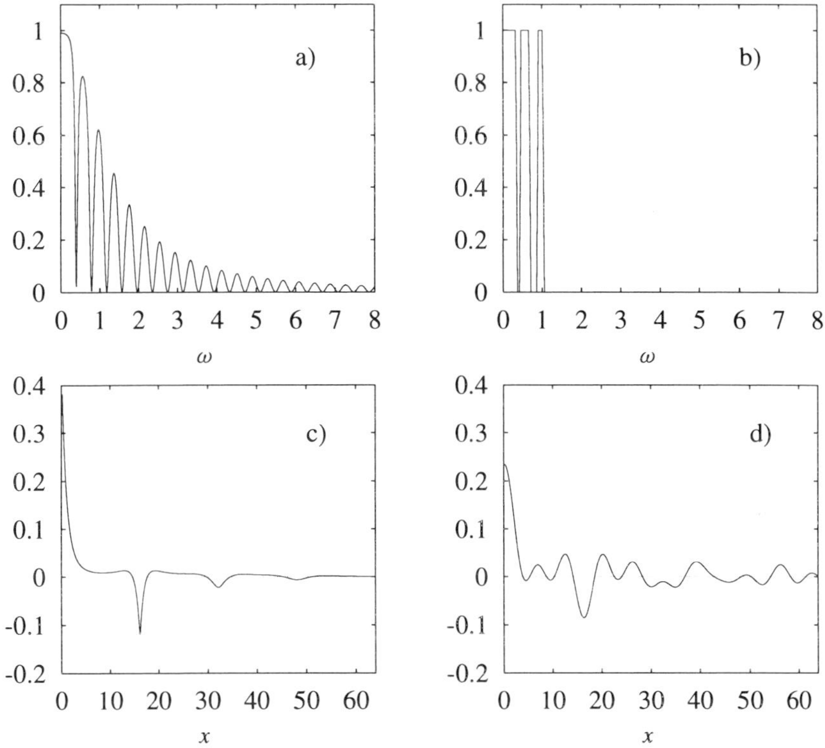

Figure 5.13. (a) The Tikhonov window function with $\mu = 10^{-2}$ for the linear motion blur in the case $s = 16$; (b) the truncated window function with the same value of μ and for the same problem; (c) the the global PSF corresponding to the Tikhonov window; (d) the global PSF corresponding to the truncated window.

As a first remark we observe that, in such a case, the truncated window is certainly less convenient than the Tikhonov window with the same value of the regularization parameter. The reason is that the modulus of the TF (5.71) at high frequencies is bounded by a factor $|\omega|^{-1}$ which tends to zero rather slowly. As a consequence, the use of a truncated window removes a large part of the frequency domain where the TF is still important. This is not true for the Gaussian blur. The reader can compare figure 5.10(a) with figure 5.13(a) and (b).

If we consider the global PSF we also see that the PSF corresponding to the truncated window is much more irregular than that corresponding to the Tikhonov window. The behaviour of the latter is rather interesting. It has a rather narrow central peak, whose FWHM (Full-Width at Half Maximum) is about 1.6 while the translation parameter of the motion blur is $s = 16$. This means that in the restored image it is possible to identify much more details than in the blurred one. Moreover, a reduction of the value of μ (which is possible if the noise is reduced), produces a reduction of the width of the central peak, hence a improvement in resolution. Indeed the width of this central peak is related to the high frequency cut-off introduced by the Tikhonov window.

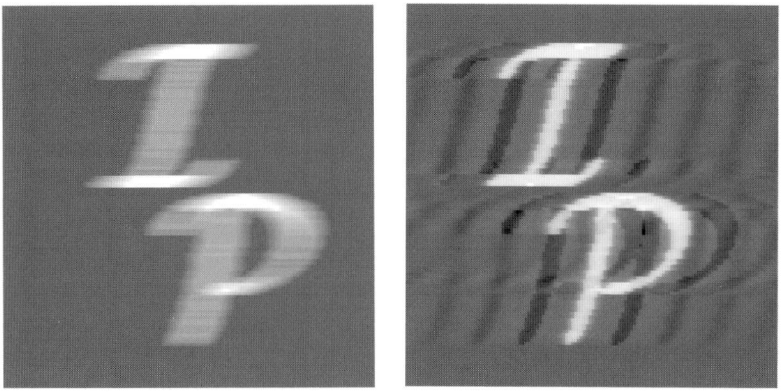

(a) (b)

Figure 5.14. Restoration by the Tikhonov method, with $\mu = 10^{-2}$, of an image blurred by motion ($S = 16$ pixels). In order to make the effect of the approximation errors evident, the image, shown to the left, was not corrupted by noise. In the restoration, shown to the right, the ghosts discussed in the text clearly appear. In the case of a noisy image these ghosts are perturbed by the noise contribution.

The central peak is flanked by a secondary and negative peak at a distance $d_0 = 16$, equal to the translation distance. The position of this peak does not depend on the value of μ but depends on the value of s because it depends on the zeros of the TF given in equation (5.71). If we reduce μ, the peak becomes narrower but its position does not change. Its effect on the restored image is the appearance of two ghosts (one on the right and one on the left) reproducing translate and negative versions of the original object. Figure 5.13(c) also suggests the appearance of other weaker ghosts at distances that are multiples of that of the first one.

This effect can be understood if we observe that the Tikhonov window function of figure 5.13 approximates the profile of a modulated grating with period equal to the distance between the zeros of the TF given in equation (5.71), i.e. $\omega_0 = 2\pi/s$. Then the inverse FT of the profile of the grating has peaks spaced by the reciprocal period $d_0 = 2\pi/\omega_0 = s$.

The appearance of ghosts is shown in figure 5.14. A picture 128×128 of the calligraphic initials of the words Inverse Problems was blurred by uniform motion corresponding to a horizontal translation of 16 pixels. In the restored image at least four ghosts are visible on both sides of the two letters.

We point out that this effect cannot be removed by the linear filtering methods considered in this chapter because it is due to the lack of information in the neighbourhoods of the zeros of the TF. In some particular cases it can be removed by some of the nonlinear methods which will be discussed in the next chapters.

5.6 Choice of the regularization parameter

The choice of the value of the regularization parameter is a crucial and difficult problem in the theory of regularization. This point has been widely discussed in the mathematical literature. No precise recipe has been discovered which could be used for any problem. In this section we consider the case of Tikhonov regularization algorithm, i.e. the regularized solution $f_\mu(\boldsymbol{x})$ is given by equation (5.12), and we summarize the main methods which are used in practice.

As we know from the discussion of section 5.4, for any image g there exists an optimum value, μ_{opt}, of the regularization parameter. For such value of μ, the corresponding regularized solution $f_\mu(\boldsymbol{x})$ has minimal distance from the true object $f^{(0)}(\boldsymbol{x})$. The problem is that the determination of this optimal value should imply the knowledge of $f^{(0)}(\boldsymbol{x})$. The methods described in section 5.1 and section 5.2 can also be considered as methods for estimating μ_{opt} when $f^{(0)}(\boldsymbol{x})$ is not known. We repropose here these methods and others which can be used in the analysis of real images.

In order to compare the various methods we also give the results obtained in the following numerical experiment. A 256×256 picture of the bell-tower of the San Donato church in Genova, given in figure 5.15, is blurred by linear motion corresponding to a horizontal translation of 8 pixels. In such a case the discrete TF, given by equation (3.22), is zero for $m = Nk/S = 32k$ $(k = 1, 2, \ldots, 8)$. However, as a consequence of round-off errors, in our experiment the condition number (given by equation (4.25)) is not infinite but of the order of the inverse of the machine precision, i.e. $\alpha = 9.68 \times 10^7$. Finally the blurred image is perturbed by white Gaussian noise (see section 7.2) with a variance whose square root is equal to 0.8% of the maximum value of the noise-free image (the procedure is described in section 12.2). The values of the square roots of the energy and of the discrepancy of the object are the following

$$E = 2.91 \times 10^4, \quad \varepsilon = 513. \tag{5.72}$$

Since we have $\|g\| = 2.78\ 10^4$, the signal-to-noise ratio, as given by equation (3.5), is SNR = 35 (dB).

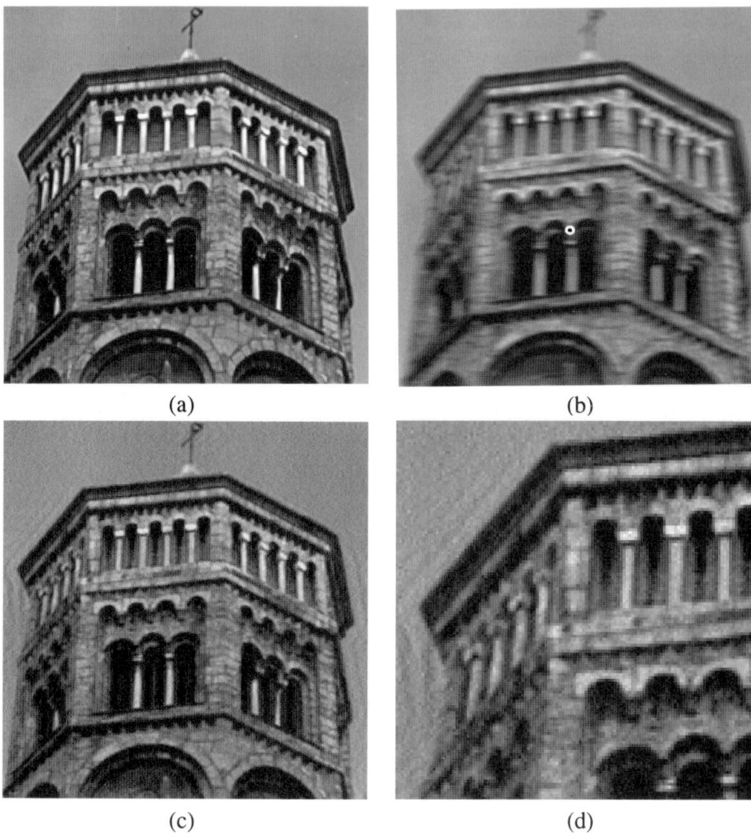

Figure 5.15. (a) The object; (b) the image blurred by linear motion (corresponding to a translation of 8 pixels) and perturbed by white Gaussian noise with SNR= 35 (dB); (c) the optimal restoration corresponding to the minimum of the restoration error; (d) detail of the restoration in (c) showing the appearance of ghosts both in the sky and in the space between the columns.

The behaviour of the relative restoration error as a function of μ is given in figure 5.6 and the optimal value of μ, corresponding to the minimum, is $\mu_{opt} = 7.0 \times 10^{-3}$. The restored image is given in figure 5.15 together with a detail showing the appearance of the ghosts discussed in the previous section.

The restoration error is about 13.2%, corresponding to a MSEIF equal to 8.7 dB (the MSEIF is defined in equation (4.29)). The restored image contains some negative values which are not very important because if we replace these values by zero (projection onto the set of positive images), the restoration error does not decrease significantly (it becomes 12.9%).

(a) Regularized solution with prescribed energy.
If we do not know $f^{(0)}(\boldsymbol{x})$ but do know its energy E^2, then the constrained least-squares problem discussed in section 5.1 can be considered as a method for estimating the optimal value of the regularization parameter. The estimate is the value $\mu_1 = \mu_1(E)$ such that the corresponding regularized solution has the same energy as the true object, i.e.

$$\|f_{\mu_1}\| = E. \tag{5.73}$$

As one can easily deduce from figure 5.2, this value is a decreasing function of E. Therefore, if we overestimate the energy of the object, we obtain a value of the regularization parameter smaller than that corresponding to the exact energy of the object. In such a case the restored image will show a higher noise contamination.

In the numerical experiment described above, using the value of E given in equation (5.72), we find $\mu_1 = 1.6 \times 10^{-3}$. This value is smaller than the optimum value by a factor of 4. However the increase in the restoration error is not dramatic, because we find a relative error of about 15.1%. The visualization (not reproduced here) of the restored image confirms that the result is still satisfactory.

(b) Regularized solution with prescribed discrepancy.
If we know a precise estimate ε of the energy of the noise, then the method discussed in section 5.2 can be considered as another method for estimating μ_{opt}. The estimate is the value $\mu_2 = \mu_2(\varepsilon)$ such that the discrepancy of the corresponding regularized solution is just equal to ε^2, i.e.

$$\|Af_{\mu_2} - g\| = \varepsilon. \tag{5.74}$$

This method is also known as the *discrepancy principle* [10]. From figure 5.4 we deduce that $\mu_2(\varepsilon)$ is an increasing function of ε. Therefore, if we overestimate the energy of the noise, we get a value of the regularization parameter which is larger than that corresponding to the exact energy of the noise. In such a case the restored image will show a smaller deblurring effect.

The value found in our numerical experiment, using the value of ε given in equation (5.72), is $\mu_2 = 1.2 \times 10^{-2}$. This value is greater than the optimum value and closer to μ_{opt} than μ_1. The increase in the restoration error is quite small since we find a relative error of about 13.5% (13.2% in the case of μ_{opt}). We point out that the range of the values of μ between μ_1 and μ_2 is rather broad and that all restorations corresponding to this interval are satisfactory. Therefore, in this example, the choice of the value of the regularization parameter is not very critical.

In the many numerical experiments we have performed, using the exact value of ε, we have found values of μ_2 sometimes greater and sometimes smaller than μ_{opt}. In all cases, however, the value of μ_2 was closer to μ_{opt} than the value of μ_1. Therefore the choice of μ based on the knowledge of the discrepancy seems better than that based on the knowledge of the energy.

(c) The Miller method

An approach to ill-posed problems proposed by Miller [8] can also be considered as a method for estimating the value of the regularization parameter. In this approach it is assumed that one knows both a bound on the energy and a bound on the discrepancy of the unknown object $f^{(0)}(\boldsymbol{x})$. Then the set of all objects $f(\boldsymbol{x})$ satisfying the two conditions

$$\|Af - g\|^2 \leq \varepsilon^2 \quad , \quad \|f\|^2 \leq E^2 \tag{5.75}$$

is called the set of *admissible approximate solutions*. This set corresponds to the shaded region of figure 5.7. It is the intersection of the ball of the objects with energy smaller than E^2 and of the ellipsoid of the objects with discrepancy smaller than ε^2. If this intersection is not empty, then the pair $\{\varepsilon, E\}$ is said to be *permissible.*

The following question arises: is it possible to establish whether a certain pair $\{\varepsilon, E\}$ is permissible for a given image $g(\boldsymbol{x})$? The answer to this question is positive and rather simple. Given $\tilde{f}_1$ or $\tilde{f}_2$, i.e. the solutions introduced in section 5.1 and section 5.2 respectively, we can check whether these functions belong or not to the set (5.75). If they do then the set is obviously not empty. If they do not, then the set is empty thanks to the following result: *if the set defined by the conditions (5.75) is not empty, then it contains both $\tilde{f}_1$ and $\tilde{f}_2$.*

The proof is easy. We give it in the case of $\tilde{f}_1$. We observe that $\tilde{f}_1$ is the object which minimizes the discrepancy functional in the set of all objects whose energy does not exceed E^2. Since there exist, by hypothesis, elements of this set whose discrepancy does not exceed ε^2, it follows that also the discrepancy of $\tilde{f}_1$ must be less than ε^2, i.e. $\tilde{f}_1$ is an admissible approximate solution. A similar argument applies to $\tilde{f}_2$ by exchanging the rolе of the two functionals.

It is obvious that this condition can be checked numerically. If the computed discrepancy of $\tilde{f}_1$ is greater than ε^2, then the set of the approximate solutions is empty. On the other hand, if the discrepancy of $\tilde{f}_1$ is smaller than ε^2, then also $\tilde{f}_2$ is an admissible approximate solution. Moreover the previous arguments also imply that, if the pair $\{\varepsilon, E\}$ is permissible, then the following inequalities hold true

$$\|\tilde{f}_2\| \leq \|\tilde{f}_1\| \quad , \quad \|A\tilde{f}_1 - g\| \leq \|A\tilde{f}_2 - g\|. \tag{5.76}$$

The first is the inequality of equation (5.36).

Finally it is possible to characterize the set of all permissible pairs. Consider in the plane $\{\varepsilon, E\}$ the curve described by the points $\{\varepsilon(f_\mu; g), E(f_\mu)\}$ when μ

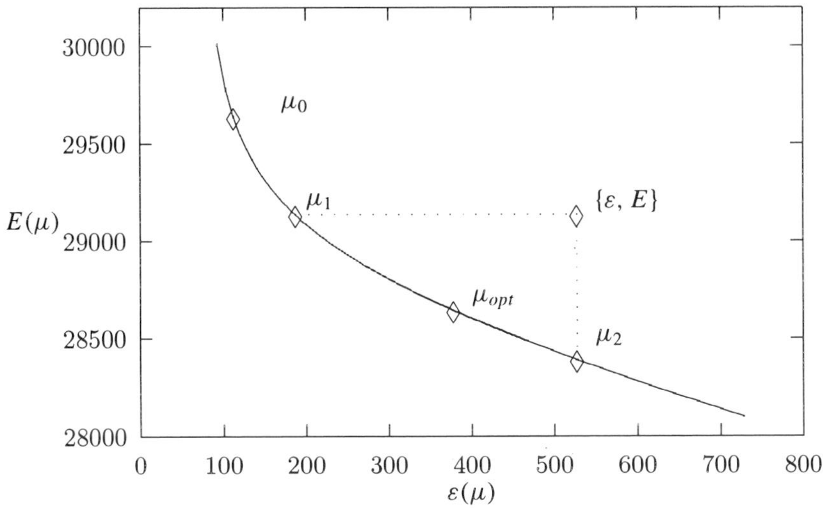

Figure 5.16. Plot of the L-curve and of the pair $\{\varepsilon, E\}$ for the example of figure 5.15. We also indicate the points on the curve corresponding to the various values of the regularization parameter obtained by means of the various methods.

varies from 0 to $+\infty$. This curve is a plot of $E(f_\mu)$ versus $\varepsilon(f_\mu; g)$ and it is called the L-curve because, in most cases, its log–log plot has the shape of the letter L. Now, if $\{\varepsilon, E\}$ is an admissible pair, we have shown that $\varepsilon \geq \|A\tilde{f}_1 - g\| = \varepsilon(f_{\mu_1}; g)$ and $E \geq \|\tilde{f}_2\| = E(f_{\mu_2})$. Therefore the point $\{\varepsilon, E\}$ is to the right and above the L-curve. Conversely any such a point is a permissible pair. In figure 5.16 we give the L-curve for the example of figure 5.15 and we indicate the pair $\{\varepsilon, E\}$ corresponding to this example.

Now, if the pair $\{\varepsilon, E\}$ is permissible, it has been shown by Miller [8] that the regularized solution corresponding to the following value of the regularization parameter

$$\mu_0 = \left(\frac{\varepsilon}{E}\right)^2 \tag{5.77}$$

satisfies the conditions (5.75) except for a factor of $\sqrt{2}$ and therefore it is essentially an admissible approximate solution.

In our numerical example we find $\mu_0 = 3.1 \times 10^{-4}$ and this value is much smaller than the values provided by the other two methods. Also the corresponding restoration error is much higher, about 21%. The visualization of the restored image shows an important contribution due to the noise propagation. Indeed all numerical experiments seem to indicate that, when the correct values of ε and E are used, the Miller method tends to underestimate the value of the regularization parameter.

(d) Generalized cross-validation

The methods considered previously require the knowledge of ε or of E or of both. In many cases one does not have a sufficiently accurate estimate of these quantities and therefore it is important to have methods which do not require this kind of information. One such method is that of *cross-validation* [11] which can only be used in problems with discrete data and is based on the idea of letting the data themselves choose the value of the regularization parameter. In other words it is required that a good value of the regularization parameter should predict missing data values.

In order to simplify the notation, let us assume that a discrete image g, represented by an array of numbers $g_{m,n}$, has been rearranged, for instance by means of lexicographic ordering, in order to obtain a vector with N^2 components, denoted by g_m. Then also the imaging matrix $\mathbf{A}$ is replaced by a matrix with only two entries and its matrix elements are denoted by $A_{m,n}$. We still use the notation $\mathbf{A}$ for this matrix.

Consider now the problem where the kth data component is missing. We denote by $f_{\mu,k}$ the regularized solution for this problem, i.e. the vector which minimizes the functional

$$\Phi_{\mu,k}(f, g) = \sum_{m \neq k} |(\mathbf{A}f)_m - g_m|^2 + \mu \sum_{n=1}^{N^2} |f_n|^2. \tag{5.78}$$

By means of $f_{\mu,k}$ one can compute the missing component, i.e. $(\mathbf{A}f_{\mu,k})_k$. If μ is a good choice, the quantity $(\mathbf{A}f_{\mu,k})_k - g_k$ should, on average, be small. The *cross-validation function* $V_0(\mu)$ is precisely the mean quadratic error we commit when we repeat this procedure for all components of the image

$$V_0(\mu) = \frac{1}{N^2} \sum_{k=1}^{N^2} |(\mathbf{A}f_{\mu,k})_k - g_k|^2 \tag{5.79}$$

and the *cross-validation method* consists of determining the value of μ which minimizes $V_0(\mu)$.

The computation of $V_0(\mu)$ does not require the solution of N^2 minimization problems. Indeed, if we denote by $\mathbf{A}(\mu)$ the following matrix

$$\mathbf{A}(\mu) = \mathbf{A}\mathbf{A}^*(\mathbf{A}\mathbf{A}^* + \mu\mathbf{I})^{-1}, \tag{5.80}$$

where $\mathbf{A}^*$ is the adjoint matrix defined in equation (2.69), and by f_μ the regularized solution (5.16), then the following relation holds true [12], [13]

$$V_0(\mu) = \frac{1}{N^2} \sum_{k=1}^{N^2} \frac{|(\mathbf{A}f_\mu)_k - g_k|^2}{|1 - A_{kk}(\mu)|^2} \tag{5.81}$$

which can be used for the computation of $V_0(\mu)$.

The main disadvantage of the method is that it is not invariant with respect to a linear transformation of the components of the image and of the object as that provided, for instance, by the DFT. The DFT, indeed, transforms the matrix $\mathbf{A}$ into a diagonal one but, in the case of a diagonal matrix, the cross-validation function is a constant, i.e. $V_0(\mu) = N^{-1}\|g\|^2$.

For this reason, the minimization of the cross-validation function, in spite of its clear meaning, is replaced by the minimization of the *generalized cross-validation function* (GCV function) defined by

$$V(\mu) = \frac{\|\mathbf{A}f_\mu - g\|^2}{(Tr[\mathbf{I} - \mathbf{A}(\mu)])^2} \tag{5.82}$$

where

$$Tr[\mathbf{I} - \mathbf{A}(\mu)] = \sum_{n=1}^{N^2} \left[1 - A_{n,n}(\mu)\right], \tag{5.83}$$

the matrix $\mathbf{A}(\mu)$ being defined in equation (5.80). This function is invariant with respect to rotations of the data vector and is obtained from equation (5.81) by replacing the denominators with their arithmetic mean.

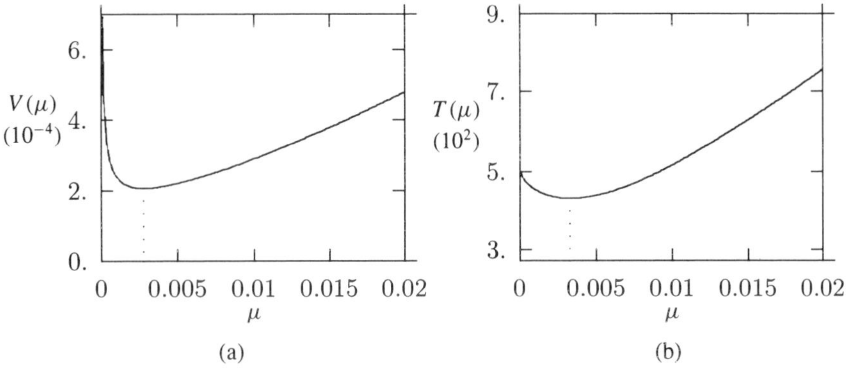

Figure 5.17. Plot of the GCV function, (a), and of the predictive mean square error, (b), in the case of the example of figure 5.15.

The GCV function has another important property. As it has been shown [12], [13], the minimizer of $V(\mu)$ provides an estimate of the minimizer of the so-called *predictive mean square error* defined by

$$T(\mu) = \|\mathbf{A}f_\mu - g^{(0)}\|, \tag{5.84}$$

where $g^{(0)}$ is the noise-free image, i.e. $g^{(0)} = \mathbf{A}f^{(0)}$.

Also this function has a unique minimum as the norm of the restoration error investigated in section 5.3, i.e. $\|f_\mu - f^{(0)}\|$. The minimization of $T(\mu)$ provides another criteria of optimality. We will denote by μ'_{opt} the minimizer of $T(\mu)$ and by μ_{GCV} the minimizer of $V(\mu)$.

In figure 5.17 we plot the GCV function $V(\mu)$ and the predictive mean square error $T(\mu)$ as functions of μ in the case of the example of figure 5.15. The minimum of $T(\mu)$ is $\mu'_{opt} = 3.3 \times 10^{-3}$ and therefore is smaller, by a factor of 2, than the optimal value μ_{opt} but larger than μ_1. The restoration error is about 13.7%. On the other hand the minimum of the GCV function is $\mu_{GCV} = 2.8 \times 10^{-3}$ and provides a satisfactory approximation of the minimum of $T(\mu)$. In this case the restoration error is 13.9%.

In table 5.1 we summarize all results obtained in the case of the example of figure 5.15.

Table 5.1. The values of the regularization parameter and the corresponding restoration errors obtained by the various methods in the example of figure 5.15.

	μ_{opt}	μ_1	μ_2	μ_0	μ'_{opt}	μ_{GCV}
μ	7.0×10^{-3}	1.6×10^{-3}	1.2×10^{-2}	3.1×10^{-4}	3.3×10^{-3}	2.8×10^{-3}
rest. error	13.2%	15.1%	13.5%	21.0%	13.7%	13.9%

(e) L-curve method

This graphically motivated method, introduced by Hansen [14], is another method which does not require information about the energy of the noise or of the true object. The starting point is that the L-curve, introduced in connection with Miller method, has, in many cases, a rather characteristic L-shaped behaviour in a log–log plot. This plot is given in figure 5.18 for the numerical example of figure 5.15.

A qualitative explanation of this behaviour is the following. We recall that $E(f_\mu)$ is large for small μ and small for large μ while $\varepsilon(f_\mu; g)$ has an opposite behaviour. Therefore $E(f_\mu)$ is large when $\varepsilon(f_\mu; g)$ is small, and conversely. This is the trade-off between noise-propagation error and approximation error already discussed in section 5.3. Now the vertical part of the L-curve corresponds to values of the regularization parameter such that f_μ is dominated by the noise propagation error. As a consequence $E(f_\mu)$ is very sensitive to variations of μ while $\varepsilon(f_\mu; g)$ is not. Analogously the horizontal part of the L-curve corresponds to values of the regularization parameter such that f_μ is dominated by the approximation error. As a consequence $\varepsilon(f_\mu; g)$ is very sensitive to variations of μ while $E(f_\mu)$ is not.

Now the L-curve method consists of taking as an estimate of the regularization parameter the value of μ, which we denote by μ_L, corresponding to the corner of the L-curve. In fact this point should correspond to the best compromise between approximation error and noise-propagation error. From

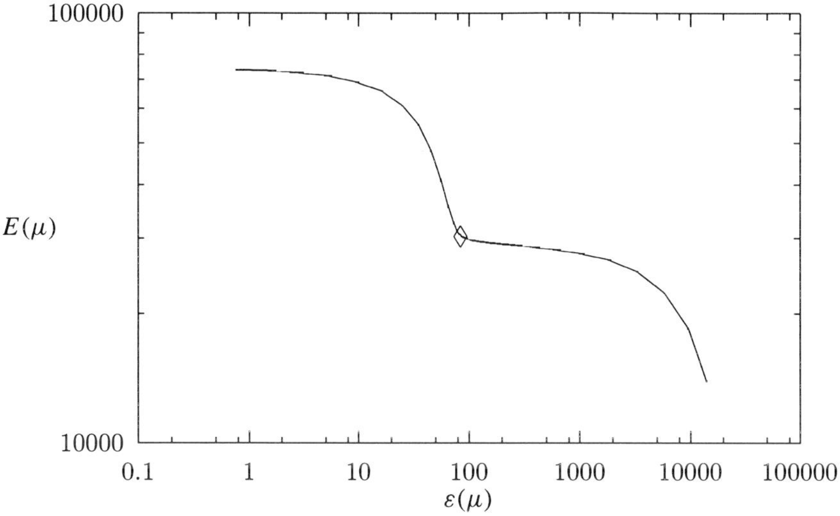

Figure 5.18. Log–log plot of the L-curve of figure 5.16

the computational point of view a convenient definition of the L-curve corner is the point with maximum curvature. Good performance of the method has been reported in the case of numerical experiments [15]. However in our example we find $\mu_L = 2.3 \times 10^{-7}$ and this is an unacceptable value.

Indeed the L-curve method, even if it can be useful in some cases, does not work in all cases and has some theoretical and practical inconveniences. It has been recently shown [16], [17] that, in certain cases, it does not provide a regularized solution converging to the exact one when the noise tends to zero. Moreover, examples can be found where the L-curve does not even have an L-shape so that the method cannot be used.

(f) The interactive method

The speed and versatility of modern digital computers allows us to restore images interactively: the user controls the restorations obtained by means of several values of the regularization parameter and, by tuning μ, he selects the best restoration on the base of his intuition or of the attainment of some specific purpose. Figure 5.8 clearly shows that excellent results can be obtained by means of this method.

References

[1] Ivanov V K 1962 *Soviet Math. Dokl.* **3** 981

[2] Phillips D L 1962 *J. Assoc. Comp. Mach.* **9** 84

[3] Tikhonov A N 1963 *Soviet Math. Dokl.* **4** 1035

[4] Bertero M 1986 *Inverse Problems* (ed G Talenti) Lecture Notes in Mathematics, vol 1225 (Berlin: Springer) 52

[5] Ivanov V K 1966 *USSR Comp. Math. Phys.* **6** 197

[6] Tikhonov A N and Arsenin V Y 1977 *Solutions of ill-posed problems* (Washington: Winston/Wiley)

[7] Bertero M 1989 *Advances in Electronics and Electron Physics* (ed P W Hawkes) vol 75 (London: Academic Press) p 1

[8] Miller K 1970 *SIAM J. Math. Anal.* **1** 52

[9] Kunt M 1986 *Digital Signal Processing* (Norwood: Artech House)

[10] Morozov V A 1984 *Methods for Solving Incorrectly Posed Problems* (Berlin: Springer)

[11] Wahba G 1977 *SIAM J. Numer. Anal.* **14** 651

[12] Craven P and Wahba G 1979 *Numer. Math.* **31** 377

[13] Golub G H, Heath M and Wahba G 1979 *Technometric* **21** 215

[14] Hansen P C 1992 *SIAM Rev.* **34** 561

[15] Hanke M and Hansen P C 1993 *Surveys Math. Indust.* **3** 253

[16] Engl H W and Grever W 1994 *Numer. Math.* **69** 25

[17] Vogel C R 1996 *Inverse Problems* **12** 535

Chapter 6

Iterative regularization methods

Some iterative methods, which have been introduced to solve linear algebraic systems, can also be used to regularize the solution of linear ill-posed problems. The basic feature is that the number of iterations plays the role of a regularization parameter because semiconvergence holds true in the case of noisy images: when the number of iterations increases, the iterates first approach the unknown object and then go away.

The simplest iterative method having this property is the so-called Landweber or successive-approximation method. Another one, which is more used in practice, is the conjugate gradient method. The main practical difference between the two methods is that the convergence to a sensible approximate solution is faster in the case of the conjugate gradient than in the case of the Landweber method.

However, an interesting feature of the Landweber method is that it can be modified in order to take into account additional *a priori* information about the solution. The resulting projected Landweber method can be used for solving some of the constrained least-squares problems discussed in section 4.7.

6.1 The van Cittert and Landweber methods

A very simple iterative method for approximating the least-squares solutions of first-kind integral equations has been introduced independently by Landweber [1] and Fridman [2]. For this reason, the method is usually called the *Landweber* or the *Landweber–Fridman method*, at least in the Western literature concerning ill-posed problems. In the Russian literature the denomination of *successive-approximations method* is preferred. It is probably more correct to call it the *Jacobi method* because the basic idea is the same as in an iterative method introduced by Jacobi for solving linear algebraic systems.

The convergence of the method was investigated by Bialy [3] in the general case of a linear and ill-posed operator equation. A particular version of the same method was proposed independently by Gerchberg [4] and Papoulis [5] for the problem of extrapolating a bandlimited signal. This algorithm will be discussed in chapter 11. It can also be shown that the Cimmino method is a particular case of the Landweber method [6]. Finally the *simultaneous iterative reconstruction technique* (SIRT), which is frequently used in seismic tomography, can also be reduced to the Landweber method if suitable weights are introduced in the data and solution space [7].

In the problem of image deconvolution the Landweber method applies to the least-squares equation introduced in section 4.5, i.e.

$$A^*Af = A^*g \tag{6.1}$$

where A^*A is a convolution operator whose TF is $|\hat{K}(\omega)|^2$, if $\hat{K}(\omega)$ is the TF of the imaging system. However, in the field of image deconvolution a method introduced by van Cittert [8] is also considered. The iterative scheme is the same of the Landweber method but it is applied to the original imaging equation

$$Af = g \tag{6.2}$$

and not to the least-squares equation (6.1). In this context the method of Landweber is also called a *reblurring method* [9] because equation (6.1) can be obtained from equation (6.2) by applying to both members the blurring operator A^*. As we will see, the van Cittert method can be used only in the case of imaging systems having some particular properties, while the method of Landweber can be applied to any imaging system. In order to treat simultaneously the two methods we consider a convolution equation of the following form

$$\bar{A}f = \bar{g}. \tag{6.3}$$

If $\bar{A} = A^*A$ and $\bar{g} = A^*g$ we get equation (6.1) while if $\bar{A} = A$ and $\bar{g} = g$ we get equation (6.2).

A very simple way to introduce both methods is the following: consider the operator

$$T(f) = f + \tau(\bar{g} - \bar{A}f) \tag{6.4}$$

where τ is the so-called *relaxation parameter* whose choice will be discussed in the following. Notice that this operator is nonlinear because of the inhomogeneous term $\tau\bar{g}$. Then any solution of equation (6.3) is also a *fixed point* of the operator T and conversely, i.e. equation (6.3) is equivalent to the following fixed point equation

$$f = T(f). \tag{6.5}$$

If we use the well-known *method of successive approximations*, which is the most natural way to approximate the fixed points of a contraction mapping, we obtain the following iterative scheme

$$f_{k+1} = T(f_k) \tag{6.6}$$

or, in a more explicit form,

$$f_{k+1} = f_k + \tau(\bar{g} - \bar{A} f_k). \tag{6.7}$$

The main difficulty in the case of an ill-posed problem is that the operator (6.4) is not a contraction mapping even if it is non-expansive (see appendix F), so that the convergence of the method is not ensured.

In order to investigate the convergence of the iterative method, it is convenient to write equation (6.7) as follows

$$f_{k+1} = \tau\bar{g} + (1 - \tau\bar{A}) f_k. \tag{6.8}$$

If we denote by $\hat{H}(\omega)$ the TF of the convolution operator $\bar{A}$ (notice that $H(\omega) = K(\omega)$ if $\bar{A} = A$ and $H(\omega) = |K(\omega)|^2$ if $\bar{A} = A^*A$), then this equation implies the following equation for the Fourier transforms

$$\hat{f}_{k+1}(\omega) = \tau\hat{\bar{g}}(\omega) + (1 - \tau\hat{H}(\omega))\hat{f}_k(\omega). \tag{6.9}$$

Let us assume again, for generality, that the imaging system has a band $\mathcal{B}$ while $\bar{\mathcal{B}}$ is the set of the out-of-band frequencies. As already remarked, this is not restrictive since $\bar{\mathcal{B}}$ can be empty, if $\mathcal{B}$ coincides with the whole frequency plane.

If $\hat{f}_0(\omega)$ is the FT of the initial approximation f_0, it is easy to show by induction that

$$\begin{aligned} \hat{f}_k(\omega) &= (1 - \tau\hat{H}(\omega))^k \hat{f}_0(\omega) \\ &+ \tau\left\{1 + (1 - \tau\hat{H}(\omega)) + (1 - \tau\hat{H}(\omega))^2 + \ldots + (1 - \tau\hat{H}(\omega))^{k-1}\right\}\hat{\bar{g}}(\omega). \end{aligned} \tag{6.10}$$

When ω is in $\bar{\mathcal{B}}$ from this equation we obtain

$$\hat{f}_k(\omega) = \hat{f}_0(\omega) + k\tau\hat{\bar{g}}(\omega). \tag{6.11}$$

Therefore if $\hat{\bar{g}}(\omega) = 0$ in $\bar{\mathcal{B}}$, then $\hat{f}_k(\omega)$ coincides with the initial approximation $\hat{f}_0(\omega)$ for any k. Otherwise the sequence does not converge. We reconsider this point in the following. On the other hand, for spatial frequencies such that $\hat{H}(\omega) \neq 0$, from the well-known relation

$$1 + \xi + \xi^2 + \ldots + \xi^{k-1} = \frac{1 - \xi^k}{1 - \xi} \quad (\xi \neq 1), \tag{6.12}$$

with $\xi = 1 - \tau\hat{H}(\omega)$, and from equation (6.10) we obtain

$$\hat{f}_k(\omega) = (1 - \tau\hat{H}(\omega))^k \hat{f}_0(\omega) + \{1 - (1 - \tau\hat{H}(\omega))^k\}\frac{\hat{\bar{g}}(\omega)}{\hat{H}(\omega)}. \tag{6.13}$$

From inspection of this equation it follows that, for any given ω such that $\hat{H}(\omega) \neq 0$, the limit of $\hat{f}_k(\omega)$, for $k \to \infty$, exists and is precisely $\hat{\bar{g}}(\omega)/\hat{H}(\omega)$ if and only if

$$|1 - \tau \hat{H}(\omega)| < 1. \tag{6.14}$$

We discuss the implications of the previous results by considering separately the van Cittert and the Landweber method.

(A) *The van Cittert method*

This is the iterative method (6.7) applied to equation (6.2), i.e. $\bar{A} = A$ and $\bar{g} = g$. If the imaging system is bandlimited and the image is affected by out-of-band noise, then equation (6.11) implies that the method does not converge. Convergence can only be obtained if the out-of-band noise is suppressed by means of some filtering. On the other hand, when ω is in $\mathcal{B}$, equation (6.13), with $\hat{H}(\omega) = \hat{K}(\omega)$ and $\hat{\bar{g}}(\omega) = \hat{g}(\omega)$, takes the following form

$$\hat{f}_k(\omega) = (1 - \tau \hat{K}(\omega))^k \hat{f}_0(\omega) + \{1 - (1 - \tau \hat{K}(\omega))^k\} \frac{\hat{g}(\omega)}{\hat{K}(\omega)} \tag{6.15}$$

and condition (6.14) becomes

$$|1 - \tau Re\hat{K}(\omega)|^2 + \tau^2 |Im\hat{K}(\omega)|^2 < 1. \tag{6.16}$$

If $Re\hat{K}(\omega)$ does not have a definite sign, then it is impossible to find a value of τ such that condition (6.16) is satisfied for any ω. If $Re\hat{K}(\omega)$ has a definite sign, then it is not restrictive to assume that

$$Re\hat{K}(\omega) > 0. \tag{6.17}$$

In such a case τ must be positive and must also satisfy, for any ω, the following condition derived from (6.16)

$$\tau |\hat{K}(\omega)|^2 - 2Re\hat{K}(\omega) < 0. \tag{6.18}$$

Such a value of τ exists if and only if

$$\hat{K}_+ = \sup_\omega \frac{|\hat{K}(\omega)|^2}{Re\hat{K}(\omega)} < \infty; \tag{6.19}$$

then the values of τ ensuring the convergence of $\hat{f}_k(\omega)$ for any ω in $\mathcal{B}$ are given by

$$0 < \tau < \frac{2}{\hat{K}_+}. \tag{6.20}$$

Condition (6.17) however is rather restrictive. It is not satisfied, for instance, by the uniform motion blur, by the out-of-focus blur, etc. In the case of a blur whose TF is real valued and satisfies condition (6.17) (for instance a Gaussian blur), condition (6.20) becomes

$$0 < \tau < \frac{2}{\hat{K}_{max}} \tag{6.21}$$

where $\hat{K}_{max}$ is the maximum value of $\hat{K}(\omega)$. In such a case the van Cittert method can be used and has a regularization effect analogous but not equivalent to that of the Landweber method, which is discussed in the following.

(B) *The Landeweber method*

This is the iterative method (6.7) applied to equation (6.1), i.e. $\bar{A} = A^*A$ and $\bar{g} = A^*g$. In such a case, if the system is bandlimited, then $\hat{\bar{g}}(\omega) = 0$ when ω is in $\bar{\mathcal{B}}$ and therefore from equation (6.11) one obtains $\hat{f}_k(\omega) = \hat{f}_0(\omega)$, for any k. On the other hand, when ω is in $\mathcal{B}$, equation (6.13), with $\hat{H}(\omega) = |\hat{K}(\omega)|^2$ and $\hat{\bar{g}}(\omega) = \hat{K}^*(\omega)\hat{g}(\omega)$, takes the following form

$$\begin{aligned} \hat{f}_k(\omega) &= (1 - \tau|\hat{K}(\omega)|^2)^k \hat{f}_0(\omega) \\ &\quad + \left\{1 - (1 - \tau|\hat{K}(\omega)|^2)^k\right\} \frac{\hat{g}(\omega)}{\hat{K}(\omega)} \end{aligned} \tag{6.22}$$

and condition (6.14) becomes

$$-1 < 1 - \tau|\hat{K}(\omega)|^2 < 1 \tag{6.23}$$

or also

$$0 < \tau < \frac{2}{|\hat{K}(\omega)|^2}. \tag{6.24}$$

This condition is satisfied for any ω in $\mathcal{B}$ if the relaxation parameter is such that

$$0 < \tau < \frac{2}{\hat{K}_{max}^2} \tag{6.25}$$

where now $\hat{K}_{max}$ is the maximum value of $|\hat{K}(\omega)|$. In such a case, equation (6.22) has a limit for any ω and any kind of blur, when $k \to \infty$.

For any ω in $\mathcal{B}$ the limit is given by

$$\lim_{k\to\infty} \hat{f}_k(\omega) = \frac{\hat{g}(\omega)}{\hat{K}(\omega)}; \tag{6.26}$$

while, when ω is in $\bar{\mathcal{B}}$, from equation (6.11) we get that $\hat{f}_k(\omega)$ is always equal to $\hat{f}_0(\omega)$. It follows that, if the generalized solution $f^\dagger(x)$ exists, i.e. the image

$g(\boldsymbol{x})$ satisfies condition (4.35), then the limit of the iterates is the following least-squares solution

$$f(\boldsymbol{x}) = f^{\dagger}(\boldsymbol{x}) + \frac{1}{(2\pi)^q}\int_{\bar{\mathcal{B}}} \hat{f}_0(\boldsymbol{\omega})e^{i\boldsymbol{x}\cdot\boldsymbol{\omega}}d\boldsymbol{\omega}. \tag{6.27}$$

More precisely, from general results on the convergence of the Landweber method [3] one can prove that the limit is in the sense of the L^2-norm, i.e. $\|f_k - f\| \to 0$ when $k \to \infty$. If $f_0(\boldsymbol{x}) = 0$, then the limit is just the generalized solution $f^{\dagger}(\boldsymbol{x})$.

The choice $f_0(\boldsymbol{x}) = 0$ is the most simple one for obtaining approximations of the generalized solution and, for this reason, it is the most frequently used. We mainly consider this case.

As we know from chapter 4, the generalized solution in general does not exist for noisy data or, if it exists as in the case of discrete images, it is deprived of any physical meaning due to the effect of noise propagation from data to solution. In such a case the Landweber method can be used as a regularization algorithm, in the sense defined in section 5.3. Indeed, from equation (6.22) with $\hat{f}_0(\boldsymbol{\omega}) = 0$ we find that the result of the kth iteration can be written as follows

$$f_k(\boldsymbol{x}) = (R^{(k)}g)(\boldsymbol{x}) = \frac{1}{(2\pi)^q}\int_{\mathcal{B}} \hat{W}^{(k)}(\boldsymbol{\omega})\frac{\hat{g}(\boldsymbol{\omega})}{\hat{K}(\boldsymbol{\omega})}e^{i\boldsymbol{x}\cdot\boldsymbol{\omega}}d\boldsymbol{\omega} \tag{6.28}$$

where

$$\hat{W}^{(k)}(\boldsymbol{\omega}) = 1 - (1 - \tau|\hat{K}(\boldsymbol{\omega})|^2)^k. \tag{6.29}$$

Notice that, if $\hat{K}(\boldsymbol{\omega}_0) = 0$ for a certain frequency $\boldsymbol{\omega}_0$ in the band $\mathcal{B}$, then we must put $\hat{W}^{(k)}(\boldsymbol{\omega}_0)/\hat{K}(\boldsymbol{\omega}_0) = 0$, as follows from equation (6.10).

Equation (6.29) has the same structure of equation (5.50) and it is easy to show that the functions $\hat{W}^{(k)}(\boldsymbol{\omega})$ satisfy the conditions, stated in section 5.4, which characterize the window functions. Therefore equations (6.28) and (6.29) define a regularization algorithm, the parameter being now the number of iterations.

It is interesting to remark that a given number of iterations produces a filtering effect which is similar to that produced by a value of the regularization parameter μ, proportional to the inverse of the number of iterations. Indeed, from equation (5.51) with $\hat{P}(\boldsymbol{\omega}) = 1$, we find that the Tikhonov window function, for small values of $\hat{K}(\boldsymbol{\omega})$, is approximately given by

$$\hat{W}_{\mu}(\boldsymbol{\omega}) \simeq \frac{1}{\mu}|\hat{K}(\boldsymbol{\omega})|^2. \tag{6.30}$$

Similarly $\hat{W}^{(k)}(\boldsymbol{\omega})$, for small values of $\hat{K}(\boldsymbol{\omega})$, is given by

$$\hat{W}^{(k)}(\boldsymbol{\omega}) \simeq \tau k|\hat{K}(\boldsymbol{\omega})|^2 \tag{6.31}$$

and therefore the two windows take approximately the same values if $\mu = 1/\tau k$. We call $\hat{W}^{(k)}(\boldsymbol{\omega})$ the *Landweber window function*.

In practice, the Tikhonov and Landweber window functions provide rather similar results. The comparison can be made as follows. Without loss of generality we can assume $|\hat{K}_{max}| = 1$ so that we can choose $\tau = 1$. Then the two window functions can be expressed in terms of the following functions

$$\hat{W}_\mu(\lambda) = \frac{\lambda}{\lambda + \mu}, \qquad \hat{W}^{(k)}(\lambda) = 1 - (1 - \lambda)^k \tag{6.32}$$

defined for $0 \leq \lambda \leq 1$. These functions are plotted in figure 6.1, as functions of $x = -\log_{10} \lambda$, in the case $\mu = 10^{-3}$ and $k = 10^3$, respectively. As we see, they are rather similar even if $\hat{W}^{(k)}(\lambda)$ is always above $\hat{W}_\mu(\lambda)$ and is practically equal to 1 up to $\lambda \simeq 10^{-2}$. This explains why one obtains rather similar results by means of the Tikhonov and Landweber methods. It is also clear that, if one uses the Landweber method as a true iterative method and not simply as a filtering method then one in general needs a very large number of iterations to obtain reliable results. This remark applies to the case $\tau = 1$. An acceleration of the method can be obtained by increasing the value of τ. Since the product τk is equivalent to the inverse of the regularization parameter μ, it is obvious that, by increasing τ we can get the same regularizing effect with a smaller number of iterations.

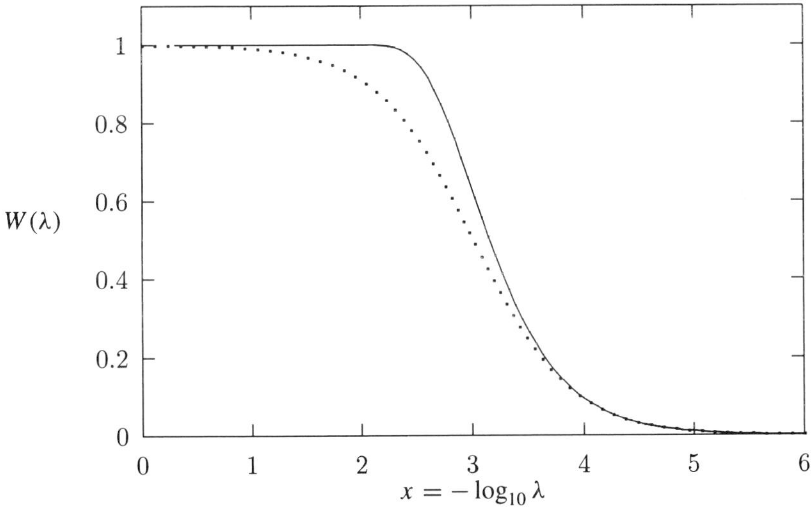

Figure 6.1. Comparison of the behaviour of the Tikhonov window function with $\mu = 10^{-3}$ (dotted line) and of the Landweber window function with $\tau = 1$ and $k = 10^3$ (full line).

Both the energy of f_k, $E^2(f_k) = \|f_k\|^2$, and the discrepancy of f_k, $\varepsilon(f_k; g) = \|Af_k - g\|^2$ can be easily computed from equation (6.28). We find that

$E(f_k)$ is an increasing function of k while $\varepsilon(f_k; g)$ is a decreasing function of k. In particular $E(f_k)$ increases from 0 to $+\infty$ (or to $E(f^\dagger)$ if $E(f^\dagger) < \infty$) while $\varepsilon(f_k; g)$ decreases from $\|g\|$ to $\|g_{out}\|$ (defined in section 5.2). The behaviour of these functions is plotted in figure 6.2 in the case of the numerical example of section 5.6. The value of the relaxation parameter is $\tau = 1$.

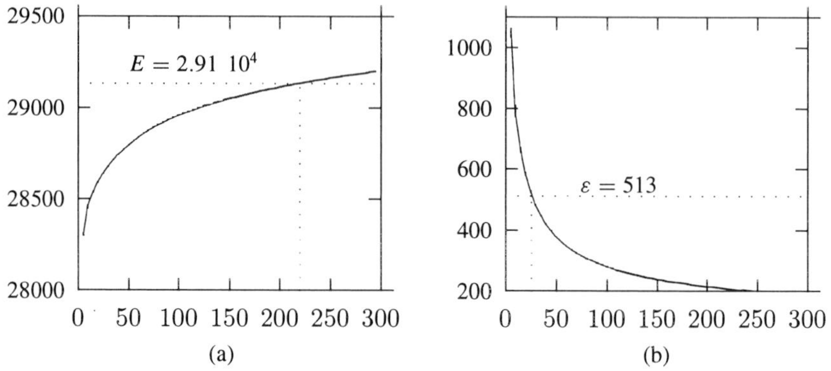

Figure 6.2. Behaviour of $E(f_k)$, (a) and of $\varepsilon(f_k; g)$, (b), as functions of the number of iterations k, in the case of the numerical example of figure 5.15. We also indicate the values of $E = \|f^{(0)}\|$ and $\varepsilon = \|Af^{(0)} - g\|$ corresponding to this numerical example as well as the numbers of iterations such that $\|f_k\| = E$ and $\|Af_k - g\| = \varepsilon$.

We conclude by investigating the behaviour of the restoration error as a function of k when the image g is noisy. As in the case of equation (5.32) the error consists of two terms

$$R^{(k)}g - P^{(\mathcal{B})}f^{(0)} = (R^{(k)}Af^{(0)} - P^{(\mathcal{B})}f^{(0)}) + R^{(k)}w. \tag{6.33}$$

The first is the approximation error while the second is the error due to noise propagation.

The square norm of the approximation error is given by

$$\|R^{(k)}Af^{(0)} - P^{(\mathcal{B})}f^{(0)}\|^2 = \frac{1}{(2\pi)^q}\int_{\mathcal{B}} |\hat{W}^{(k)}(\omega) - 1|^2|\hat{f}^{(0)}(\omega)|^2 d\omega$$
$$= \frac{1}{(2\pi)^q}\int_{\mathcal{B}} (1 - \tau|\hat{K}(\omega)|^2)^{2k}|\hat{f}^{(0)}(\omega)|^2 d\omega \tag{6.34}$$

and therefore this error is a decreasing function of k, when the relaxation parameter satisfies the inequalities (6.25): it takes the value $\|P^{(\mathcal{B})}f^{(0)}\|$ for $k = 0$ and tends to zero for $k \to \infty$. The behaviour is plotted in figure 6.3(a) also in the case of the numerical example discussed in section 5.6.

On the other hand the square norm of the error due to the noise propagation

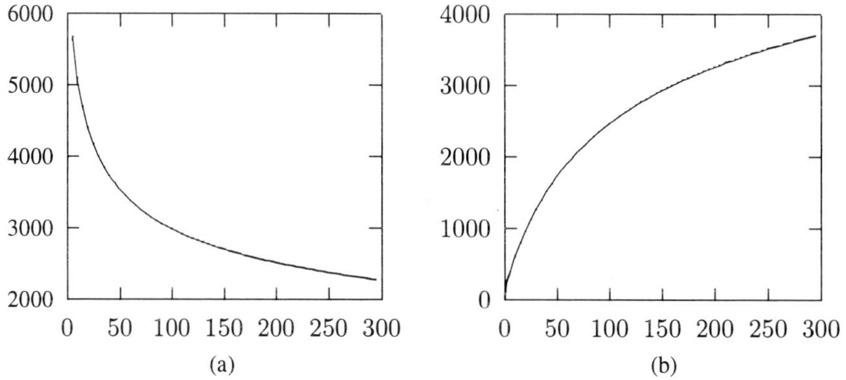

Figure 6.3. (a) Behaviour of the norm of the approximation error; (b) behaviour of the norm of the noise-propagation error, both as functions of k. The figures correspond to the numerical example of figure 5.15.

is given by

$$\|R^{(k)}w\|^2 = \frac{1}{(2\pi)^q}\int_{\mathcal{B}} |1-(1-\tau|\hat{K}(\omega)|^2)^k|^2 \frac{|\hat{w}(\omega)|^2}{|\hat{K}(\omega)|^2} d\omega \qquad (6.35)$$

and this is an increasing function of k, which is zero for $k = 0$ and tends to infinity (or to a very large number) when $k \to \infty$ (see figure 6.3(b)).

By combining the behaviour of the approximation error and that of the noise-propagation error, we deduce that the norm of the restoration error has a minimum for a certain value of k. This behaviour implies that, when we increase the number of iterations, the iterates first approach $P^{(\mathcal{B})}f^{(0)}$ and then go away, i.e. the Landweber method has the *semiconvergence property* which was already stressed in the case of the Tikhonov regularization algorithm. Therefore there exists an optimum value of the number of iterations, k_{opt}, corresponding to an iterate which has minimal distance from $P^{(\mathcal{B})}f^{(0)}$.

In figure 6.4 we plot the behaviour of the relative norm of the restoration error, i.e. the quantity $\|f_k - f^{(0)}\|/\|f^{(0)}\|$, as a function of k, for the numerical example of section 5.6. Since in this example $\hat{K}_{max} = 1$, the range of variation of the relaxation parameter is: $0 < \tau < 2$. In figure 6.4 the curves correspond to the values $\tau = 1$ and $\tau = 1.8$. In the case $\tau = 1.8$ the minimum of the restoration error is reached after a number of iterations smaller than that corresponding to the case $\tau = 1$. This result is in agreement with the previous remark that the product τk should be related to the inverse of the regularization parameter. Indeed for $\tau = 1$ we have $k_{opt} = 80$ and for $\tau = 1.8$ we have $k_{opt} = 45$. The product τk_{opt} is essentially the same in the two cases: 80 for $\tau = 1.$ and 81 for $\tau = 1.8$. Since in the case of the Tikhonov method we found

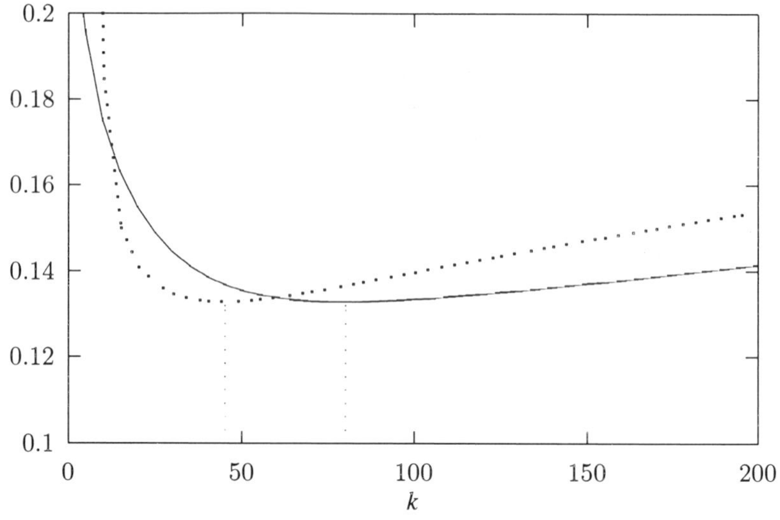

Figure 6.4. Behaviour of the relative norm of the restoration error, as a function of the number of iterations k, for two values of the relaxation parameter: $\tau = 1$ (full line) and $\tau = 1.8$ (dotted line). The curves have been computed for the numerical example of figure 5.15. The optimal numbers of iterations are also indicated.

$\mu_{opt} = 7 \times 10^{-3}$ (see section 5.6), it follows that $1/\mu_{opt} \simeq 143$ and this value is considerably greater than τk_{opt}. In other words the relationship $\tau k = 1/\mu$ can be used for comparing the high-frequency cut-off of the two methods but not the restoration errors.

The minimum restoration error is essentially the same for the two values of τ and precisely 13.2%. This is also the minimum error provided by the Tikhonov method (see section 5.6) and therefore the two methods, Tikhonov and Landweber, are basically equivalent in this particular example. As concerns the choice of the value of τ, we observe that if we increase τ, then $k_{opt} = k_{opt}(\tau)$ decreases up to a certain value of τ (with a restoration error approximately constant) and then suddenly increases, in agreement with the fact that for $\tau = 2$ the method does not converge. This behaviour may be observed in many examples. It implies that, in general, there exists an optimum value of the relaxation parameter, and that, in practice, τ must be chosen not too close to the upper limit of the interval of allowed values.

In the case of real images it is not possible to compute k_{opt} and therefore one needs methods for estimating the optimum number of iterations. These methods are called *stopping rules*.

The most simple stopping rule can be obtained by an extension of the *discrepancy principle* which is discussed in section 5.6: let $b \geq 1$ be a given number; then, if there exists a value $\tilde{k}$ of k such that

$$\|Af_{\tilde{k}} - g\| > b\varepsilon \quad , \quad \|Af_{\tilde{k}+1} - g\| \leq b\varepsilon \qquad (6.36)$$

then $\tilde{k} = \tilde{k}(\varepsilon)$ can be taken as an estimate of k_{opt} and $\tilde{f} = f_{\tilde{k}}$ can be taken as an estimate of the best approximation to $P^{(\mathcal{B})}f^{(0)}$ provided by the Landweber method. It is possible to prove in a rather simple way [10] that, if $b > 1$ and if the noise on the data tends to zero, so that $\varepsilon \to 0$, then $\|\tilde{f} - P^{(\mathcal{B})}f^{(0)}\|$ also tends to zero.

This method does not always work well in practice. In the numerical example of section 5.6 we obtain: $\tilde{k} = 25$ and a restoration error of 15.8% (somewhat too large with respect to the minimum restoration error of 13.2%) if we take $b = 1$ and $\tau = 1$; $\tilde{k} = 20$ and a restoration error of 14.1% (not too bad) if we take $b = 1$ and $\tau = 1.8$. If we use larger values of b, then the results are worse as follows from the behaviour of $\varepsilon(f_k; g)$ given in figure 6.2(b). However, the condition $b > 1$ is essentially introduced for technical reasons, i.e. in order to prove convergence. In practice, we can take $b = 1$.

If we look at figure 6.4 we see that the curve of the restoration error is rather steep before and rather flat after the minimum. Therefore the stopping rules which tend to underestimate the number of iterations can provide restorations which are not sufficiently accurate while methods which tend to overestimate the number of iterations can provide satisfactory restorations. For instance, in the numerical example of section 5.6, the number of iterations such that $\|f_k\| = E$ is rather large ($k = 220$—see figure 6.2(a)), but the corresponding restoration error is 14.3%, a little better than that corresponding to $\tilde{k}$ (for $\tau = 1$).

6.2 The projected Landweber method

The method of Landweber discussed in detail in the previous section is not a genuine iterative method, when applied to the problem of image deconvolution, because it is equivalent to a filter and therefore, in practice, it is much more convenient and faster to implement the filter rather than the iterative procedure. In any case it is a good example to show that the number of iterations can play the role of a regularization parameter.

Another interesting feature of this method is that it can be used, after some appropriate modifications, for solving constrained least-squares problems such as those introduced in section 4.7. As shown there, many physical constraints on the unknown object can be expressed by requiring that it belong to some given closed and convex set $\mathcal{C}$; then the constrained least-squares problem is the minimization of the discrepancy functional over this set, i.e.

$$\|Af - g\| = minimum, \quad f \in \mathcal{C}. \qquad (6.37)$$

This problem is still ill-posed if the set $\mathcal{C}$ does not satisfy additional conditions. A general discussion of this problem is beyond the scope of this book. We only observe that, from some general mathematical results [11] it follows that if the imaging system is not bandlimited and if the set C is interior to the set of objects with a prescribed energy E^2 then there exists a unique solution of the problem (6.37) for any image g. Therefore the problem is well-posed as concerns uniqueness and existence of the solution.

An example of a constraint which is rather natural in many problems of image restoration is the positivity of the solution, i.e. the set C is the set of all non-negative functions. In such a case, as in others, the problem (6.37) may be ill-posed. Therefore what we need is not a method for obtaining the solution of problem (6.37) (which, in general, does not exist for noisy images) but a regularization algorithm, i.e. a family of approximate solutions which converge to the true object in the case of a noise-free image and has the semiconvergence property in the case of noisy images. As we will show, the Landweber method can be modified in order to provide an algorithm which has these properties. This result mainly derives from numerical experiments because the mathematical results which have been obtained on the convergence of the modified method [12] do not yet allow us to prove either the convergence or the semiconvergence property.

From now on we will use in this chapter the following notation, already used in section 5.1

$$\bar{A} = A^*A, \quad \bar{g} = A^*g. \tag{6.38}$$

Moreover we will denote by P_C the projection operator onto the set C, as defined in section 4.7 (see also appendix F). Then the modified Landweber method, also called the *projected Landweber method* is as follows

$$f_{k+1}^{(C)} = P_C\left[f_k^{(C)} + \tau(\bar{g} - \bar{A}f_k^{(C)})\right] \tag{6.39}$$

with τ satisfying conditions (6.25): at each step the method consists of a Landweber iteration followed by a projection onto the convex set C.

In the case of image restoration the method can be implemented as follows:

- compute $\hat{f}_k^{(C)}(\boldsymbol{\omega})$ from $f_k^{(C)}(\boldsymbol{x})$
- compute

$$\hat{h}_{k+1}(\boldsymbol{\omega}) = \tau\hat{K}^*\hat{g}(\boldsymbol{\omega}) + (1 - \tau|\hat{K}(\boldsymbol{\omega})|^2)\hat{f}_k^{(C)}(\boldsymbol{\omega}) \tag{6.40}$$

- compute $h_{k+1}(\boldsymbol{x})$ from $\hat{h}_{k+1}(\boldsymbol{\omega})$
- compute $f_{k+1}^{(C)}(\boldsymbol{x}) = (P_C h_{k+1})(\boldsymbol{x})$.

We see that, at each iteration step, the implementation of the method requires the computation of one direct and one inverse Fourier transform followed by the projection onto the set C. Therefore the iterates $f_k^{(C)}$ are easily computable if this projection is easily computable.

One example is provided by the case where the set C is the subspace of the functions $f(\boldsymbol{x})$ whose support is interior to a bounded domain $\mathcal{D}$. This case will be discussed in detail in chapter 11 in connection with the problem of super-resolution. Another example is provided by the set of the square-integrable functions $f(\boldsymbol{x})$ which are non-negative. In such a case the projection operator is given by equation (4.58). The constraint of positivity, however, is a particular example of a more general constraint which consists of upper and lower bounds on $f(\boldsymbol{x})$. If C is the set of all square-integrable functions $f(\boldsymbol{x})$ such that $a \le f(\boldsymbol{x}) \le b$, with a and b given real numbers, then the projection operator onto this set is given by

$$(P_C f)(\boldsymbol{x}) = \begin{cases} f(\boldsymbol{x}) & \text{if } a \le f(\boldsymbol{x}) \le b \\ a & \text{if } f(\boldsymbol{x}) < a \\ b & \text{if } f(\boldsymbol{x}) > b. \end{cases} \tag{6.41}$$

The constraint of positivity is re-obtained when $a = 0$ and $b = +\infty$. Other examples of convex and closed sets C are given in section 4.5.

As concerns the properties of this algorithm, some general results have been proved [12] which imply that, if $f^{(0)}$ is an object in C and $g = g^{(0)} = Af^{(0)}$ is the corresponding noise-free image, then the algorithm (6.39) converges weakly to a solution of the constrained least-squares problem (6.37) (the definition of weak convergence is given in remark 5.1). If the solution of the problem is unique then the limit is just $f^{(0)}$; if the solution is not unique than the limit is a constrained least-squares solution and depends on the choice of the initial approximation f_0.

If g is noisy, no result has been proved ensuring that the iterates $f_k^{(C)}$ first approach $f^{(0)}$ and then go away. However, numerical simulations strongly suggest that this result must be true, because the following properties are in general satisfied:

- for any k, $\|f_k^{(C)} - f^{(0)}\| < \|f_k - f^{(0)}\|$ where f_k is the kth iterate of the unconstrained Landweber method (6.7)
- the restoration error $\|f_k^{(C)} - f^{(0)}\|$ has one minimum corresponding to an optimal value k_{opt} of the number of iterations and therefore the algorithm has the semiconvergence property.

These properties (which, we repeat, have not yet been proved) can be verified, for instance, if we apply the projected Landweber method, with the constraint of positivity, to the numerical example of figure 5.15. The relative restoration error of the Landweber method (with $f_0 = 0$ and $\tau = 1$) is given by the full line of figure 6.4. The relative restoration error of the projected Landweber method (also with $f_0 = 0$ and $\tau = 1$) for each k is smaller than that of the non-projected method even if, in this case, the improvement is not very large: the minimum restoration error, corresponding to $k_{opt} = 82$, is 12.1% for the constrained method, while it is 13.2% for the unconstrained one. Indeed, in this example, the positivity constraint is not very important because the optimal solution provided by the Landweber method does not contain large negative values.

An example where the difference between the constrained and unconstrained method is more spectacular is that given in figure 5.14. In this example the object consists of only two levels: 1/2 for the background and 1 for the letters. As shown in figure 5.14 the restoration provided by the Tikhonov method is affected by ghosts and this is true also for the restoration provided by the Landweber method because this is a property of any linear restoration method.

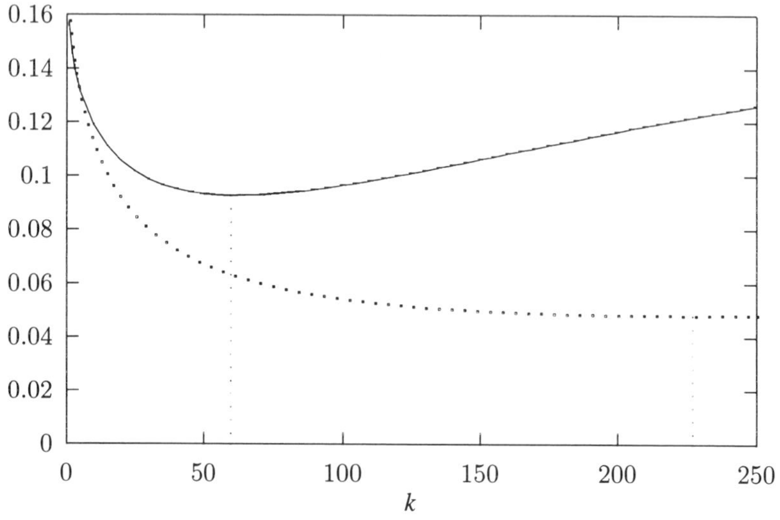

Figure 6.5. Comparison of the relative restoration errors for the unconstrained Landweber method (full line) and the projected Landweber method (dotted line), with lower-bound constraint. The curves are computed in the case of the image of figure 5.14, corrupted with white Gaussian noise (square root of the variance $s = 0.008$). Initial approximation: $f_0 = 0$; relaxation parameter: $\tau = 1$.

In order to apply the projected method we have considered two noisy versions of the image of figure 5.14(a) both corrupted by white Gaussian noise: one with square root of the variance given by $s = 0.008$ (corresponding to a relative r.m.s error of 1.5%, i.e. SNR=36.5 dB) and the other with $s = 0.055$ (corresponding to a relative r.m.s error of 10%, i.e. SNR=20 dB). In figure 6.5 we plot the behaviour of the relative restoration error, for the first noisy version, in the case $\tau = 1$ and $f_0 = 0$: the full line represents the restoration error of the non-projected Landweber method while the dotted line represents the restoration error of the projected one, the projection being on the set of the functions with values greater or equal to 1/2 (see equation (6.41) with $a = 1/2$ and $b = +\infty$). The minimum restoration error of the Landweber method is 9.3% corresponding to $k_{opt} = 60$ while the minimum restoration error of the projected method is 4.8% corresponding to $k_{opt} = 227$. The improvement is about a factor of 2, even if the minimum is reached after a rather large number of iterations, much greater than that of the usual Landweber method. It is clear however, that the minimum is very flat, so that after 100 iterations one has a restoration error which is already acceptable.

The value of k_{opt} and the shape of the minimum however, depends on the noise level. In the case $s = 0.055$ we have $k_{opt} = 20$ and the restoration error is 11.7% (comparable with the r.m.s. error of the image, 10%). The behaviour of the relative restoration error in the two cases is plotted in figure 6.6. In the case of larger noise the Landweber method also converges faster: we find $k_{opt} = 6$, while the corresponding restoration error is 14.2%.

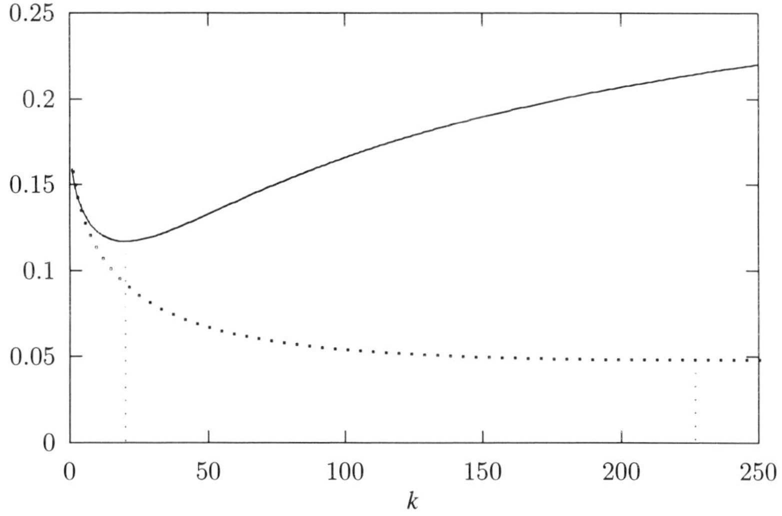

Figure 6.6. Behaviour of the relative restoration error of the projected Landweber method for two noise levels: white noise with $s = 0.008$ (dotted line) and white noise with $s = 0.055$ (full line). The object is the same as that of figure 6.5

In this case the gain provided by the projected Landweber method does not seem to be very large. However if we look at the restorations provided by the two methods, which are shown in figure 6.7, we see that the effect of the constraint is the suppression of the ghosts. This result is not in conflict with our analysis of section 5.5 because the constrained method is nonlinear. Since the presence of ghosts is due to the lack of information around the zeros of the transfer function (see the discussion of section 5.5), it follows that the constrained method provides a sort of extrapolation of the Fourier transform of the object in the regions where information is lacking.

The projected Landweber method provides rather good results in many circumstances. The difficulty is that, in general, the number of iterations required

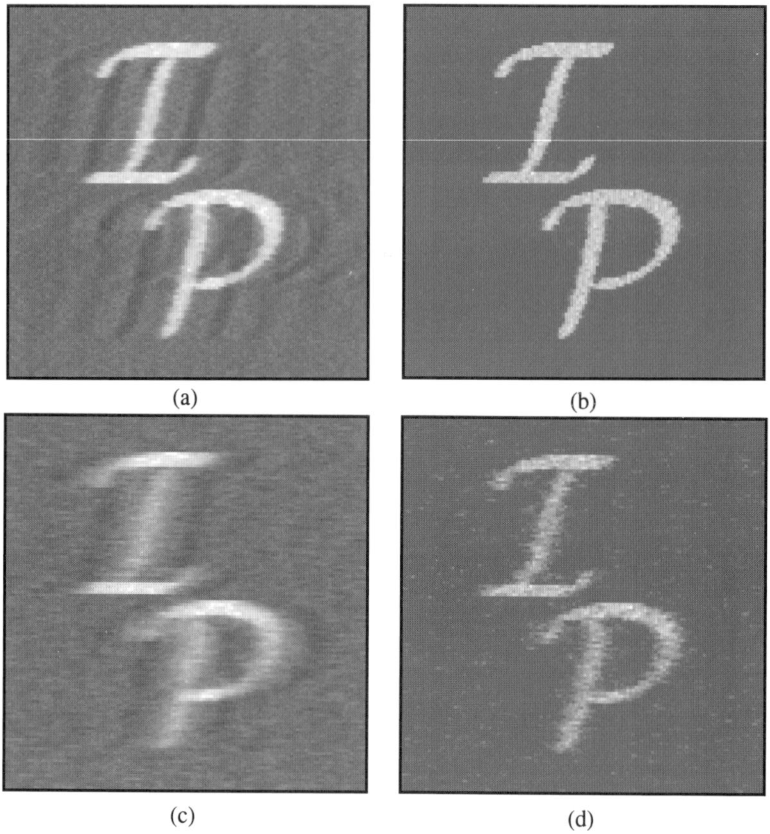

Figure 6.7. Optimal restorations of noisy versions of the image of figure 5.14 provided by the Landweber method in (a) and (c) (white noise with $s = 0.008$ in (a) and $s = 0.055$ in (c)) and by the projected Landweber method in (b) and (d) ($s = 0.008$ in (b) and $s = 0.055$ in (d))

for obtaining a satisfactory result is rather large. In the case of the unconstrained algorithm, section 6.1, a method for increasing the rate of convergence was proposed by Strand [13]. The method consists of modifying the equation for the least-squares solutions as follows

$$D\bar{A}f = D\bar{g} \tag{6.42}$$

where D is a linear and bounded operator with a bounded inverse. An additional requirement is that D commutes with $\bar{A}$. This modification does not change the set of all least-squares solutions. It is a form of *preconditioning*, a method introduced in numerical analysis [14] for improving the efficiency of the conjugate gradient method (the application of conjugate gradient to image restoration will be discussed in section 6.4 and section 6.5).

In the case of image restoration, the assumptions of Strand are satisfied if D is a convolution operator

$$(Df)(\boldsymbol{x}) = \frac{1}{(2\pi)^q} \int \hat{D}(\boldsymbol{\omega}) \hat{f}(\boldsymbol{\omega}) e^{i\boldsymbol{x}\cdot\boldsymbol{\omega}} d\boldsymbol{\omega} \tag{6.43}$$

and the function $\hat{D}(\boldsymbol{\omega})$ is a bounded positive function with a positive lower bound. Moreover $\hat{D}(\boldsymbol{\omega})$ must be chosen in such a way that:

- $\hat{D}(\boldsymbol{\omega})|\hat{K}(\boldsymbol{\omega})|^2 \leq 1$ for any $\boldsymbol{\omega}$
- $\hat{D}(\boldsymbol{\omega})|\hat{K}(\boldsymbol{\omega})|^2$ is close to 1 or, at least, not much smaller than 1, for all spatial frequencies $\boldsymbol{\omega}$ such that $|\hat{K}(\boldsymbol{\omega})|$ is greater than some suitable threshold value.

An operator satisfying these conditions will be called a *preconditioner*. A possible choice of D, which is closely related to the Tikhonov method, is the following [15]

$$\hat{D}(\boldsymbol{\omega}) = \left(|\hat{K}(\boldsymbol{\omega})|^2 + \gamma\right)^{-1} \tag{6.44}$$

with γ chosen according to the criteria used for estimating the regularization parameter—see section 5.6.

By applying the Landweber method to equation (6.42) we find

$$f_{k+1} = f_k + \tau D(\bar{g} - \bar{A} f_k) \tag{6.45}$$

with τ such that $0 < \tau < 2$ (this is a consequence of the first condition on $\hat{D}(\boldsymbol{\omega})$). Then, in the case $f_0(\boldsymbol{x}) = 0$, from equation (6.13) with $\hat{\bar{g}}(\boldsymbol{\omega}) = \hat{D}(\boldsymbol{\omega})\hat{K}^*(\boldsymbol{\omega})\hat{g}(\boldsymbol{\omega})$ and $\hat{H}(\boldsymbol{\omega}) = \hat{D}(\boldsymbol{\omega})|\hat{K}(\boldsymbol{\omega})|^2$ we find

$$\hat{f}_k(\boldsymbol{\omega}) = \left\{1 - \left[1 - \tau \hat{D}(\boldsymbol{\omega})|\hat{K}(\boldsymbol{\omega})|^2\right]^k\right\} \frac{\hat{g}(\boldsymbol{\omega})}{\hat{K}(\boldsymbol{\omega})}. \tag{6.46}$$

If we compare this equation with equation (6.22) we see that the modification introduced by Strand is equivalent to a modification of the filter. However the choice of D also has effects on the speed of convergence. For instance, if $\hat{D}(\boldsymbol{\omega})$ is given by (6.44) and $\tau = 1$ we have

$$\hat{f}_k(\boldsymbol{\omega}) = \left[1 - \left(\frac{\gamma}{|\hat{K}(\boldsymbol{\omega})|^2 + \gamma}\right)^k\right] \frac{\hat{g}(\boldsymbol{\omega})}{\hat{K}(\boldsymbol{\omega})} \tag{6.47}$$

and therefore the window function becomes approximately 1 after few iterations for those frequencies such that $|\hat{K}(\boldsymbol{\omega})|^2$ is greater than or of the order of γ.

The effect on the rate of convergence is especially important in the case of the projected Landweber method. It can be shown [16] that the algorithm obtained by applying the projected Landweber method to equation (6.42), i.e.

$$f_{k+1}^{(C)} = P_C\left[f_k^{(C)} + \tau D(\bar{g} - \bar{A} f_k^{(C)})\right] \tag{6.48}$$

is the projected Landweber method for the following least-squares problem

$$(D(Af - g), Af - g) = minimum, \quad f \in \mathcal{C}. \tag{6.49}$$

Since, in this way, the original least-squares problem, as defined in equation (6.37), has been modified, it may be interesting to investigate the effect of this modification by considering, for instance, the preconditioner (6.44). As a rule, one finds that, by decreasing γ, the number of iterations required for reaching the minimum restoration error decreases, but the minimum restoration error increases.

The statement above can be illustrated by applying the method to the simulated image of a star cluster, produced by AURA/STScI for testing the methods used for the restoration of the pre-COSTAR images of the Hubble Space Telescope. The object, a cluster of 470 stars, and the corresponding image are given in figure 6.8.

The minimum restoration error produced by the projected Landweber method, with positivity constraints and $\tau = 1.98/\hat{K}^2_{max}$, is 3.68% and is reached after about 1300 iterations. The good quality of the restoration is explained by the fact that the condition number of the problem is rather low ($\alpha = 23.74$).

If we use the preconditioner (6.44) with $\gamma = 0.04$, we find the minimum after 56 iterations and the restoration error is 5.24%. If we use $\gamma = 0.01$, the minimum is reached after 8 iterations but its value is now 7.16%, about twice the error of the method without preconditioning. The restorations obtained without preconditioner and with the preconditioner (6.44) corresponding to $\gamma = 0.04$ are given, for comparison, in figure 6.8. The difference between the two restorations is visible. Both restorations, however, are much better than the one provided by the Tikhonov regularization method since, in that case, the minimum restoration error is 10.2% (corresponding to $\mu_{opt} = 5 \times 10^{-4}$).

6.3 The projected Landweber method for the computation of constrained regularized solutions

In section 5.5 it has been shown that the linear deconvolution methods and, in particular, the Tikhonov regularization method, can introduce ringing effects. This is the case, for instance, of the restoration of an object consisting of bright spots against a black background or containing sharp intensity variations. An example has been considered in the previous section.

Ringing is not desirable first of all because the original object is known to be without these effects; secondly because ringing artifacts reduce not only the visual but also the measurable quality of the restored image.

Since ringing is not compatible with our *a priori* knowledge about the object, it should be suppressed through the use of some suitable *a priori* information. For instance, if the object is positive and the ringing manifests itself through negative values of the restored image, the positivity constraint can be used.

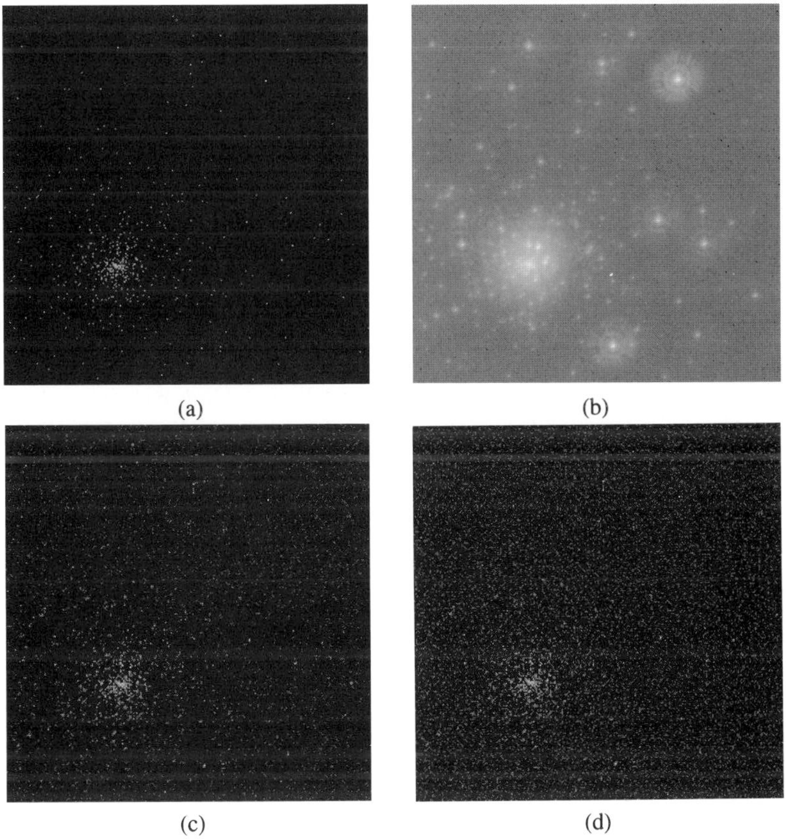

Figure 6.8. Example of a restoration obtained by means of the projected Landweber method with preconditioning. In (a) we show the object, a cluster of 470 stars, and in (b) its image (both images have been produced by AURA/STScI). In (c) we give the best restoration obtained by the projected Landweber method and in (d) the best restoration obtained by means of the same method with the preconditioner (6.44) ($\gamma = 0.04$).

The projected Landweber method discussed in the previous section is a way to achieve these results. Another way which has been proposed [17] is to compute constrained regularized solutions. If $\mathcal{C}$ is the closed and convex set expressing the constraints on the unknown object, then one has to solve the following minimization problem

$$\Phi_\mu(f; g) = \|Af - g\|^2 + \mu\|f\|^2 = minimum, \quad f \in \mathcal{C}. \tag{6.50}$$

It can be proved [18] that the problem (6.50) has a unique solution in $\mathcal{C}$, let us say $f_\mu^{(\mathcal{C})}$, for any $\mu > 0$ and any g. Also in this case there exists an optimum value of μ such that $\|f_\mu^{(C)} - f^{(0)}\|$ is a minimum. Therefore, in the case of numerical simulations, one can solve the problem (6.50) for any μ and determine the optimum value of μ by minimizing the restoration error. In the case of a real image, one can estimate a value of μ by means of one of the methods of section 5.6 and then solve the problem (6.50) with this particular value of μ. The main question is: how can we solve the problem (6.50)?

The projected Landweber method introduced in the previous section provides the answer. Indeed, let us consider the iterative scheme (6.7) for the solution of equation (6.3). If we use the notations (6.38), the Euler equation (5.23) associated with the functional $\Phi_\mu(f; g)$ has the following form

$$(\bar{A} + \mu I)f = \bar{g} \tag{6.51}$$

and therefore it is obtained from equation (6.3) by adding the term μI to $\bar{A}$. Accordingly we can modify the operator T of equation (6.4), and introduce the operator T_μ defined as follows

$$T_\mu(f) = f + \tau[\bar{g} - (\bar{A} + \mu I)f]. \tag{6.52}$$

If we require that the relaxation parameter τ satisfies condition (6.14) with $\hat{H}(\omega) = |\hat{K}(\omega)|^2 + \mu$, we find

$$0 < \tau < \frac{2}{\hat{K}_{max}^2 + \mu} \tag{6.53}$$

and for these values of τ the following quantity $\rho_\mu(\tau)$ is strictly smaller than one

$$\rho_\mu(\tau) = \min_\omega |1 - \tau(|\hat{K}(\omega)|^2 + \mu)| < 1. \tag{6.54}$$

This implies that the nonlinear operator T_μ defines a contraction mapping (see appendix F). Indeed, from the Parseval equality we get, for any pair of functions $f^{(1)}$, $f^{(2)}$

$$\begin{aligned} \|T_\mu(f^{(2)}) - T_\mu(f^{(1)})\|^2 &= \|\left[I - \tau(\bar{A} + \mu I)\right](f^{(1)} - f^{(2)})\|^2 \qquad (6.55) \\ &= \frac{1}{(2\pi)^q} \int |1 - \tau(|\hat{K}(\omega)|^2 + \mu)|^2 |\hat{f}^{(2)}(\omega) - \hat{f}^{(1)}(\omega)|^2 d\omega \\ &\leq \frac{\rho_\mu^2(\tau)}{(2\pi)^q} \int |\hat{f}^{(2)}(\omega) - \hat{f}^{(1)}(\omega)|^2 d\omega \end{aligned}$$

and therefore

$$\|T_\mu(f^{(2)}) - T_\mu(f^{(1)})\| \leq \rho_\mu(\tau)\|f^{(2)} - f^{(1)}\|. \tag{6.56}$$

If the projection operator P_C is non-expansive, the operator $P_C T_\mu$ is also a contraction mapping

$$\begin{aligned} \|P_C T_\mu(f^{(2)}) - P_C T_\mu(f^{(1)})\| &\leq \|T_\mu(f^{(2)}) - T_\mu(f^{(1)})\| \qquad (6.57) \\ &\leq \rho_\mu(\tau)\|f^{(2)} - f^{(1)}\| \end{aligned}$$

and therefore the projected Landweber method

$$f_{k+1}^{(C,\mu)} = P_C T_\mu (f_k^{(C,\mu)}), \tag{6.58}$$

which is precisely the successive approximation method for the contraction operator $P_C T_\mu$, converges, for any initial approximation f_0, to the unique fixed point of $P_C T_\mu$. Moreover it can be proved [17] that this fixed point coincides with the unique solution of the problem (6.50).

For image restoration, the implementation of the iterative algorithm (6.58) is similar to that of the projected Landweber method. It is as follows:

- compute $\hat{f}_k^{(C,\mu)}(\boldsymbol{\omega})$ from $f_k^{(C,\mu)}(\boldsymbol{x})$
- compute $\hat{h}_{k+1}^{(\mu)}(\boldsymbol{\omega}) = \tau \hat{K}^*(\boldsymbol{\omega})\hat{g}(\boldsymbol{\omega}) + [1 - \tau(|\hat{K}(\boldsymbol{\omega})|^2 + \mu)]\hat{f}_k^{(C,\mu)}(\boldsymbol{\omega})$
- compute $h_{k+1}^{(\mu)}(\boldsymbol{x})$ from $\hat{h}_{k+1}^{(\mu)}(\boldsymbol{\omega})$
- compute $f^{(C,\mu)}(\boldsymbol{x}) = (P_C h_{k+1}^{(\mu)})(\boldsymbol{x})$.

Also in this case, at each iteration step we need the computation of one direct and one inverse Fourier transform followed by the projection onto the set $\mathcal{C}$.

Since the convergence of the algorithm is rather slow, one can use the method of Strand discussed in section 6.2, for increasing the speed of convergence.

6.4 The steepest descent and the conjugate gradient method

The Landweber method is an example of the so-called *gradient methods*, i.e. of the methods where, at each step, the new approximation is obtained by modifying the old one in the direction of the gradient of the discrepancy functional. Another example is provided by the *steepest descent method.*

(A) *The steepest-descent method*

Let us remark first that the minimization of the discrepancy functional $\varepsilon^2(f; g)$, as defined in equation (5.2), is equivalent to the minimization of the functional

$$\eta(f; g) = \frac{1}{2}(\bar{A}f, f) - (\bar{g}, g) \tag{6.59}$$

where the notations (6.38) are used. Indeed, if we expand the square of the norm of $Af - g$ by the use of the properties of the scalar product, we find that

$$\varepsilon^2(f; g) = 2\eta(f; g) + \|g\|^2. \tag{6.60}$$

Now, it is easy to show that the gradient of the function $\eta(f; g)$ is just one-half the gradient of the functional $\varepsilon^2(f; g)$, derived in appendix E

$$\nabla_f \eta(f; g) = \bar{A}f - \bar{g}, \tag{6.61}$$

so that this gradient coincides with the negative residual associated with the least-squares equation

$$\bar{r} = \bar{g} - \bar{A} f. \tag{6.62}$$

Given an approximation f_k for the object, in a neighbourhood of this point the functional $\eta(f; g)$ decreases most rapidly in the direction of the negative gradient, i.e. in the direction of the residual $\bar{r}$. If we consider the Landweber method and if we put $\bar{r}_k = \bar{g} - \bar{A} f_k$, we see that it can be written as follows

$$f_{k+1} = f_k + \tau \bar{r}_k \tag{6.63}$$

and therefore it corresponds to modifying the kth iterate precisely in the direction of the negative gradient. However, if we take a value of τ satisfying condition (6.25), this value may not be the best one, in the sense that it does not minimize $\eta(f_{k+1}; g)$. By simple computations we find that

$$\eta(f_{k+1}; g) = \eta(f_k; g) + \frac{1}{2}\tau^2 \|A\bar{r}_k\|^2 - \tau \|\bar{r}_k\|^2 \tag{6.64}$$

(the relation $(\bar{A}\bar{r}_k, \bar{r}_k) = (A^* A\bar{r}_k, \bar{r}_k) = (A\bar{r}_k, A\bar{r}_k)$ must be used) and therefore the value of τ, let us say τ_k, which minimizes $\eta(f_{k+1}; g)$ is given by

$$\tau_k = \frac{\|\bar{r}_k\|^2}{\|A\bar{r}_k\|^2}. \tag{6.65}$$

The modification of the Landweber method provided by this choice of the relaxation parameter (which depends on k), i.e.

$$f_{k+1} = f_k + \tau_k \bar{r}_k \tag{6.66}$$

is the so-called *steepest-descent method*.

From equations (6.65) and (6.66) we find that the residuals $\bar{r}_k$ have the following properties

- they can be obtained by means of the iterative scheme

$$\bar{r}_{k+1} = \bar{r}_k - \tau_k \bar{A}\bar{r}_k \tag{6.67}$$

  with τ_k given by equation (6.65)
- they satisfy the orthogonality condition

$$(\bar{r}_{k+1}, \bar{r}_k) = 0. \tag{6.68}$$

Since $\bar{A}$ is a positive-definite matrix, this method has a very simple interpretation: if f_0 is the initial approximation, then one moves from f_0 in the direction orthogonal to the level surface of $\varepsilon^2(f; g)$ (an ellipsoid), passing through f_0, up to a point f_1 on the level surface tangent to this direction. Then one moves from f_1 in the direction orthogonal to the new level surface up to

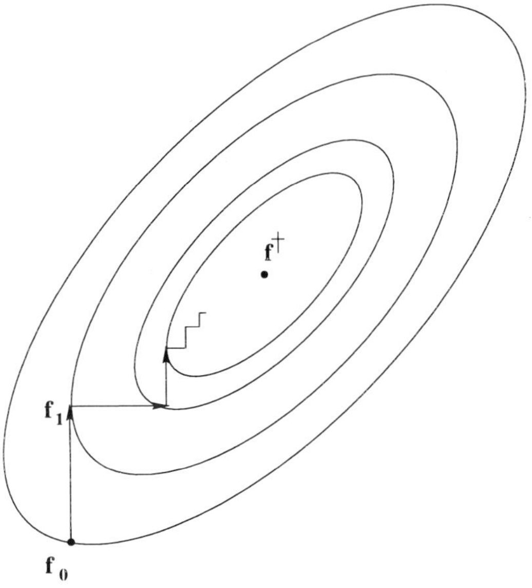

Figure 6.9. Two-dimensional representation of the steepest descent method.

another level surface tangent to this direction and so on. A two-dimensional representation is given in figure 6.9

The steepest-descent method is a nonlinear method for approximating least-squares solutions. It is also a regularization method. It has been proved [19] that, if $f_0 = 0$, then the iterates f_k converge to $f^\dagger$ in the case of a noise-free image. Moreover they have the semiconvergence property in the case of noisy images so that, also for this method, there exists an optimum number of iterations providing the best approximation of the unknown object.

If we apply the method to the numerical example of section 5.6, the minimum restoration error we obtain is 13.3% with $k_{opt} = 41$. The values of the quantities τ_k, equation (6.65), oscillate around 2, between 1.8 and 2.1. We recall that in the case of the Landweber method with $\tau = 1.8$ the minimum restoration error is 13.2% with $k_{opt} = 45$. We see in this example that the speed of convergence is not significantly improved. The reason can be understood by looking at figure 6.9. In the case of an ill-conditioned problem, as we showed in section 4.6, the level surfaces of the discrepancy functional are very elongated ellipsoids and minimization corresponds to finding the lowest point in a very flat and steep-sided valley. The various steps in the steepest-descent method correspond to move back and forth across the valley. Moreover the steps become smaller and smaller. In the *conjugate gradient* (CG) method, which was proposed by Hestenes and Stiefel [20], the steps correspond to moving down the valley and therefore the convergence is faster.

(B) *The conjugate gradient method*

The origin of the name *conjugate gradient* resides in the following definition: given two vectors (or functions or images), f and h, they are said to be *conjugate with respect to* $\bar{A}$ or also $\bar{A}$*-orthogonal* if

$$(f, \bar{A}h) = 0. \tag{6.69}$$

Then, given a level surface (ellipsoid) of the discrepancy functional and a point of this surface, the direction of the conjugate gradient at this point is the direction $\bar{A}$-orthogonal to the tangent plane. In the finite-dimensional case it is possible to prove that this direction points toward the centre of the ellipsoid, as shown in figure 6.10. In this figure we indicate both the gradient $\boldsymbol{r}$ and the conjugate gradient $\boldsymbol{q}$.

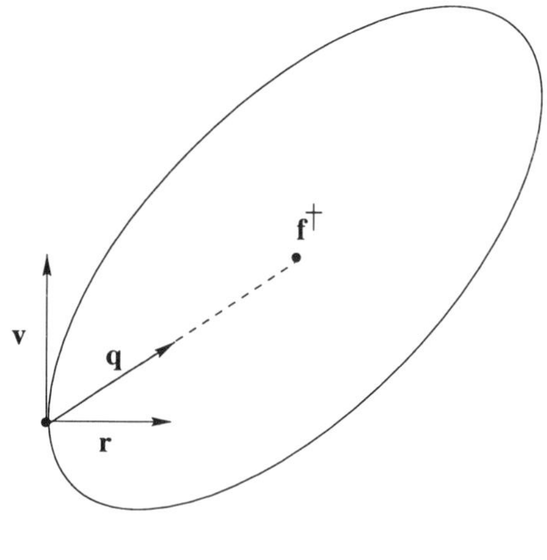

Figure 6.10. Two-dimensional representation of $\bar{A}$-orthogonal vectors $\boldsymbol{v}$ and $\boldsymbol{q}$. If $\boldsymbol{v}$ is tangent to a level line of the discrepancy functional, then $\boldsymbol{q}$ is directed toward the centre of the ellipsoid. The direction of the usual gradient $\boldsymbol{r}$ is also indicated.

The starting point of the method is the observation that methods such as Landweber or steepest descent provide, at step k, an approximation which is a linear combination of the functions $\bar{g}, \bar{A}\bar{g}, \bar{A}^2\bar{g}, \ldots, \bar{A}^{k-1}\bar{g}$ (we assume here that the initial approximation is $f_0 = 0$). Therefore, whatever the choice of the relaxation parameter is, the result of the kth iteration always lies in the subspace which is the span of these functions. This is the so-called *Krylov subspace* $\mathcal{K}^{(k)}(\bar{A}; \bar{g})$ whose dimension is k if the above-mentioned elements are linearly independent. For the sake of simplicity, we will assume that this condition is satisfied for any k.

The Landweber or the steepest-descent method does not provide, at the step k, the element of the Krylov subspace which best approximates the generalized solution, i.e. the orthogonal projection of the generalized solution onto $\mathcal{K}^{(k)}(\bar{A};\bar{g})$. In general this projection cannot be easily computed and is not computed by the CG method which is doing something else. It provides the function of $\mathcal{K}^{(k)}(\bar{A};\bar{g})$ which minimizes the discrepancy functional (or else the functional (6.59)).

This property can be obtained by proving that the CG method is a *projection method*, in the sense that, at step k, it computes the solution of the least-squares equation projected onto $\mathcal{K}^{(k)}(\bar{A};\bar{g})$. If P_k is the projection operator onto $\mathcal{K}^{(k)}(\bar{A};\bar{g})$, then the projected equation is

$$P_k\bar{A}P_kf = P_k\bar{g}, \tag{6.70}$$

and the solution provided by the CG method satisfies the condition $P_kf = f$.

This solution is the minimum of the discrepancy functional restricted to the Krylov subspace $\mathcal{K}^{(k)}(\bar{A};\bar{g})$. Indeed, if we look for a least-squares solution in $\mathcal{K}^{(k)}(\bar{A};\bar{g})$, this is a solution of the minimization problem

$$\|AP_kf - g\| = minimum, \quad P_kf = f. \tag{6.71}$$

Then equation (6.70) is the Euler equation of this problem (see appendix E), as follows from the remark that the adjoint of the operator AP_k is the operator P_kA^*. We also remember that the solution of the problem (6.71) is the centre of the level lines of the discrepancy functional in $\mathcal{K}^{(k)}(\bar{A};\bar{g})$ and that the conjugate gradient points precisely towards this centre.

The projection methods are not iterative methods in a strict sense even if the construction of the projected equation is performed recursively. The *Lanczos method* [7] is another example of a projection method. It provides the construction of a $k \times k$ tridiagonal matrix which is isomorphic to the projected matrix $P_k\bar{A}P_k$. Since the inverse of a tridiagonal matrix can be easily computed, the projected equation can be easily solved. The CG method, however, provides the same result in a more stable way, avoiding the explicit construction of the projected system. We also observe that if $\bar{A}$ is a matrix $N \times N$ and $\mathcal{K}^{(N)}(\bar{A};\bar{g})$ has dimension N, then, in exact arithmetic, the method converges necessarily in N steps. Properties of the CG method are derived in any book of numerical analysis (see, for instance, [14] or [7]). Here we sketch only the basic points.

The CG method is based on the following iterative construction of two bases $\bar{r}_k$ and $\bar{p}_k$:

-
$$\bar{r}_0 = \bar{p}_0 = \bar{g} \tag{6.72}$$

- compute

$$\bar{\alpha}_k = \frac{\|\bar{r}_k\|^2}{(\bar{r}_k, \bar{A}\bar{p}_k)}$$

- compute

$$\bar{r}_{k+1} = \bar{r}_k - \bar{\alpha}_k \bar{A}\bar{p}_k$$

- compute

$$\bar{\beta}_k = -\frac{(\bar{r}_{k+1}, \bar{A}\bar{p}_k)}{(\bar{p}_k, \bar{A}\bar{p}_k)}$$

- compute

$$\bar{p}_{k+1} = \bar{r}_{k+1} + \bar{\beta}_k \bar{p}_k.$$

Then the iterative scheme for the computation of the approximate solutions is given by

$$\begin{aligned} f_0 &= 0 \\ f_{k+1} &= f_k + \bar{\alpha}_k \bar{p}_k. \end{aligned} \tag{6.73}$$

The definition of $\bar{\alpha}_k$ and $\bar{\beta}_k$, given above, is not the usual one but it may be useful for understanding what the CG method is doing. First let us observe that the coefficients $\bar{\alpha}_k$ and $\bar{\beta}_k$ are chosen in such a way that the following orthogonality conditions are satisfied

$$(\bar{r}_{k+1}, \bar{r}_k) = 0, \quad (\bar{p}_{k+1}, \bar{A}\bar{p}_k) = 0 \tag{6.74}$$

i.e. $\bar{r}_1$ is orthogonal to $\bar{r}_0$, $\bar{r}_2$ is orthogonal to $\bar{r}_1$ and so on; similarly $\bar{p}_1$ is $\bar{A}$-orthogonal (conjugate) to $\bar{p}_0$, $\bar{p}_2$ is $\bar{A}$-orthogonal to $\bar{p}_1$ and so on.

It is not difficult to prove by induction that the following properties hold true:

- the functions $\{\bar{r}_0, \bar{r}_1, \ldots, \bar{r}_{k-1}\}$ form an orthogonal basis of $\mathcal{K}^{(k)}(\bar{A}; \bar{g})$;
- the functions $\{\bar{p}_0, \bar{p}_1, \ldots, \bar{p}_{k-1}\}$ form an $\bar{A}$-orthogonal basis of $\mathcal{K}^{(k)}(\bar{A}; \bar{g})$;
- the $\bar{r}_k$ are precisely the residuals associated with the approximations f_k, i.e.

$$\bar{r}_k = \bar{g} - \bar{A} f_k \tag{6.75}$$

- the following alternative expressions of the coefficients $\bar{\alpha}_k$, $\bar{\beta}_k$ hold true

$$\bar{\alpha}_k = \frac{\|\bar{r}_k\|^2}{(\bar{p}_k, \bar{A}\bar{p}_k)}, \quad \bar{\beta}_k = \frac{\|\bar{r}_{k+1}\|^2}{\|\bar{r}_k\|^2}, \tag{6.76}$$

  and therefore the coefficients $\bar{\alpha}_k$, $\bar{\beta}_k$ are positive.

The first property implies that $\bar{r}_k$ is orthogonal to $\mathcal{K}^{(k)}(\bar{A}; \bar{g})$, or also that $P_k \bar{r}_k = 0$. Then, from equation (6.75) we obtain equation (6.70) and we prove in this way that the CG method is a projection method.

We can see now how the CG method works. We start with $f_0 = 0$ and we move in the direction orthogonal to the level surface of the discrepancy functional, passing through f_0, up to the level surface which is tangent to this direction. Next we do not move in the direction of the gradient but we move up to the centre of the ellipse which is the intersection of this level surface with the plane spanned by the functions $\bar{r}_0$, $\bar{r}_1$ (the Krylov subspace $\mathcal{K}^{(2)}(\bar{A}; \bar{g})$). This is precisely the result we obtain by moving f_1 in the direction of the conjugate gradient $\bar{p}_1$. Next we consider the level surface through f_2 and so on. In figure 6.11 we present this process in the two-dimensional case. If we take into account that, as shown in figure 6.10, the conjugate gradient $\boldsymbol{q}$ points toward the centre of the ellipse, we see that, in this case, we reach the centre in two steps.

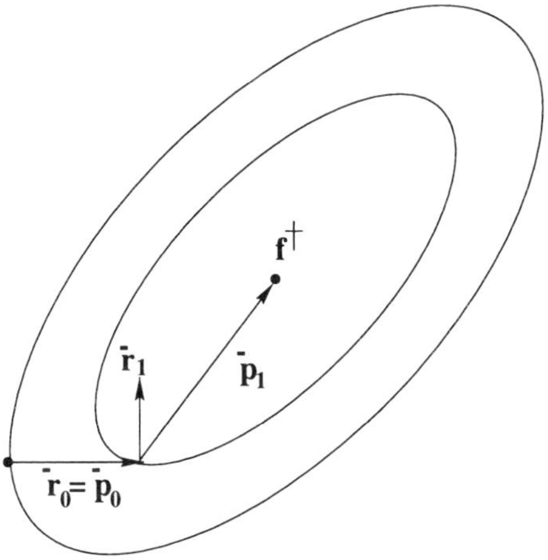

Figure 6.11. Two-dimensional representation of the conjugate gradient method.

In the case of the problem of image deconvolution we can write the iterative process in terms of the Fourier transforms. We recognize immediately that if the imaging system is bandlimited, then the FT of $\bar{r}_k$, $\bar{p}_k$ and f_k is zero outside the band $\mathcal{B}$ of the system and therefore f_k is a bandlimited approximation of the solution of the problem. Indeed we have the following algorithm, where we use the expression (6.76) of the coefficients $\bar{\alpha}_k$,$\bar{\beta}_k$

-
$$\hat{f}_0(\omega) = 0; \quad \widehat{\bar{r}}_0(\omega) = \widehat{\bar{p}}_0(\omega) = \hat{K}^*(\omega)\hat{g}(\omega) \tag{6.77}$$

- compute

$$\bar{\alpha}_k = \frac{\int_{\mathcal{B}} |\widehat{\bar{r}}_k(\omega)|^2 d\omega}{\int_{\mathcal{B}} |\hat{K}^*(\omega)|^2 |\widehat{\bar{p}}_k(\omega)|^2 d\omega} \tag{6.78}$$

- compute

$$\widehat{\bar{r}}_{k+1}(\omega) = \widehat{\bar{r}}_k(\omega) - \bar{\alpha}_k |\hat{K}(\omega)|^2 \widehat{\bar{p}}_k(\omega) \tag{6.79}$$

- compute

$$\bar{\beta}_k = \frac{\int_B |\widehat{\bar{r}}_{k+1}(\omega)|^2 d\omega}{\int_B |\widehat{\bar{r}}_k(\omega)|^2 d\omega} \tag{6.80}$$

- compute

$$\widehat{\bar{p}}_{k+1}(\omega) = \widehat{\bar{r}}_{k+1}(\omega) + \bar{\beta}_k \widehat{\bar{p}}_k(\omega) \tag{6.81}$$

- compute

$$\hat{f}_{k+1}(\omega) = \hat{f}_k(\omega) + \bar{\alpha}_k \widehat{\bar{p}}_k(\omega). \tag{6.82}$$

The convergence properties of the CG method in the case of an ill-posed operator equation are proved in [21]. From these general results one can derive the following convergence properties of the algorithm (6.77)–(6.82):

- in the case of a noise-free image $g^{(0)} = Af^{(0)}$, the iterates f_k converge to the generalized solution $f^{\dagger} = P^{(\mathcal{B})} f^{(0)}$ when $k \to \infty$, i.e. $\| f_k - f^{\dagger} \| \to 0$;
- in the case of a noisy image the iterates f_k have the semiconvergence property, i.e. the f_k first approach $f^{(0)}$ and then go away, so that the number of iterations acts as a regularization parameter and the restoration error $\| f_k - f^{\dagger} \|$ has a minimum for a certain value of k, k_{opt};
- the optimal number of iterations can be estimated by means of the discrepancy principle (6.36), which provides a finite value $\tilde{k}$ because the norm of the residual of the kth iterate of the CG method is always smaller than the norm of the corresponding residual of the Landweber method.

The second and the third property are illustrated in the case of the numerical example of section 5.6. In figure 6.12 we plot the behaviour of the relative restoration error $\| f_k - f^{(0)} \| / \| f^{(0)} \|$, as a function of k. It shows a minimum at $k_{opt} = 11$ and the corresponding restoration error is 13.6%. By applying the discrepancy principle, as given in equation (6.36) with $b = 1$, we find $\tilde{k} = 7$ and a restoration error of 14.2%.

As we see in this example, the number of iterations required by the CG method is much smaller than that required by the Landweber or the steepest-descent method. In some cases, however, it may be useful to improve the rate of convergence of the CG method and, to this purpose, preconditioning can be used. In other words one can apply the CG method to equation (6.42). In the case of deconvolution problems one can choose a preconditioner D which is also a convolution operator and an example is given in equation (6.44). In such a case we find that the TF of the operator $D\bar{A}$ is given by $|\hat{K}(\omega)|^2 (|\hat{K}(\omega)|^2 + \gamma)^{-1}$ and therefore it is approximately 1 for the space frequencies ω such that $|\hat{K}(\omega)|^2 \gg \gamma$ and approximately zero for space frequencies ω such that $|\hat{K}(\omega)|^2 \ll \gamma$. Then the effect on the rate of convergence can be understood by taking into account the filtering properties of the CG method which will be discussed in the next section.

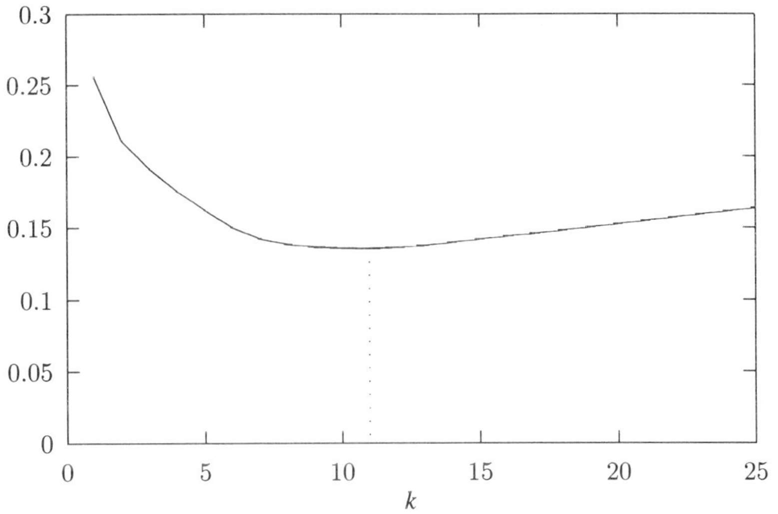

Figure 6.12. Behaviour of the relative restoration error as a function of the number of iterations of the CG method for the restoration of the image of figure 5.15.

The discrete version of the computational scheme (6.77)–(6.82) can be obtained straightforwardly and therefore it is not reported. As concerns the convergence properties, we only observe that, for the restoration of a discrete image $N \times N$, in exact arithmetic the CG method, as with any projection method, must terminate in N^2 steps because the dimension of the Krylov subspace cannot exceed N^2. The final result is the generalized solution which, as we know, is completely corrupted by noise in the case of an ill-conditioned problem. Therefore, also in the case of discrete images one must use the semiconvergence property of the algorithm for obtaining reliable restorations.

6.5 Filtering properties of the conjugate gradient method

The presentation of the CG method given in the previous section does not clarify why this method has regularization properties. In this section we investigate this point by showing that the method has some peculiar filtering properties even if these properties cannot be described in terms of a global PSF because the method is basically nonlinear.

We already know that the approximation f_k, provided by the kth iteration, is the element of the Krylov subspace $\mathcal{K}^{(k)}(\bar{A};\bar{g})$ which minimizes the discrepancy functional

$$\|Af_k - g\| = \min_{u \in \mathcal{K}^{(k)}(\bar{A};\bar{g})} \|Au - g\|. \tag{6.83}$$

Moreover, if we observe that any element of $\mathcal{K}^{(k)}(\bar{A}; \bar{g})$ can be written as the result of the application to $\bar{g}$ of a polynomial, of degree $k-1$, formed with the powers of $\bar{A}$:

$$u = P_{k-1}(\bar{A})\bar{g} = P_{k-1}(\bar{A})A^* g, \tag{6.84}$$

we find that a similar representation holds true also for f_k and we denote by Q_{k-1} the corresponding polynomial

$$f_k = Q_{k-1}(\bar{A})A^* g. \tag{6.85}$$

From equation (6.83) it follows that Q_{k-1} is the polynomial solving the following minimization problem

$$\|A Q_{k-1}(\bar{A})A^* g - g\| = \min_{P_{k-1}} \|A P_{k-1}(\bar{A})A^* g - g\|. \tag{6.86}$$

For the problem of image restoration, using the Parseval equality we have

$$\begin{aligned} &\|A P_{k-1}(\bar{A})A^* g - g\| \\ &= \frac{1}{(2\pi)^q} \int |\hat{g}(\omega)|^2 \left|1 - |\hat{K}(\omega)|^2 P_{k-1}(|\hat{K}(\omega)|^2)\right|^2 d\omega \end{aligned} \tag{6.87}$$

and therefore, if $Q_{k-1}(|\hat{K}(\omega)|^2)$ is the polynomial minimizing this functional, we see that $|\hat{K}(\omega)|^2 Q_{k-1}(|\hat{K}(\omega)|^2)$ is of the order of 1 for those spatial frequencies such that $|\hat{g}(\omega)|$ is large. On the other hand the function $|\hat{K}(\omega)|^2 Q_{k-1}(|\hat{K}(\omega)|^2)$ is a filter acting on the generalized solution since we can write

$$\begin{aligned} \hat{f}_k(\omega) &= Q_{k-1}(|\hat{K}(\omega)|^2)\hat{K}^*(\omega)\hat{g}(\omega) \\ &= |\hat{K}(\omega)|^2 Q_{k-1}(|\hat{K}(\omega)|^2)\frac{\hat{g}(\omega)}{\hat{K}(\omega)}. \end{aligned} \tag{6.88}$$

This filter is nonlinear because, even if it is not indicated explicitly, the polynomial Q_{k-1} depends on the image g. This point has been clarified when we have observed that $|\hat{K}(\omega)|^2 Q_{k-1}(|\hat{K}(\omega)|^2)$ is of the order of 1 when $|\hat{g}(\omega)|$ is large. This also means that the CG method provides a very clever filter. Indeed this filter not only tends to pick up the frequencies corresponding to large values of $|\hat{K}(\omega)|$ (as the Tikhonov and Landweber filters do) but also frequencies corresponding to large values of the product $|\hat{K}(\omega)\hat{f}^{(0)}(\omega)|$ where $f^{(0)}$ is, as usual, the unknown object.

The filter (6.88) is often written in the following form

$$\hat{f}_k(\omega) = \left\{1 - R_k(|\hat{K}(\omega)|^2)\right\} \frac{\hat{g}(\omega)}{\hat{K}(\omega)} \tag{6.89}$$

where R_k, the so-called *Ritz polynomial*, is a polynomial of degree k related to the polynomial Q_{k-1} by

$$R_k(t) = 1 - t Q_{k-1}(t). \tag{6.90}$$

The zeros of the Ritz polynomial are the so-called *Ritz values*, indicated by $\theta_1^{(k)}, \ldots, \theta_k^{(k)}$. Since $R_k(0) = 1$, we have the following representation

$$R_k(t) = \prod_{j=1}^{k} \left(1 - \frac{t}{\theta_j^{(k)}}\right). \tag{6.91}$$

It is possible to prove (see [7] for a discussion and references) that the Ritz values are the eigenvalues of the projected matrix $P_k \bar{A} P_k$. They can be computed by means of the Lanczos method, since this method provides a tridiagonal representation of the matrix, which can be easily diagonalized.

References

[1] Landweber L 1951 *Am. J. Math.* **73** 615
[2] Fridman V 1956 *Usp. Math. Nauk.* **11** 232
[3] Bialy H 1959 *Arch. Rat. Mech. Anal.* **4** 166
[4] Gerchberg R W 1974 *Opt. Acta* **21**
[5] Papoulis A 1975 *IEEE Trans Circuits and Systems* **CAS-22** 735
[6] Kammerer W J and Nashed M Z 1972 *J. Math. Anal. Appl.* **40** 547
[7] van der Sluis and van der Vorst 1987 *Seismic Tomography* ed G Nolet (Dordrecht: Reidel) 49
[8] van Cittert P H 1931 *Z. Phys.* **69** 298
[9] Biemond J, Lagendijk R L and Mersereau R M 1990 *Proc. IEEE* **78** 856
[10] Defrise M and De Mol C 1987 *Inverse Problems: An interdisciplinary Study* ed P C Sabatier *Advances Electr. Elect. Physics* Supplement 19 (London: Academic Press) 261
[11] Ivanov V K 1962 *Soviet Math. Dokl.* **3** 981
[12] Eicke B 1992 *Num. Funct. Anal. Optimiz.* **13** 413
[13] Strand O N 1974 *SIAM J. Num. Anal.* **11** 798
[14] Golub G H and van Loan C F 1989 *Matrix Computations* (Baltimore, MD: Johns Hopkins University Press)
[15] Sanz J L C and Huang T S 1983 *J. Opt. Soc. Am.* **73** 1455
[16] Piana M and Bertero M 1997 *Inverse Problems* **13** 441
[17] Lagendijk R L, Biemond J and Bockee D E 1988 *IEEE. Trans. Acoust. Speech Signal Processing* **ASSP-36** 1874
[18] Neubauer A 1988 *J. Approx. Theory* **53** 304
[19] Gilyazov S F 1977 *Moscow Univ. Comput. Math. Cybernet.* **3** 78
[20] Hestenes M R and Stiefel E 1952 *J. Nat. Bur. Standards* **B-49** 409
[21] Engl H W, Hanke M and Neubauer A 1996 *Regularization of Inverse Problems* (Dordrecht: Kluwer)

Chapter 7

Statistical methods

The basic feature of the statistical methods is that they take into account the random nature of noise which implies that the recorded image is the realization of a random process.

Among the many methods which have been proposed it is possible to identify two main approaches. The first consists of assuming that the object is deterministic so that it plays the role of a set of parameters characterizing the probability distribution of the image. Then methods of parameter estimation such as the *method of maximum likelihood* can be used for image deconvolution.

In the second approach, the object is also assumed to be a realization of a random process with a given probability distribution. This is a form of *a priori information* about the object and its use is the feature characterizing the so-called *Bayesian methods*. In the case of Gaussian processes one obtains, as a particular example, the method of the *Wiener filter*, which shows interesting analogies with the method of Tikhonov regularization.

Maximum likelihood methods, in general, lead to a reformulation of the problem which is still ill-posed because no *a priori* information, or only a rather weak version, is used. On the other hand Bayesian methods usually provide stable solutions.

7.1 Maximum likelihood (ML) methods

The methods considered in the previous chapters are basically deterministic because they do not take into account the random nature of noise. If some statistical property of the noise, such as the expectation value or the correlation function or the probability distribution, is known, then one can develop methods where this information is used for image deconvolution.

In this chapter we only consider, for simplicity, the case of discrete images. The required mathematical background consists essentially in basic facts about probability and random variables, as treated, for instance in the first part of the book by Papoulis [1].

The starting point is the discrete model of image formation introduced in section 3.1

$$g = \mathbf{A} f^{(0)} + w \tag{7.1}$$

where g,$f^{(0)}$ and w are $N \times N$ arrays and $\mathbf{A}$ is a cyclic matrix with four entries. However, also in this chapter it is convenient to simplify the notations by rearranging the $N \times N$ arrays into vectors depending on only one index (using, for instance, a lexicographic ordering of the arrays) so that $\mathbf{A}$ becomes a matrix with two entries. Accordingly we denote by $A_{m,n}$ the matrix elements of $\mathbf{A}$ and by g_m, w_m and $f_n^{(0)}$ the components of the image, noise and object, respectively.

Since the noise term w is a realization of a random vector, this is true also for the noisy image g, as follows from equation (7.1). In problems where the recorded values of the image have the physical meaning of numbers of counts of particles (most frequently photons) emitted by the unknown sources, the random nature of the image is mainly due to the statistical fluctuations in the counting process. This is precisely the case which will be considered in section 7.3.

We assume that the notion of a random variable (r.v.) [1] is known. Greek letters are used for denoting such variables which are functions defined on a probability space with values in a set of possible outcomes of an experiment.

The first basic assumption we make is that the components w_m of the noise vector w are the values (realizations) of r.v. ν_m which are the components of a random vector ν, i.e. of a vector-valued random variable.

From equation (7.1) it follows that also the components g_m of the noisy image g are the values of r.v. which are denoted by η_m and are the components of a random vector η. Then the following relation holds true between the random vectors η and ν

$$\eta = \mathbf{A} f^{(0)} + \nu. \tag{7.2}$$

The second basic assumption is that the expectation value of the random vector η is just $\mathbf{A} f^{(0)}$.

Let us denote by $p_\eta(g)$ the density function of η. In the case of continuous random variables, $p_\eta(g)$ is a non-negative function such that its integral with respect to g is one; in the case of discrete random variables, $p_\eta(g)$ is a distribution, more precisely a linear combination, with positive coefficients, of delta distributions concentrated on the values of the random variable. In such a case the sum of the coefficients is one. In any case the *expectation value* of η is defined by

$$E\{\eta\} = \int g p_\eta(g) dg \tag{7.3}$$

so that the second basic assumption is

$$E\{\eta\} = \mathbf{A} f^{(0)}. \tag{7.4}$$

This expectation value is not known because $\boldsymbol{f}^{(0)}$ is not known. However the random vector $\boldsymbol{\eta}$ is a member of a family $\boldsymbol{\eta}(\boldsymbol{f})$ defined by

$$\boldsymbol{\eta}(\boldsymbol{f}) = \mathbf{A}\boldsymbol{f} + \boldsymbol{\nu} \tag{7.5}$$

where $\boldsymbol{f}$ is an element of the class of admissible objects containing $\boldsymbol{f}^{(0)}$.

Then the third basic assumption is that the density functions of the random vectors $\boldsymbol{\eta}(\boldsymbol{f})$ are known. In other words, these density functions depend on $\boldsymbol{f}$ as a (vector-valued) parameter and when $\boldsymbol{f} = \boldsymbol{f}^{(0)}$ we obtain the density function of $\boldsymbol{\eta}$, equation (7.2), i.e. $\boldsymbol{\eta}(\boldsymbol{f}^{(0)}) = \boldsymbol{\eta}$.

Summarizing all previous assumptions, the starting point of a statistical approach to image restoration can be restated as follows: *a family of density functions, denoted by $p_{\eta}(\boldsymbol{g}|\boldsymbol{f})$, is given, which has the following properties:*

- *$p_{\eta}(\boldsymbol{g}|\boldsymbol{f})$ is the density function of the r.v. $\boldsymbol{\eta}(\boldsymbol{f})$ defined by equation (7.5);*
- *the expectation value of $\boldsymbol{\eta}(\boldsymbol{f})$ is $\mathbf{A}\boldsymbol{f}$;*
- *when $\boldsymbol{f} = \boldsymbol{f}^{(0)}$, $p_{\eta}(\boldsymbol{g}|\boldsymbol{f}^{(0)})$ is the density function of the r.v. $\boldsymbol{\eta}$, equation (7.2), corresponding to the unknown object $\boldsymbol{f}^{(0)}$.*

We observe that the notation $p_{\eta}(\boldsymbol{g}|\boldsymbol{f})$ is similar to the notation used, in general, for a conditional density function. The justification of this choice will be clear in section 7.4.

An example can be useful for illustrating the previous assumptions. Assume that $\boldsymbol{\nu}$ describes additive noise with zero expectation value and known density function $p_{\nu}(\boldsymbol{w})$. Then the density function of $\boldsymbol{\eta}(\boldsymbol{f})$ is given by

$$p_{\eta}(\boldsymbol{g}|\boldsymbol{f}) = p_{\nu}(\boldsymbol{g} - \mathbf{A}\boldsymbol{f}) \tag{7.6}$$

and it is easily verified that this density function satisfies the conditions stated above.

We have now the following problem: given a noisy image $\boldsymbol{g}$, which is an observed value (realization) of the random vector $\boldsymbol{\eta}$, equation (7.2), which member of the family $\boldsymbol{\eta}(\boldsymbol{f})$ is the most likely candidate to represent $\boldsymbol{\eta}$? This is a typical estimation problem in statistics whose solution provides an estimate of $\boldsymbol{f}^{(0)}$.

The *method of maximum likelihood* (*ML method*) is most useful for solving this kind of problems [2]. In our particular case it can be formulated as follows. Let $\boldsymbol{g}$ be an observed value of the random vector $\boldsymbol{\eta}$, equation (7.2), i.e. a noisy image of the object $\boldsymbol{f}^{(0)}$. Then the density function of the random vector $\boldsymbol{\eta}(\boldsymbol{f})$ takes in $\boldsymbol{g}$ the following value

$$L(\boldsymbol{f}) = p_{\eta}(\boldsymbol{g}|\boldsymbol{f}). \tag{7.7}$$

Once the observed image $\boldsymbol{g}$ is given, $L(\boldsymbol{f})$ contains only one unknown, i.e. the parameter (or object) $\boldsymbol{f}$. As a function of $\boldsymbol{f}$, $L(\boldsymbol{f})$ is called the *likelihood* (or *likelihood function*) and is defined over the class of all admissible objects. In the case of $N \times N$ objects, it is a function of N^2 variables. If the objects are non-negative then the domain of definition of $L(\boldsymbol{f})$ is the first quadrant in the vector space with dimension N^2.

The maximum-likelihood estimate of $\boldsymbol{f}^{(0)}$ *is the object* $\tilde{\boldsymbol{f}}$ *which maximizes* $L(\boldsymbol{f})$ *on its domain of definition.* In other words the maximum-likelihood estimate is the object which maximizes the probability of obtaining the observed image. This is the basic argument of the method.

In order to show how the method works we consider two simple examples which mimic two important cases considered in the next sections.

Example 7.1. Assume that η is a Gaussian random variable with known variance s^2 and unknown expectation value λ. If a given experiment provides the value g of η, then the likelihood function associated with this experiment is given by

$$L(\lambda) = (2\pi s^2)^{-1/2} \exp\left[-\frac{1}{2s^2}(g-\lambda)^2\right]. \tag{7.8}$$

It is obvious that the maximum of the likelihood function is reached at $\lambda = g$ and therefore the maximum-likelihood estimate of the parameter coincides with the result of the experiment.

Example 7.2. Assume that η is a Poisson random variable with unknown parameter λ. Therefore η is a discrete random variable which can take only integer values. If the result of the experiment is the integer value g of η, then the likelihood function is

$$L(\lambda) = e^{-\lambda}\frac{\lambda^g}{(g)!}. \tag{7.9}$$

By deriving with respect to λ the logarithm of $L(\lambda)$ we find that the maximum is reached at $\lambda = g$. Also in this case the maximum-likelihood estimate of the parameter coincides with the result of the experiment.

In these very simple examples the likelihood function has only one maximum and this can be easily computed. In general this is not the case. In image restoration the likelihood is a function of N^2 variables and it can happen that it has several local maxima. Of course one can prefer the global one, i.e. that with the largest value of the likelihood function, but, in any case, the computations can become very complex.

To determine the position of the maximum (maxima) of $L(\boldsymbol{f})$ one can use the standard methods of differential calculus, i.e. compute the first derivatives of $L(\boldsymbol{f})$ with respect to the components of $\boldsymbol{f}$ and set them equal to zero; look at the Hessian matrix, etc. However, when the components of $\eta(\boldsymbol{f})$ are independent random variables, the density function of $\eta(\boldsymbol{f})$ is the product of N^2 functions.

Since the derivatives of a product with many factors are quite cumbersome to handle, it is useful to consider the logarithm of the likelihood function

$$l(\boldsymbol{f}) = \ln L(\boldsymbol{f}) = \ln p_\eta(\boldsymbol{g}|\boldsymbol{f}). \tag{7.10}$$

This function is called the *logarithmic likelihood function* and its maxima obviously coincide with the maxima of $L(\boldsymbol{f})$.

We conclude by pointing out a formal analogy between the ML method and the least-squares method discussed in chapter 4. In the least-squares method the basic rule is to minimize the discrepancy functional $\varepsilon^2(f; g) = \|\mathbf{A}f - g\|^2$. In the ML method the basic rule is to maximize the functional defined on the space of the objects by the logarithmic likelihood function $l(\boldsymbol{f})$. Therefore both are variational methods and in the next section we show that, at least in one case, the connection between the two methods is much deeper than a purely formal analogy.

7.2 The ML method in the case of Gaussian noise

In many circumstances it is reasonable to assume that the noise term $\boldsymbol{w}$ in equation (7.1) is the realization of a vector-valued Gaussian random variable $\boldsymbol{\nu}$, i.e. the components ν_m of $\boldsymbol{\nu}$ are jointly normal random variables.

The usual assumption is that the expectation value of $\boldsymbol{\nu}$ is zero

$$E\{\boldsymbol{\nu}\} = \mathbf{0} \tag{7.11}$$

and this assumption implies equation (7.4). Then, if we consider the general case where the components ν_m are correlated, the joint density function of the normally distributed r.v. ν_m is the following

$$p_\nu(\boldsymbol{w}) = [(2\pi)^{N^2}|\mathbf{S}_\nu|]^{-1/2} \exp\left[-\frac{1}{2}(\mathbf{S}_\nu^{-1}\boldsymbol{w}\cdot\boldsymbol{w})\right] \tag{7.12}$$

where $\mathbf{S}_\nu$ is the *covariance matrix* of the ν_m, $|\mathbf{S}_\nu|$ is the determinant of $\mathbf{S}_\nu$ and the scalar product is the canonical one.

We recall that, in the case of r.v. ν_m with zero expectation value, the covariance matrix is defined by

$$\left(\mathbf{S}_\nu\right)_{m,m'} = E\{\nu_m \nu_{m'}^*\}, \tag{7.13}$$

the complex conjugation being used in the case of complex-valued random variables. This matrix is positive semi-definite because its definition implies that

$$\left(\mathbf{S}_\nu\boldsymbol{w}\cdot\boldsymbol{w}\right) = E\left\{|(\boldsymbol{\nu}\cdot\boldsymbol{w})|^2\right\} \geq 0 \tag{7.14}$$

for any vector $\boldsymbol{w}$. Moreover, this matrix is positive definite, and hence invertible as required in equation (7.12), if and only if no component of the noise is zero with probability one. We make this assumption in the following. We also observe that the diagonal elements of $\mathbf{S}_\nu$ are the variances s_m^2 of the components ν_m of the noise. When the matrix is diagonal

$$\left(\mathbf{S}_\nu\right)_{m,m'} = s_m^2 \delta_{m,m'}, \tag{7.15}$$

the components of the noise are uncorrelated. Since they are normally distributed, from equation (7.12) it follows that they are independent r.v.

The particular case where $\mathbf{S}_\nu$ is diagonal and all variances are equal, i.e. $s_m^2 = s^2$, so that

$$\mathbf{S}_\nu = s^2 I \tag{7.16}$$

is the case of the so-called *white Gaussian noise*, which is often used as a simplified model of noise. It is also a particular case of stationary noise because it is invariant with respect to translations.

Remark 7.1. The more general case of stationary and Gaussian noise (for 2D discrete and finite images) is provided by a covariance matrix which is a cyclic matrix with four entries:

$$(\mathbf{S}_\nu)_{m,n;m',n'} = (\bar{\mathbf{S}}_\nu)_{m-m',n-n'} \tag{7.17}$$

where the periodic array $(\bar{\mathbf{S}})_{m,n}$ (with period N in both indices) is a discrete version of the correlation function of the noise. In particular we observe that, if we put $s^2 = (\bar{\mathbf{S}}_\nu)_{0,0}$, then from equation (7.17) we obtain $(\mathbf{S}_\nu)_{m,n;m,n} = s^2$, i.e. all components of the noise have the same variance. The case of white noise corresponds to the following four-entries covariance matrix

$$(\mathbf{S}_\nu)_{m,n;m',n'} = s^2 \delta_{m,m'} \delta_{n,n'}. \tag{7.18}$$

Finally, the DFT of the array $(\bar{\mathbf{S}}_\nu)_{m,n}$, i.e. $(\hat{\bar{\mathbf{S}}}_\nu)_{k,l}$, is an array with non-negative elements, as follows from equation (7.14). It represents the power spectrum of the noise. In the case of white noise the power spectrum is constant, i.e. $(\hat{\bar{\mathbf{S}}}_\nu)_{k,l} = s^2$.

If the covariance matrix $\mathbf{S}_\nu$ of the noise is known, then the density function (7.12) is completely determined and the density function of the random vector $\eta(\boldsymbol{f})$, equation (7.5), can be obtained from equation (7.6) so that

$$p_\eta(g|\boldsymbol{f}) = [(2\pi)^{N^2} |\mathbf{S}_\nu|]^{-1/2} \exp\left[-\frac{1}{2}\left(\mathbf{S}_\nu^{-1}(g - \mathbf{A}\boldsymbol{f}) \cdot (g - \mathbf{A}\boldsymbol{f})\right)\right]. \tag{7.19}$$

It also follows that, once the noisy image g is given, the logarithmic likelihood function (7.10) is the following

$$l(\boldsymbol{f}) = -\frac{1}{2}(\mathbf{S}_\nu^{-1}(g - \mathbf{A}\boldsymbol{f}) \cdot (g - \mathbf{A}\boldsymbol{f})) - \frac{1}{2}\ln\left[(2\pi)^{N^2}|\mathbf{S}_\nu|\right]. \tag{7.20}$$

We can conclude that *the maximization of the likelihood function is equivalent to the minimization of the following generalized discrepancy functional*

$$\varepsilon_\nu^2(f;g) = \left(\mathbf{S}_\nu^{-1}(\mathbf{A}f - g)\cdot(\mathbf{A}f - g)\right). \tag{7.21}$$

In the particular case of white noise, equation (7.16), we reobtain the discrepancy functional which was basic for all methods discussed in chapters 4–6. Therefore, *in the case of white Gaussian noise the maximum-likelihood estimates coincide with the least-squares solutions of the problem. As a consequence the ML method is affected by all difficulties of the least-squares method such as non-uniqueness and numerical instability of the solution.*

The ML method can also lead to constrained least-squares problems. Indeed, since one has to maximize the likelihood function over its domain of definition, one can restrict this domain by the introduction of constraints defining the set of permissible objects. For instance, one can assume that the energy of the object cannot exceed a given quantity E^2 or that the objects must be non-negative and so on. One easily realizes that in this way one re-obtains all constrained problems considered above. In other words one can inject *a priori* information into the ML method by restricting in a suitable way the domain of definition of the likelihood function. This remark provides a statistical foundation to some of the regularization methods.

If the noise is not white, i.e. $\mathbf{S}_\nu$ is not a multiple of the identity matrix, the solutions of the ML problem may not coincide with the solutions of the standard least-squares problem. For discussing this point we observe that the functional (7.21) is associated with the following weighted scalar product defined in the vector space of the $N \times N$ arrays

$$(g^{(1)} \cdot g^{(2)})_\nu = (\mathbf{S}_\nu^{-1} g^{(1)} \cdot g^{(2)}), \tag{7.22}$$

the scalar product in the second member being the canonical one. This kind of scalar product is investigated in appendix C, where the weighting matrix is denoted by $\mathbf{C}$. In appendix C it is required that $\mathbf{C}$ must be positive definite and this condition is satisfied by $\mathbf{S}_\nu^{-1}$, because $\mathbf{S}_\nu$ is positive definite.

The functional (7.21) is now the norm, induced by the scalar product (7.22), of the vector $\mathbf{A}f - g$. From the results proved in appendix E, it follows that the Euler equation associated with the minimization of this functional is given by

$$\mathbf{A}^*\mathbf{S}_\nu^{-1}\mathbf{A}f = \mathbf{A}^*\mathbf{S}_\nu^{-1}g. \tag{7.23}$$

This equation is a generalization of equation (4.36).

If the inverse matrix $\mathbf{A}^{-1}$ exists, then the unique solution $f^\dagger$ of the equation $\mathbf{A}f = g$, as given by equation (4.17), is also the unique solution of equation (7.23). Indeed, it is easy to verify that $f^\dagger$ is a solution. The uniqueness results from the following remark: if f is such that $\mathbf{A}^*\mathbf{S}_\nu\mathbf{A}f = \mathbf{0}$, then, from the relation

$$0 = \left(\mathbf{A}^*\mathbf{S}_\nu^{-1}\mathbf{A}f \cdot f\right) = \left(\mathbf{S}_\nu^{-1}\mathbf{A}f \cdot \mathbf{A}f\right) \tag{7.24}$$

and the fact that $\mathbf{S}_\nu^{-1}$ is positive definite, it follows that $\mathbf{A}\boldsymbol{f} = \mathbf{0}$. Since $\mathbf{A}$ is invertible, we obtain $\boldsymbol{f} = \mathbf{0}$.

A consequence of the previous remark is that the solutions of equation (7.23) can be different from those of equation (4.36) only in the case where $\mathbf{A}^{-1}$ does not exist (bandlimited systems). We briefly discuss this point.

If the Gaussian noise is stationary, the covariance matrix is cyclic (see remark 7.1) and therefore commutes with the cyclic matrix $\mathbf{A}$. If we multiply both sides of equation (7.23) by $\mathbf{S}_\nu$, we re-obtain equation (4.36). It follows that, *in the case of stationary Gaussian noise the set of the maximum-likelihood estimates coincides with the set of the least-squares solutions.* This is a generalization of the result obtained in the case of white noise.

The situation is different in the case of non-stationary noise. Since $\mathbf{S}_\nu^{-1}$ may not commute with $\mathbf{A}$, the set of the solutions of equation (7.23) does not necessarily coincide with the set of the solutions of equation (4.36). This happens, for instance, when the matrix $\mathbf{S}_\nu$ is diagonal and the variances of the noise affecting the different pixels are different. An example of such a situation is provided by the Gaussian approximation of the Poisson noise considered in the next section. We also observe that, since the matrix $\mathbf{A}^*\mathbf{S}_\nu^{-1}\mathbf{A}$ is not cyclic, equation (7.23) cannot be solved by means of the DFT. It can nevertheless be solved by means of the more general techniques discussed in chapter 9.

7.3 The ML method in the case of Poisson noise

As already mentioned in section 7.1, in some applications to astronomy and microscopy the recorded values g_m of the image have the physical meaning of numbers of detected photons. In image restoration the following model can be used to describe this situation.

The component $f_n^{(0)}$ of the object represents the mean intensity of the light emitted by the source located at the pixel n of the object domain. Analogously the component $(\mathbf{A}\boldsymbol{f}^{(0)})_m$ of the noise-free image represents the mean intensity of the light arriving at the pixel m of the image domain. The matrix element $A_{m,n}$ is the fraction of light emitted at the pixel n (in the object domain) and arrived at pixel m (in the image domain). It follows that all quantities $f_n^{(0)}$, $(\mathbf{A}\boldsymbol{f}^{(0)})_m$ and $A_{m,n}$ are non-negative.

Now, g_m is the number of photons detected at pixel m. If the mean intensity $(\mathbf{A}\boldsymbol{f}^{(0)})_m$ is time-independent, the fluctuations in the process of photon counting obey to Poisson statistics and therefore g_m is the value of a Poisson r.v. η_m whose expectation value is just $(\mathbf{A}\boldsymbol{f}^{(0)})_m$. The difference between g_m and $(\mathbf{A}\boldsymbol{f}^{(0)})_m$, i.e. the term w_m in equation (7.1), is the so-called photon noise or quantum noise. The term Poisson noise is also used. This model can be applied, for instance, to the images of the Hubble Space Telescope (see section 3.4) even if, in such a case, other sources of noise are present.

If we put $\lambda_m = (\mathbf{A}f^{(0)})_m$, the probability distribution of η_m is given by

$$p_{\eta_m}(g_m) = e^{-\lambda_m}\frac{(\lambda_m)^{g_m}}{(g_m)!} \tag{7.25}$$

where g_m can take the values $0, 1, 2, \cdots$. We also recall that λ_m is both the expectation value and the variance of η_m

$$E\{\eta_m\} = \lambda_m, \quad E\{(\eta_m - \lambda_m)^2\} = \lambda_m. \tag{7.26}$$

Remark 7.2. If the value of λ_m is large, for values of g_m close to λ_m the Poisson distribution (7.25) can be approximated by a Gaussian distribution with expectation and variance given by equation (7.26)

$$p_{\eta_m}(g_m) \simeq \frac{1}{\sqrt{2\pi\lambda_m}}\exp\left[-\frac{1}{2\lambda_m}(g_m - \lambda_m)^2\right]. \tag{7.27}$$

This approximation justifies the least-squares method where weights proportional to g_m^{-1} are used, i.e. the method based on the following minimization problem

$$\sum_{m=1}^{N^2}\frac{1}{g_m}\left|(\mathbf{A}f)_m - g_m\right|^2 = minimum. \tag{7.28}$$

Indeed, if the r.v. η_m are independent, then from equation (7.27) and the results of the previous section it follows that the ML method is equivalent to the least-squares problem (7.21) with $\mathbf{S}_\nu$ given by

$$\left(\mathbf{S}_\nu\right)_{m,m'} = \lambda_m\delta_{m,m'}. \tag{7.29}$$

The parameters λ_m, however, are unknown. Then the standard procedure consists of replacing the λ_m by their ML estimates. As shown in example 7.2, given g_m, the ML estimate of λ_m is just g_m. From equation (7.21) and equation (7.29) with λ_m replaced by g_m we obtain equation (7.28).

It is obvious that, in such a case, the noise is not stationary, so the solution of problem (7.28) cannot be obtained by means of DFT based methods.

In imaging problems it is reasonable to assume that the components g_m of the image are the values of independent Poisson r.v. with expectation values $\lambda_m = (\mathbf{A}f^{(0)})_m$. It follows that the family of random vectors $\eta(f)$, introduced in section 7.1, can be defined as a family of random vectors whose components are independent Poisson r.v. with expectations $(\mathbf{A}f)_m$. The corresponding density functions are given by

$$p_\eta(g|f) = \prod_{m=1}^{N^2} e^{-(\mathbf{A}f)_m}\frac{(\mathbf{A}f)_m^{g_m}}{(g_m)!} \tag{7.30}$$

and the domain of the permissible objects is the first quadrant since, as discussed above, all their components must be non-negative. As a consequence, once the image $\boldsymbol{g}$ (which is a vector whose components are integer numbers) is given, the logarithmic likelihood function has the following expression

$$l(\boldsymbol{f}) = \sum_{m=1}^{N^2} \{g_m \ln(\mathbf{A}\boldsymbol{f})_m - (\mathbf{A}\boldsymbol{f})_m - \ln(g_m)!\} \tag{7.31}$$

and is defined on the first quadrant of the object space. Its maximum or maxima in this domain can possibly lie on the boundary, i.e. some of the components of the ML estimates can be zero.

We prove now that the function (7.31) has at least one maximum in its domain of definition. By elementary computations we find that the gradient and the Hessian matrix of $l(\boldsymbol{f})$ are given by

$$(\nabla l)_n(\boldsymbol{f}) = \frac{\partial l(\boldsymbol{f})}{\partial f_n} = -\alpha_n + \left(\mathbf{A}^T \frac{\boldsymbol{g}}{\mathbf{A}\boldsymbol{f}}\right)_n, \tag{7.32}$$

$$\left[\mathbf{H}(\boldsymbol{f})\right]_{n,n'} = \frac{\partial^2 l(\boldsymbol{f})}{\partial f_n \partial f_{n'}} = -\sum_{m=1}^{N^2} \frac{A_{m,n} A_{m,n'} g_m}{(\mathbf{A}\boldsymbol{f})_m^2}. \tag{7.33}$$

In equation (7.32) the following notations are used: α_n is the sum of the elements of the nth column of the matrix $\mathbf{A}$

$$\alpha_n = \sum_{m=1}^{N^2} A_{m,n}; \tag{7.34}$$

the division of the two vectors $\boldsymbol{g}$ and $\mathbf{A}\boldsymbol{f}$ is defined as a division component by component (and therefore it is still a vector)

$$\left(\frac{\boldsymbol{g}}{\mathbf{A}\boldsymbol{f}}\right)_m = \frac{g_m}{\sum_{n=1}^{N^2} A_{m,n} f_n}; \tag{7.35}$$

and finally $\mathbf{A}^T$ denotes the transposed matrix of $\mathbf{A}$, i.e. $(\mathbf{A}^T)_{n,m} = A_{m,n}$.

From equation (7.33) it follows that the Hessian matrix is negative semidefinite, because for any vector $\boldsymbol{\xi}$ we have

$$(\mathbf{H}(\boldsymbol{f})\boldsymbol{\xi} \cdot \boldsymbol{\xi}) = -\sum_{m=1}^{N^2} g_m \frac{(\mathbf{A}\boldsymbol{\xi})_m^2}{(\mathbf{A}\boldsymbol{f})_m^2}. \tag{7.36}$$

Therefore $l(\boldsymbol{f})$, equation (7.31), is a *concave function*. Moreover from the example 7.2 of section 7.1, we derive that, for given g_m, the upper bound of the Poisson distribution (7.25), as a function of λ_m is reached when $\lambda_m = g_m$. It follows that $l(\boldsymbol{f})$ is bounded by

$$l(\boldsymbol{f}) \leq \sum_{m=1}^{N^2} \{g_m \ln g_m - g_m - \ln(g_m)!\} \tag{7.37}$$

(when $g_m = 0$, the contribution of g_m to the upper bound is zero). Since $l(\boldsymbol{f})$ is concave and bounded, we conclude that *all maxima of $l(\boldsymbol{f})$ are global and at least one maximum exists.*

Remark 7.3. The bound (7.37) is reached, i.e. it is the maximum value of $l(\boldsymbol{f})$, if and only if the basic equation

$$\mathbf{A}f = g \tag{7.38}$$

*has at least one non-negative solution. In such a case, the properties of $l(\boldsymbol{f})$ imply that the set of its global maxima coincides with the set of the non-negative solutions of equation (7.38). However it is unlikely that this condition is satisfied in practice. As we know, the matrix **A** is highly ill-conditioned and therefore non-negative solutions of the equation, in general, do not exist when the data are noisy. In such a case the ML method certainly provides non-negative estimates but it is not yet clear whether they are physically sound or not. Numerical experiments indicate that, in most cases, they are not.*

The problem is now to find methods for computing or approximating the maxima of $l(\boldsymbol{f})$. To this purpose we can observe that, since $l(\boldsymbol{f})$ is concave, the *Kuhn–Tucker conditions* [3] are not only necessary but also sufficient conditions for an object $\tilde{\boldsymbol{f}}$ to be a maximum of $l(\boldsymbol{f})$. Since the domain of definition of $l(\boldsymbol{f})$ is the first quadrant, they can be written in the following way:

$$f_n \frac{\partial l(\boldsymbol{f})}{\partial f_n}\bigg|_{f=\tilde{f}} = 0; \quad n = 1, 2, \cdots, N^2 \tag{7.39}$$

$$\frac{\partial l(\boldsymbol{f})}{\partial f_n}\bigg|_{f=\tilde{f}} \leq 0, \quad \text{if } \tilde{f}_n = 0\,. \tag{7.40}$$

In particular, the first condition is necessary for any maximum and also sufficient when the maximum is interior to the domain of definition, i.e. all components of $\tilde{\boldsymbol{f}}$ are positive. The second condition applies to the maxima which lie on the boundary of the domain and, when combined with the first condition, implies that the extreme point is a maximum.

In any case, condition (7.39) is a necessary condition. By combining this equation with equation (7.32) we deduce that the maxima of $l(\boldsymbol{f})$ must be solutions of the following nonlinear equation

$$\alpha_n \tilde{f}_n = \tilde{f}_n \left(\mathbf{A}^T \frac{g}{\mathbf{A}\tilde{f}}\right)_n ; \quad n = 1, 2, \cdots, N^2. \tag{7.41}$$

Remark 7.4. By summing both members of equation (7.41) with respect to n we find that

$$\sum_{n=1}^{N^2}\left(\sum_{m=1}^{N^2} A_{m,n}\right)\tilde{f}_n = \sum_{n=1}^{N^2}\tilde{f}_n \sum_{m=1}^{N^2} A_{m,n}\frac{g_m}{\sum_{n'=1}^{N^2} A_{m,n'}\tilde{f}_{n'}} \tag{7.42}$$

and, by exchanging the summation order, we get

$$\sum_{m=1}^{N^2} (\mathbf{A}\tilde{f})_m = \sum_{m=1}^{N^2} g_m. \tag{7.43}$$

The physical meaning of this equation is the following: the total mean intensity of the noise-free image of $\tilde{f}$ coincides with the total number of counts of the noisy image of $f^{(0)}$. The mathematical meaning is that all maxima of $l(f)$ lie in the intersection of the first quadrant with the hyperplane defined by equation (7.43). This is a bounded domain of the object space.

If we introduce the nonlinear operator

$$\mathbf{M}(f) = \frac{f}{\alpha}\left(\mathbf{A}^T \frac{g}{\mathbf{A}f}\right) \tag{7.44}$$

where α is the vector with components α_n and the division of two vectors is defined as in equation (7.35), then equation (7.41) implies that the ML estimates are fixed points of this operator, i.e.

$$\tilde{f} = \mathbf{M}(\tilde{f}). \tag{7.45}$$

The converse, however, is not true: not all fixed points of $\mathbf{M}(f)$ are maxima of $l(f)$ even if they satisfy condition (7.43), as follows from remark 7.4.

Remark 7.5. It is easy to show, using property (7.43), that the operator (7.44) has at least N^2 fixed points $f^{(n)} (n = 1, 2, \cdots, N^2)$ given by

$$(f^{(n)})_n = \frac{1}{\alpha_n}\sum_{m=1}^{N^2} g_m; \quad (f^{(n)})_{n'} = 0; \; n' \neq n. \tag{7.46}$$

These fixed points are objects with only one component different from zero. In general, they are not ML estimates.

If we apply the method of successive approximations to the fixed-point equation (7.45), we obtain the following iterative algorithm

$$\tilde{f}_{k+1} = \mathbf{M}(\tilde{f}_k), \tag{7.47}$$

which is known as the *Richardson–Lucy method* (*RL-method*) or also the *Expectation–Maximization method* (*EM-method*). It was introduced by Richardson [4] and Lucy [5] in image restoration and by Shepp and Vardi [6] in emission tomography. Shepp and Vardi derived this iterative algorithm as a particular case of the EM-algorithm introduced by Dempster, Laird and Rubin [7] for the computation of maximum likelihood estimates from incomplete data.

For the interested reader we give this derivation in appendix G. Shepp and Vardi [6] also proved that the *iterates* $\tilde{f}_k$ *always converge to a maximum of* $l(\boldsymbol{f})$ if all components of the initial guess $\tilde{\boldsymbol{f}}_0$ are positive, so that $\tilde{\boldsymbol{f}}_0$ cannot be a fixed point of the operator (7.44) without being a maximum of the likelihood function. The limit depends on the initial approximation $\tilde{f}_0$ if $l(\boldsymbol{f})$ has several maxima. We do not give the proof of the convergence of the iterates.

We summarize the main properties of the algorithm:

- each approximation $\tilde{\boldsymbol{f}}_k$ is non-negative;
- each approximation $\tilde{\boldsymbol{f}}_k$ satisfies condition (7.43) i.e. the total mean intensity of $\mathbf{A}\tilde{\boldsymbol{f}}_k$ coincides with the total number of counts of the image g;
- the logarithmic likelihood function of $\tilde{\boldsymbol{f}}_k$ is non-decreasing, in the sense that $l(\tilde{\boldsymbol{f}}_{k+1}) \geq l(\tilde{\boldsymbol{f}}_k)$ (the proof is given in [6]);
- the sequence $\tilde{\boldsymbol{f}}_k$ converges to a maximum of $l(\boldsymbol{f})$ (the proof is given in [8]).

In the case of image restoration, the implementation of the algorithm is rather simple. First we observe that, for cyclic matrices, all coefficients α_n, equation (7.34), are equal because the columns are obtained by means of permutations of the components of a vector (in 1D case) or of an array (in 2D or 3D case). In particular in the 2D case we have

$$\alpha_{m,n} = \sum_{k,l=1}^{N^2} K_{m-k,n-l} = \sum_{k,l=1}^{N^2} K_{k,l} = \hat{K}_{0,0} \tag{7.48}$$

i.e. the coefficients coincide with the zero-frequency component of the TF. If we normalize the PSF in such a way that $\hat{K}_{0,0} = 1$, then the EM algorithm for image restoration can be written as follows:

$$(\tilde{\boldsymbol{f}}_{k+1})_{m,n} = (\tilde{\boldsymbol{f}}_k)_{m,n} \left(\mathbf{K}^T * \frac{g}{\mathbf{K} * \tilde{\boldsymbol{f}}_k} \right)_{m,n} . \tag{7.49}$$

The implementation of the algorithm requires the computation of four FFT at each iteration step: two for the computation of $\mathbf{K} * \tilde{\boldsymbol{f}}_k$ and two for the computation of the convolution of $\mathbf{K}^T$ with $g/(\mathbf{K} * \tilde{\boldsymbol{f}}_k)$. Therefore the computational cost of one step of this algorithm is higher than that of one step of the projected Landweber method, which requires the computation of only two FFT. Moreover, from numerical experience it follows that also the convergence of the EM-method is rather slow.

As already remarked, the limit of $\tilde{\boldsymbol{f}}_k$ may not be a sensible estimate of $\boldsymbol{f}^{(0)}$. Indeed, numerical simulations seem to indicate that the EM-algorithm has the semiconvergence property typical of the regularization methods discussed in the previous chapters.

In figure 7.1 we plot the relative norm of the restoration error, defined by $\|\tilde{f}_k - f^{(0)}\|/\|f^{(0)}\|$, as a function of k in the case of the example of figure 5.15. The function has a minimum for $k_{opt} = 53$ and the corresponding restoration error is 12.1%. The minimum restoration error is just the same as that provided by the projected Landweber method with positivity constraint (see section 6.2) and it is reached after a smaller number of iterations (53 instead of 82). The computation time, however, is greater because, as we already pointed out, each iteration step of the EM-method requires the computation of four DFT, against the two of the projected Landweber method.

The coincidence of the minimum value of the restoration error in the two cases is not surprising because the EM-method contains, in natural way, the constraint of positivity. This property, however, is significant only when this constraint may have important effects.

In order to explain the meaning of this remark let us consider the application of the EM-method to the example of figure 5.14, the image being corrupted by additive white noise with variance $s^2 = (0.008)^2$ (see also figure 6.7). This is

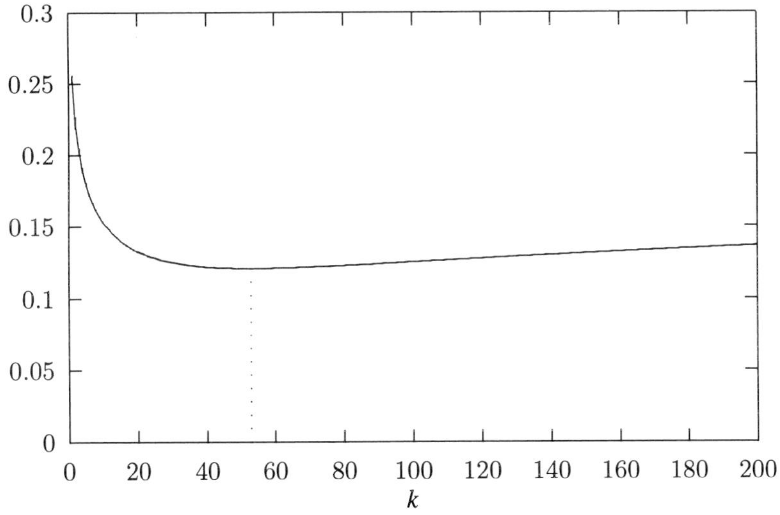

Figure 7.1. Behaviour of the relative norm of the restoration error as a function of the number of iterations of the EM method, in the case of the image of figure 5.14.

the image of an object which consists of two values: 0.5 and 1. The minimum restoration error provided by the EM-method is 8.5% and is reached after 59 iterations. These values are close to those corresponding to the Landweber method: restoration error 9.3%, number of iterations 60. Moreover, if we look at the corresponding restored images, we find that both exhibit the presence of ghosts. In other words, in this case, the EM-method behaves as the linear filtering methods. The reason is that, in this case, the positivity constraint is not effectual while it is the constraint consisting in a lower bound 0.5. As shown in section 6.2, the restoration error of the projected Landweber method with this constraint is 4.8% (even if 227 iterations are needed).

Different results are obtained if we consider now the image of the same object with the background 0.5 replaced by the background zero. The image is corrupted again by additive white noise with variance $s^2 = (0.008)^2$. The application of the previous methods to this new example provides now the following results for the minimum restoration error: 31.4% for the Landweber method, 12.4% for the projected Landweber method with positivity, 15% for the EM-method. The corresponding restorations are shown in figure 7.2. In this example, where the constraint of positivity is really significant, the linear method provides the worst restoration, while the restoration of the EM-method is close to that of the projected Landweber method.

The last example, however, is not yet completely appropriate because the noise added to the image is white and Gaussian. Therefore we have also generated an image corrupted by Poisson noise. The starting point is the noise-free image of the object of the previous example, but with the levels 0 and 50 instead of 0 and 1. This image has been corrupted by means of a generator of Poisson noise based on the rejection method [9]. The norm of the resulting noise is about 19.5% of the norm of the noise-free image and this means a rather large Poisson noise. The minimum restoration error for the two methods in competition is: 39.5% for the projected Landweber method and 33.3% for the EM-method. The number of iterations is 33 in the first case and 15 in the second. This last numerical experiment seems to indicate that EM is the appropriate regularization method when both Poisson noise and positivity constraint are significant.

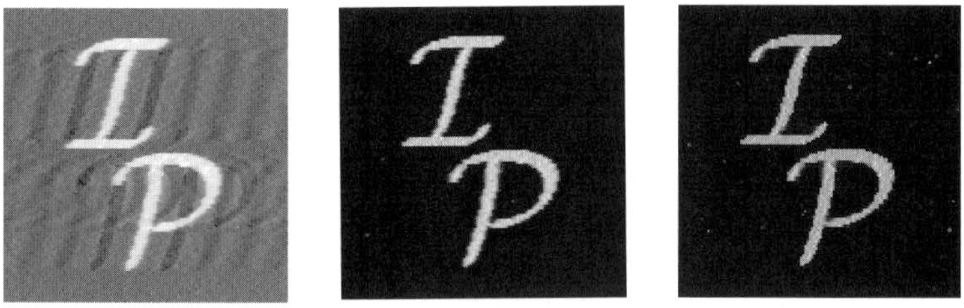

(a) (b) (c)

Figure 7.2. Optimal restorations of the image of figure 5.14, with background zero (noise variance, $s^2 = (0.008)^2$) provided by the Landweber method, (a), by the projected Landweber method with lower bound zero, (b) and by the EM-method, (c).

7.4 Bayesian methods

As we pointed out several times, some *a priori* information about the unknown object is required for obtaining sensible approximate solutions of the problem of image deconvolution. In the methods considered in the previous chapters, the information used takes the form of rigid constraints on the solution. Indeed they essentially consist of requiring that the approximate solutions belong to some fixed domain in the object space, for instance the set of all objects whose energy does not exceed a fixed quantity E^2. Similar constraints can be introduced also in maximum likelihood methods by restricting the domain of definition of the likelihood function. Rigid, or deterministic, constraints, however, do not represent the unique kind of *a priori* information which can be used.

Indeed it is also possible to introduce *a priori* information about the unknown object by giving a probability distribution, i.e. by assuming that the unknown object $\boldsymbol{f}^{(0)}$ is the value of a random vector ϕ whose density function is given. This can be a way, for instance, to allow a complete use of previous experience about the possible objects to be restored. Even if no such experience exists, the probabilistic approach can be used to formulate assumptions about the unknown objects.

The use of probabilistic *a priori* information is the basic feature of the so-called *Bayesian methods* and also the feature which characterizes these methods compared to ML methods.

Let us assume that the unknown object is the value of a random vector ϕ whose density function $p_\phi(\boldsymbol{f})$ is given. In such a case all vectors related by equation (7.1) are values of vector valued r.v. η, ϕ and ν which therefore are related by

$$\eta = \mathbf{A}\phi + \nu. \tag{7.50}$$

This relationship implies that a complete probabilistic description of the problem is provided by the joint density function of two of the three random vectors, for instance ϕ and η. Let us denote by $p_{\phi\eta}(\boldsymbol{f}, \boldsymbol{g})$ this density function. Then the *a priori* density function of ϕ, $p_\phi(\boldsymbol{f})$, must be considered as the *marginal density function* [2] of ϕ, i.e.

$$p_\phi(\boldsymbol{f}) = \int p_{\phi\eta}(\boldsymbol{f}, \boldsymbol{g}) d\boldsymbol{g}. \tag{7.51}$$

Moreover the density function $p_\eta(\boldsymbol{g}|\boldsymbol{f})$, introduced in section 7.1 as the density function of a random vector η depending on the parameter $\boldsymbol{f}$, must be interpreted as the conditional density function of η, being given a value $\boldsymbol{f}$ of ϕ. If both $p_\phi(\boldsymbol{f})$ and $p_\eta(\boldsymbol{g}|\boldsymbol{f})$ are given, this information is equivalent to a complete knowledge of $p_{\phi\eta}(\boldsymbol{f}, \boldsymbol{g})$. Indeed, from the Bayes formula for the conditional density function [2], it follows that

$$p_{\phi\eta}(\boldsymbol{f}, \boldsymbol{g}) = p_\eta(\boldsymbol{g}|\boldsymbol{f}) p_\phi(\boldsymbol{f}). \tag{7.52}$$

We see that the joint density function is obtained as the product of the likelihood function and of the *a priori* density function of ϕ.

The result of an imaging experiment is a value $\boldsymbol{g}$ of the random vector η. Therefore the obvious question is the following: what can we say about the random vector ϕ when a value of η is known? The answer is provided by the conditional density function of ϕ which, again using the Bayes formula, is given by

$$\begin{aligned} p_\phi(\boldsymbol{f}|\boldsymbol{g}) &= \frac{p_{\phi\eta}(\boldsymbol{f},\boldsymbol{g})}{p_\eta(\boldsymbol{g})} \\ &= \frac{p_\eta(\boldsymbol{g}|\boldsymbol{f})p_\phi(\boldsymbol{f})}{\int p_\eta(\boldsymbol{g}|\boldsymbol{f})p_\phi(\boldsymbol{f})d\boldsymbol{f}}. \end{aligned} \tag{7.53}$$

This conditional density function is also called the *a posteriori* (or *posterior*) density function of the random vector ϕ.

The density function $p_\phi(\boldsymbol{f}|\boldsymbol{g})$ provides all information we can have about the unknown object when we know its *a priori* probability distribution and its image $\boldsymbol{g}$. Indeed, in this way, we do not have a unique estimate of the unknown object but rather a set of possible estimates with different probabilities. Since, in practical applications, the usual requirement is to provide a restoration of the unknown object, we can use $p_\phi(\boldsymbol{f}|\boldsymbol{g})$ for computing various significant estimates of $\boldsymbol{f}^{(0)}$. The most used in practice are the following:

- the estimate provided by the *a posteriori* expectation value of ϕ:

$$\tilde{\boldsymbol{f}} = E\{\phi|\boldsymbol{g}\} = \int \boldsymbol{f}\,p_\phi(\boldsymbol{f}|\boldsymbol{g})d\boldsymbol{f}; \tag{7.54}$$

- the estimate (or estimates) provided by the global maximum (or maxima) of the *a posteriori* density function $p_\phi(\boldsymbol{f}|\boldsymbol{g})$.

If $p_\phi(\boldsymbol{f}|\boldsymbol{g})$ has a unique global maximum and is concentrated in a small neighbourhood of this maximum, then the two estimates indicated above are very close.

Bayesian methods are very flexible and allow the introduction of rather sophisticated constraints. The difficulty is that they are, in general, very expensive from the computational point of view. The treatment of these methods goes beyond the scope of this book. We will only consider in the next section the most simple case which leads to the so-called Wiener filter.

7.5 The Wiener filter

The determination of the Bayesian estimates is rather simple in the case of Gaussian processes. The basic assumptions are the following:

- the noise ν is a Gaussian random vector with zero expectation value and covariance matrix $\mathbf{S}_\nu$, so that its density function is given by equation (7.12);

- the object ϕ is also a Gaussian random vector with zero expectation value and covariance matrix $\mathbf{S}_\phi$, so that its density function is given again by equation (7.12) with $\mathbf{S}_\nu$ replaced by $\mathbf{S}_\phi$;
- the noise ν and the object ϕ are independent random vectors so that their *cross-covariance matrix* $\mathbf{S}_{\phi\nu}$ defined by

$$(\mathbf{S}_{\phi\nu})_{n,m} = E\{\phi_n \nu_m^*\} \tag{7.55}$$

  is zero.

From equation (7.19) it follows that the conditional density function of η, for a given value f of ϕ is Gaussian, with expectation value $E\{\eta|f\} = \mathbf{A}f$ and covariance matrix (independent of f) given by $\mathbf{S}_{\eta|\phi} = \mathbf{S}_\nu$. Then, from equation (7.52) it follows that

$$\begin{aligned} p_{\phi\eta}(f, g) &= p_\nu(g - \mathbf{A}f)p_\phi(f) \\ &= \left[(2\pi)^{N^2}|\mathbf{S}_\nu||\mathbf{S}_\phi|\right]^{-1/2} \exp\left[-\frac{1}{2}\Phi(f, g)\right] \end{aligned} \tag{7.56}$$

where

$$\Phi(f, g) = \left(\mathbf{S}_\nu^{-1}(\mathbf{A}f - g) \cdot (\mathbf{A}f - g)\right) + \left(\mathbf{S}_\phi^{-1} f \cdot f\right). \tag{7.57}$$

These equations imply that the marginal density function of η is also Gaussian, so that it is completely characterized by the expectation value and the covariance matrix of η, which can be obtained directly from equation (7.50) and the assumptions on ν and ϕ. Indeed, since both ν and ϕ have zero expectation value, from equation (7.50) we obtain that also the expectation value of η is zero

$$E\{\eta\} = \mathbf{0}. \tag{7.58}$$

Moreover, from equation (7.50) and the statistical independence of ν and ϕ we can also obtain the covariance matrix of η. The result is

$$\mathbf{S}_\eta = \mathbf{A}\mathbf{S}_\phi\mathbf{A}^* + \mathbf{S}_\nu. \tag{7.59}$$

Remark 7.6. The proof of equation (7.59) is the following. The independence of the r.v. ϕ_n and ν_m implies $E\{\phi_n \nu_m^*\} = E\{\nu_m \phi_n^*\} = 0$, so that from equation (7.50) we obtain

$$\begin{aligned} (\mathbf{S}_\eta)_{m,m'} &= E\{\eta_m \eta_{m'}^*\} \\ &= \sum_{n,n'=1}^{N^2} E\{A_{m,n}\phi_n A_{m',n'}^* \phi_{n'}^*\} + E\{\nu_m \nu_{m'}^*\} \\ &= \sum_{n,n'=1}^{N^2} A_{m,n}(\mathbf{S}_\phi)_{n,n'} A_{m',n'}^* + (\mathbf{S}_\nu)_{m,m'} \end{aligned} \tag{7.60}$$

and this is precisely equation (7.59), where $\mathbf{A}^*$ *is the adjoint matrix, defined by* $(\mathbf{A}^*)_{n,m} = A^*_{m,n}$.

Equations (7.58) and (7.59) imply that the marginal density function of η is given by

$$p_\eta(g) = \left[(2\pi)^{N^2}|\mathbf{S}_\eta|\right]^{-1/2} \exp\left[-\frac{1}{2}\left(\mathbf{S}_\eta^{-1} g \cdot g\right)\right]. \tag{7.61}$$

By inserting this equation and equation (7.56) in equation (7.53), we obtain the conditional density function of ϕ

$$p_\phi(f|g) = \left[(2\pi)^{N^2}\frac{|\mathbf{S}_\nu||\mathbf{S}_\phi|}{|\mathbf{S}_\eta|}\right]^{-1/2} \exp\left[-\frac{1}{2}\Phi(f,g) + \frac{1}{2}\left(\mathbf{S}_\eta^{-1} g \cdot g\right)\right]. \tag{7.62}$$

It is obvious that this conditional density function is also Gaussian and this remark has a first important consequence: the estimate provided by the *a posteriori* expectation value of ϕ and that provided by the maximum of the *a posteriori* density function of ϕ coincide. If we denote by $\tilde{f}$ this Bayesian estimate and by $\mathbf{S}_{\phi|\eta}$ the *a posteriori* covariance matrix, equation (7.62) must take the following form

$$p_\phi(f|g) = \left[(2\pi)^{-N^2}|\mathbf{S}_{\phi|\eta}|\right]^{-1/2} \exp\left[-\frac{1}{2}\left(\mathbf{S}_{\phi|\eta}^{-1}(f-\tilde{f})\cdot(f-\tilde{f})\right)\right]. \tag{7.63}$$

We can obtain $\tilde{f}$ and $\mathbf{S}_{\phi|\eta}$ by algebraic manipulations of the exponent of equation (7.62). A much simpler method is the following:

- compute $\mathbf{S}_{\phi|\eta}$ by making equal the terms quadratic in f at the exponents of equation (7.62) and (7.63);
- compute $\tilde{f}$ by minimizing the exponent of equation (7.62).

Indeed, by making equal the quadratic terms we get

$$\left(\mathbf{S}_{\phi|\eta}^{-1} f \cdot f\right) = \left(\mathbf{S}_\nu^{-1}\mathbf{A}f \cdot \mathbf{A}f\right) + \left(\mathbf{S}_\phi^{-1} f \cdot f\right). \tag{7.64}$$

Since this equality must hold for any f, we obtain

$$\mathbf{S}_{\phi|\eta} = \left(\mathbf{A}^*\mathbf{S}_\nu^{-1}\mathbf{A} + \mathbf{S}_\phi^{-1}\right)^{-1}. \tag{7.65}$$

As concerns the determination of $\tilde{f}$, we first observe that the minimization of the exponent of equation (7.62) is equivalent to the minimization of the functional (7.57). This functional is just the basic functional of regularization theory, as considered in chapter 5, when both $\mathbf{S}_\nu$ and $\mathbf{S}_\phi$ are proportional to the identity matrix. When this condition is not satisfied, it has the same structure of more general regularizing functional considered in chapter 10 and appendix E.

Indeed the two terms in the r.h.s. of equation (7.57) can be viewed as norms in weighted Euclidean spaces, as defined in appendix C.

From equation (E.20) of appendix E, with $\mathbf{C} = \mathbf{S}_\nu^{-1}$, $\mathbf{B} = \mathbf{S}_\phi^{-1}$ and $\mu = 1$, we obtain that $\tilde{f}$ is the solution of the following equation

$$\left(\mathbf{A}^*\mathbf{S}_\nu^{-1}\mathbf{A} + \mathbf{S}_\phi^{-1}\right)\tilde{f} = \mathbf{A}^*\mathbf{S}_\nu^{-1}g \tag{7.66}$$

so that it is given by

$$\tilde{f} = \mathbf{S}_{\phi|\eta}\mathbf{A}^*\mathbf{S}_\nu^{-1}g. \tag{7.67}$$

In conclusion, we can summarize the main results of the previous analysis as follows:

- the *a posteriori* density function of ϕ is given in equation (7.63) with $\tilde{f}$ and $\mathbf{S}_{\phi|\eta}$ given respectively by equation (7.67) and equation (7.65);
- the two Bayesian estimates, introduced at the end of the section 7.4, coincide and are given by $\tilde{f}$;
- the conditional covariance matrix $\mathbf{S}_{\phi|\eta}$ is independent of the value g of η;
- the diagonal elements of $\mathbf{S}_{\phi|\eta}$ give a measure of the reliability of the estimates of the various components of the object as provided by $\tilde{f}$;
- the estimate $\tilde{f}$ is obtained by applying to the image g the matrix

$$\mathbf{R}_0 = (\mathbf{A}^*\mathbf{S}_\nu^{-1}\mathbf{A} + \mathbf{S}_\phi{}^{-1})^{-1}\mathbf{A}^*\mathbf{S}_\nu^{-1}. \tag{7.68}$$

If we use the following identity

$$\mathbf{A}^*\mathbf{S}_\nu^{-1}\left(\mathbf{A}\mathbf{S}_\phi\mathbf{A}^* + \mathbf{S}_\nu\right) = \left(\mathbf{A}^*\mathbf{S}_\nu^{-1}\mathbf{A} + \mathbf{S}_\phi^{-1}\right)\mathbf{S}_\phi\mathbf{A}^* \tag{7.69}$$

we find that the matrix $\mathbf{R}_0$ can also be written in the following form

$$\mathbf{R}_0 = \mathbf{S}_\phi\mathbf{A}^*\left(\mathbf{A}\mathbf{S}_\phi\mathbf{A}^* + \mathbf{S}_\nu\right)^{-1}. \tag{7.70}$$

The matrix $\mathbf{R}_0$ is precisely the *Wiener filter.*

Remark 7.7. In the case of white processes, i.e. if we assume that

$$\mathbf{S}_\nu = \varepsilon^2\mathbf{I}, \quad \mathbf{S}_\phi = E^2\mathbf{I} \tag{7.71}$$

then from equation (7.68) we obtain

$$\mathbf{R}_0 = (\mathbf{A}^*\mathbf{A} + \left(\frac{\varepsilon}{E}\right)^2\mathbf{I})^{-1}\mathbf{A}^* \tag{7.72}$$

and this is just the Tikhonov regularizer with the choice of the regularization parameter provided by the Miller method—see section 5.6 .

As it is known, the Wiener filter provides the optimal solution of a linear estimation problem. Assume that we would like to estimate the random vector ϕ by applying a matrix $\mathbf{R}$ to the random vector η. The reliability of the estimate can be measured by the quantity

$$\varrho(\mathbf{R}) = E\{\|\mathbf{R}\eta - \phi\|^2\}; \tag{7.73}$$

then the optimal estimate is provided by the matrix $\mathbf{R}_0$ which minimizes $\varrho(\mathbf{R})$.

In order to solve this problem we do not need the joint density function of the random vectors ϕ, η; we only need their covariance matrices $\mathbf{S}_\phi$ and $\mathbf{S}_\eta$ as well as their cross-covariance matrix $\mathbf{S}_{\phi\eta} = \mathbf{S}^*_{\eta\phi}$.

By means of simple computations we find that

$$\begin{aligned} E\{\|\mathbf{R}\eta - \phi\|^2\} &= E\{(\mathbf{R}\eta \cdot \mathbf{R}\eta) - (\mathbf{R}\eta \cdot \phi) - (\phi \cdot \mathbf{R}\eta) + (\phi \cdot \phi)\} \\ &= Trace\{\mathbf{R}\mathbf{S}_\eta\mathbf{R}^* - \mathbf{R}\mathbf{S}_{\eta\phi} - \mathbf{S}_{\phi\eta}\mathbf{R}^* + \mathbf{S}_\phi\}. \end{aligned} \tag{7.74}$$

If we introduce the matrix

$$\mathbf{R}_0 = \mathbf{S}_{\phi\eta}\mathbf{S}_\eta^{-1} \tag{7.75}$$

the following identity can be easily proved

$$E\{\|\mathbf{R}\eta - \phi\|^2\} = Trace\{(\mathbf{R} - \mathbf{R}_0)\mathbf{S}_\eta(\mathbf{R} - \mathbf{R}_0) + \mathbf{S}_\phi - \mathbf{S}_{\phi\eta}\mathbf{S}_\eta^{-1}\mathbf{S}_{\eta\phi}\} \tag{7.76}$$

and therefore the minimum is reached when $\mathbf{R} = \mathbf{R}_0$. The matrix $\mathbf{R}_0$ is called the *optimum linear filter* or also the *Wiener filter* .

Now, in the case of random vectors ϕ, η related by equation (7.50), assuming that $\mathbf{S}_{\phi\nu} = \mathbf{0}$, we find that $\mathbf{S}_\eta$ is given by equation (7.59) and that $\mathbf{S}_{\phi\eta}$ is given by

$$\mathbf{S}_{\phi\eta} = \mathbf{S}_\phi\mathbf{A}^*. \tag{7.77}$$

We conclude that $\mathbf{R}_0$, as defined by equation (7.75), coincides with $\mathbf{R}_0$ as defined by equation (7.70).

References

[1] Papoulis A 1965 *Probability, Random Variables and Stochastic Processes* (New York: Mc Graw-Hill)

[2] Rahman N A 1968 *A Course in Theoretical Statistics* (London: Griffin)

[3] Zangwill W I 1969 *Nonlinear Programming* (Englewood Cliffs, NJ: Prentice-Hall)

[4] Richardson W H 1972 *J. Opt. Soc. Am.* **62** 55

[5] Lucy L B 1974 *Astron. J.* **79** 745

[6] Shepp L A and Vardi Y 1982 *IEEE Trans. Med. Imaging* **MI-1** 113

[7] Dempster A P, Laird N M and Rubin D B 1977 *J. Royal Stat. Soc.* **B39** 1

[8] Vardi Y, Shepp L A and Kaufman L 1984 *J. Am. Stat. Assoc.* **80** 8

[9] Press W H, Teukolsky S A, Vetterling W T and Flannery B P 1992 *Numerical Recipes* (Cambridge: Cambridge University Press)

PART 2

LINEAR INVERSE IMAGING PROBLEMS

Chapter 8

Examples of linear inverse problems

In the first part of the book the problem of image restoration has been investigated in the case of a space-invariant imaging system. In this second part more general inverse problems in imaging will be considered and a few significant examples of them are provided in this chapter.

8.1 Space-variant imaging systems

The imaging systems considered in the first part are space-invariant. However, as already observed, this property is seldom satisfied in practice. Therefore, if the system is linear, the general relationship between the object $f^{(0)}(\boldsymbol{x})$ and the noise-free image $g^{(0)}(\boldsymbol{x})$ is described by a linear integral operator of the following type

$$g^{(0)}(\boldsymbol{x}) = \int K(\boldsymbol{x}, \boldsymbol{x}') f^{(0)}(\boldsymbol{x}') d\boldsymbol{x}'. \tag{8.1}$$

The function $K(\boldsymbol{x}, \boldsymbol{x}')$ is the *space-variant PSF* of the system, already defined in section 3.1. In the theory of integral equations this function is also called the *kernel* (or the integral kernel) of the integral operator of equation (8.1).

The problem of image restoration consists then of determining an estimate $f(\boldsymbol{x})$ of $f^{(0)}(\boldsymbol{x})$ when a noisy version $g(\boldsymbol{x})$ of $g^{(0)}(\boldsymbol{x})$ and the space-variant PSF $K(\boldsymbol{x}, \boldsymbol{x}')$ are given. The starting point is an equation like (8.1) with $g(\boldsymbol{x})$, $f(\boldsymbol{x})$ in place of $g^{(0)}(\boldsymbol{x})$, $f^{(0)}(\boldsymbol{x})$. Such an equation is called a *Fredholm integral equation of the first kind* and its solution is a classical example of an ill-posed problem.

Equation (8.1) can be discretized by the method already used for convolution integrals. Here we consider the most frequent case of 2D images and we assume that both the object and the image are defined in the same domain, a square of side $2X$.

This domain can be divided into $N_0 = N^2$ square subdomains (pixels) with size $\delta = 2X/N$ and area δ^2—see figure 8.1 in the case of $N = 4$. If the area of the pixels is sufficiently small both the object and the image are approximately constant over a single pixel, i.e. they are approximately equal to a value taken in some particular point of the pixel, for instance the vertex at its bottom left (see figure 8.1).

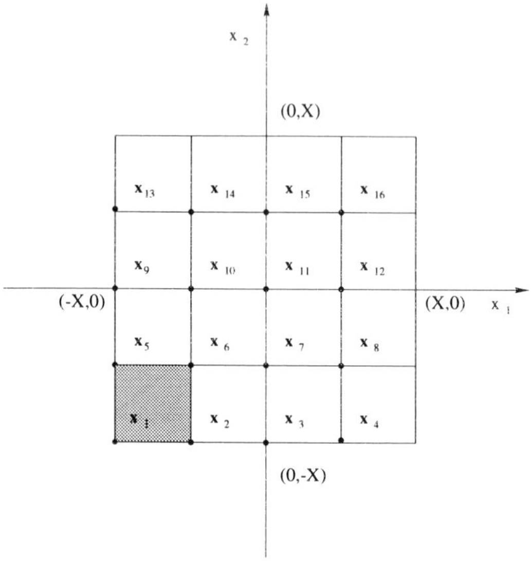

Figure 8.1. Partition of a square into 16 pixels. Each pixel is associated with the vertex at bottom left. For instance, the shaded pixel is associated with $\boldsymbol{x}_1$.

If these points are taken in the order indicated in figure 8.1 and denoted by $\boldsymbol{x}_1, \boldsymbol{x}_2, \cdots, \boldsymbol{x}_{N^2}$, then the integral in equation (8.1) can be approximated by the following integral sum

$$g^{(0)}(\boldsymbol{x}_m) = \sum_{n=1}^{N^2} K(\boldsymbol{x}_m, \boldsymbol{x}_n) f^{(0)}(\boldsymbol{x}_n)\delta^2 \quad ; \quad m = 1, \cdots, N^2. \tag{8.2}$$

Therefore, in the discrete case, the effect of the PSF is described by the matrix

$$A_{m,n} = K(\boldsymbol{x}_m, \boldsymbol{x}_n)\delta^2 \tag{8.3}$$

which can be called the *space-variant discrete PSF*.

In the case of a space-invariant imaging system, i.e. $K(\boldsymbol{x}_m, \boldsymbol{x}_n) = K(\boldsymbol{x}_m - \boldsymbol{x}_n)$, the matrix (8.3) can be replaced by the 2D cyclic matrix associated with the N^2 values $K(\boldsymbol{x}_n)$ of the space-invariant PSF ($n = 1, 2, \cdots, N^2$). In the general case of a space-variant system one needs N^4 values of the space-variant PSF. This number can be exceedingly large. For instance, in the case of an image 512 × 512, we have $N^4 = (512)^4 \simeq 6.9 \times 10^{10}$. If the matrix (8.3) does not have some particular structure, then too many matrix elements must be stored (even when the matrix is sparse) and the required storage space may be prohibitive. Also the number of operations required for computing the matrix-vector product of equation (8.2) is of the order of N^4 and therefore it is also too large.

This remark implies that the numerical treatment of equation (8.2) is not easy in the case of large images. For many optical systems, however, the space-variant PSF can be locally approximated by a space-invariant PSF. The meaning of this statement is the following.

Let us consider a sub-domain $\mathcal{D}_1$ of the image domain and a point $\boldsymbol{x}_1$ of $\mathcal{D}_1$. Typically $\mathcal{D}_1$ is a square with size d_1 and $\boldsymbol{x}_1$ is the centre of this square. In all practical cases the PSF $h_1(\boldsymbol{x}') = K(\boldsymbol{x}_1, \boldsymbol{x}')$ corresponding to $\boldsymbol{x}_1$ is significantly different from zero over a domain which is much smaller than the domain of the object so that the contributions to the value of the image in $\boldsymbol{x}_1$ do not come from the whole object domain but only from a limited portion of it. We assume that its size is also smaller than the size of $\mathcal{D}_1$. Then in $\mathcal{D}_1$ the PSF is approximately space-invariant if there exists a function $K_1(\boldsymbol{x})$ such that, for any $\boldsymbol{x}$ in $\mathcal{D}_1$

$$K(\boldsymbol{x}, \boldsymbol{x}') \simeq K_1(\boldsymbol{x} - \boldsymbol{x}' - \boldsymbol{x}_1). \tag{8.4}$$

The function $K_1(\boldsymbol{x})$ can be obtained, for instance, from the relation $K_1(-\boldsymbol{x}') = K(\boldsymbol{x}_1, \boldsymbol{x}')$ $(= h_1(\boldsymbol{x}'))$. If we replace $-\boldsymbol{x}'$ with $\boldsymbol{x} - \boldsymbol{x}' - \boldsymbol{x}_1$ in this relation, we see that the approximation (8.4) holds true if $K(\boldsymbol{x}, \boldsymbol{x}')$ satisfies the condition $K(\boldsymbol{x}, \boldsymbol{x}') \simeq K(\boldsymbol{x}_1, \boldsymbol{x} - \boldsymbol{x}' + \boldsymbol{x}_1)$, for any $\boldsymbol{x}$ in $\mathcal{D}_1$. It should be obvious that this is a rather restrictive property of the function $K(\boldsymbol{x}, \boldsymbol{x}')$.

Let us assume that it is satisfied in P non-overlapping subdomains $\mathcal{D}_1, \mathcal{D}_2, \cdots, \mathcal{D}_P$, for instance squares with centres at the points $\boldsymbol{x}_1, \boldsymbol{x}_2, \cdots, \boldsymbol{x}_P$, and sizes $d_1, d_2, \cdots, d_P$. Then there exist P functions $K_1(\boldsymbol{x}), K_2(\boldsymbol{x}), \cdots, K_P(\boldsymbol{x})$ such that $K(\boldsymbol{x}, \boldsymbol{x}') \simeq K_i(\boldsymbol{x} - \boldsymbol{x}' - \boldsymbol{x}_i)$ for any $\boldsymbol{x}$ in $\mathcal{D}_i$. If follows that the PSF can be approximated everywhere by

$$K(\boldsymbol{x}, \boldsymbol{x}') \simeq \sum_{i=1}^{P} \chi_i(\boldsymbol{x}) K_i(\boldsymbol{x} - \boldsymbol{x}' - \boldsymbol{x}_i) \tag{8.5}$$

where $\chi_i(\boldsymbol{x})$ is the characteristic function of the domain $\mathcal{D}_i$.

This approximation was used, for instance, for the PSF of the wide-field/planetary camera (WF/PC) of the Hubble Space Telescope (HST), before correction of the spherical aberration. The PSF of this system was space-variant because of vignetting in the internal repeater optics [1]. In this application a 5 × 5 grid of space-invariant PSF was typically used both for the computation of synthetic images and for image restoration. An example of two PSF from one of these grids is shown in figure 8.2. Since these PSF are rather close, the discontinuities introduced by the approximation (8.5) are not very important. Indeed the image of an object $f^{(0)}(\boldsymbol{x})$ provided by this PSF can be obtained by computing the P images provided by the P space-invariant PSF and by attributing each one of these images to the appropriate subdomain.

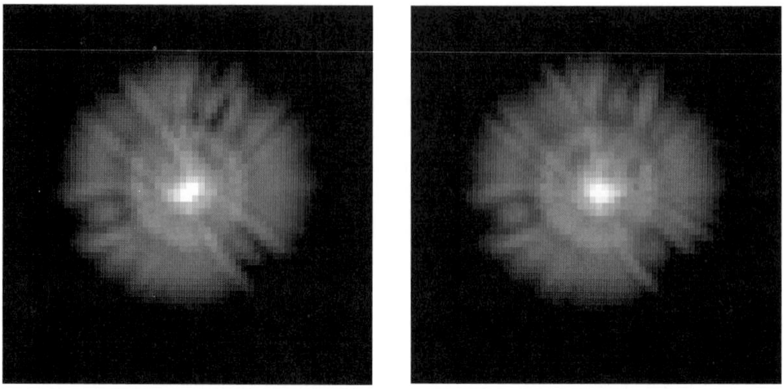

Figure 8.2. A sample of two PSF taken from a 5×5 grid of 25 PSF used for the restoration of images of the of wide-field/planetary camera of the HST. The PSF reproduced here corresponds to two corners of the grid (images produced by AURA/STScI).

As concerns the restoration problem, if the PSF can be approximated by a space-invariant PSF in a domain $\mathcal{D}_i$, then it is possible to use the methods developed in part I for restoring the image in this domain. It should be obvious, however, that a satisfactory restoration of the complete image cannot be obtained by a collection of restorations in the non-overlapping domains $\mathcal{D}_i$ because, in such a way, one introduces boundary effects which degrade the quality of the restored image. Therefore some form of matching must be used, considering, for instance, suitable overlapping sub-domains.

8.2 X-ray tomography

The word *tomography* comes from the Greek and means cross-sectional representation of a 3D object, i.e. representation of a 3D object by means of its 2D cross-sections or slices. In this sense the word only indicates a particular way of representing 3D objects and therefore can also be applied, for instance, to the case of 3D confocal microscopy, discussed in section 3.4. However, with reference to the first case where this name was introduced, it is now generally used for denominating imaging methods which are based on the knowledge of integrals of the object over lines or surfaces or other manifolds.

The first important example is the X-ray tomography, introduced by Hounsfield in 1971, which is usually called *computed tomography* (CT). A machine producing CT images is called a *CT scanner*. Several generations of CT scanners have been already designed, corresponding essentially to different ways of collecting data. A short description of the schemes of the various machines can be found in [2].

A CT image can be described as follows: it looks as though a planar slice of the body had been physically removed and then radiographed by passing X-rays through it in a direction perpendicular to its plane. Such an image shows the human anatomy with a spatial resolution of about 1 mm and a density discrimination of about 1%. In figure 8.3 an example of a CT image is shown.

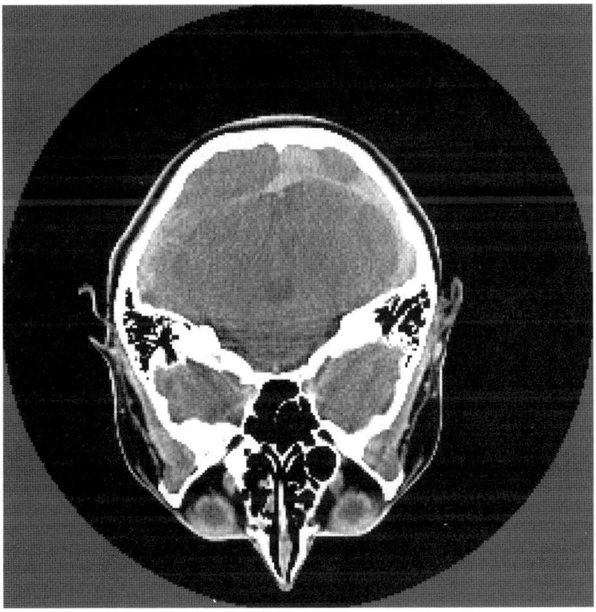

Figure 8.3. CT image of a slice of the head of one of the authors of this book.

The elementary process at the basis of CT is the X-ray absorption by the tissues of the human body. This absorption is described by a function whose value at point x is the *linear attenuation coefficient* of the tissue at that point. This function, which will be denoted $f^{(0)}(x)$ and called linear attenuation function, is roughly proportional to the density of the body and is just the function to be imaged, i.e. it is the object of the CT imaging system.

Let us consider a finely collimated source S emitting a pencil beam of X-rays which propagates through the body along a straight line L up to a well collimated detector R (see figure 8.4(a)). If $I(x)$ is the intensity of the beam at the point x of the straight line L, then the loss of intensity in the corresponding element dl of L is given by

$$dI(x) = -f^{(0)}(x)I(x)dl. \tag{8.6}$$

By integrating this equation we obtain

$$I = I_0 \exp\left(-\int_L f^{(0)}(\boldsymbol{x})dl\right) \tag{8.7}$$

where I_0 is the intensity of the beam at the exit of the source S and I is the intensity of the beam detected by the receiver R. Therefore, if we know both I_0 and I, by taking the logarithm of the ratio I_0/I we obtain the integral of the linear attenuation function $f^{(0)}(\boldsymbol{x})$ along the line L.

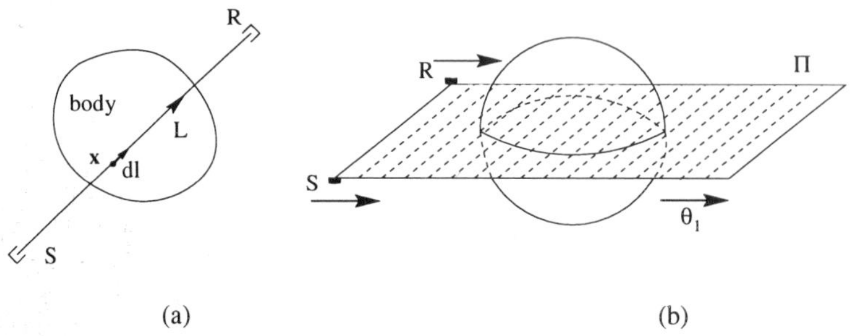

Figure 8.4. (a) Schematic representation of the elementary process of X-ray tomography; (b) the planar slice defined by the movement of the source–receiver pair.

By repeating the measurement outlined above for several different positions of the source–receiver pair, the basic information we obtain is a set of line integrals of $f^{(0)}(\boldsymbol{x})$ and therefore the problem is the recovery of $f^{(0)}(\boldsymbol{x})$ from these data. To this purpose the strategy followed in the collection of the data is important. We describe that chosen in the first CT scanner, which is also basic for understanding other methods of data collection.

As shown in figure 8.4(b), the source and the receiver are moved simultaneously along two parallel straight lines which define a plane Π, in practice a planar slice of the body under consideration. The common direction of the two straight lines is denoted by $\boldsymbol{\theta}_1$. By measuring the ratio I_0/I for all positions of the source–receiver pair, we obtain the integrals of $f^{(0)}(\boldsymbol{x})$ along all parallel lines defined by these positions. These integrals provide the so-called projection of $f^{(0)}(\boldsymbol{x})$ in the direction $\boldsymbol{\theta}_1$, whose precise definition will be given below.

Once these data have been collected, the source–receiver system is rotated by an angle $\Delta\phi$ around an axis orthogonal to the plane Π and then moved along two parallel straight lines with direction $\boldsymbol{\theta}_2$. As shown in figure 8.5, $\Delta\phi$ is the angle between the directions $\boldsymbol{\theta}_1$ and $\boldsymbol{\theta}_2$. The intersection O between the rotation axis and the plane Π is also indicated in the same figure. By measuring the ratio I_0/I for all positions of the source–receiver pair along the direction $\boldsymbol{\theta}_2$, we obtain the projection of $f^{(0)}(\boldsymbol{x})$ in this direction. By repeating this measure

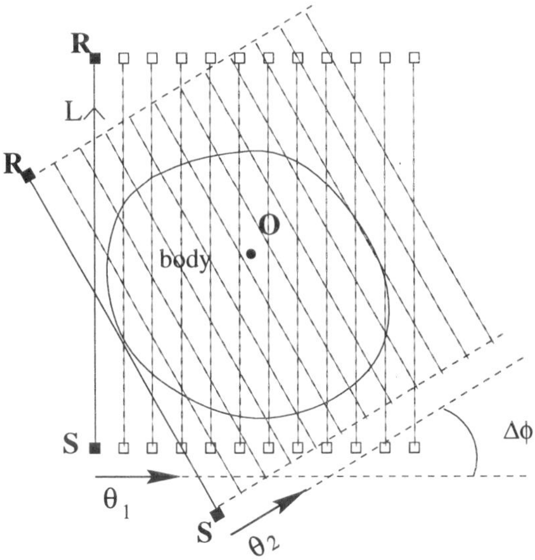

Figure 8.5. Scheme of the scanning used for measuring various projections of $f^{(0)}(\boldsymbol{x})$

for several directions $\boldsymbol{\theta}_1, \boldsymbol{\theta}_2, \cdots, \boldsymbol{\theta}_p$, we obtain the corresponding projections of $f^{(0)}(\boldsymbol{x})$.

The procedure outlined above is a double scanning consisting of an angular scanning and of a linear scanning for each angle. This is the origin of the name CT scanner used for a machine producing CT images. Moreover, the geometry of the scanning procedure outlined above is the so-called *parallel beam* geometry because one collects data corresponding to bunches of parallel lines. This was used in the first-generation scanners. In more recent scanners, the scanning procedure is based on the so-called *fan beam* geometry. The name derives from the fact that, in this case, a single source illuminates a ring of detectors with a broad fan beam of X-rays, so that one collects data corresponding to bunches of lines emanating from a point source. The difference between the parallel beam and the fan beam geometry is shown in figure 8.6. The important point is that in both cases one collects data which provide the integrals of $f^{(0)}(\boldsymbol{x})$ along lines lying in a well-defined plane Π.

The problem is now the recovery of the function $f^{(0)}(\boldsymbol{x})$ in the plane Π from the knowledge of its line integrals. Since, as shown above, the data provided by a CT scanner can be grouped into projections of the function $f^{(0)}(\boldsymbol{x})$, this problem is usually called *image reconstruction from projections* [3] while in the mathematical literature, the name *Radon transform inversion* is preferred. In this book we consider the problem in the case of parallel beam geometry. The modifications required in the case of fan beam geometry can be found in [4].

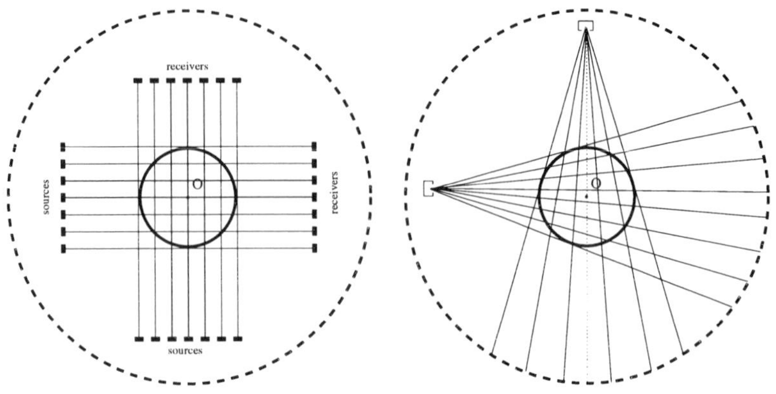

Figure 8.6. Schematic representation of the parallel beam geometry (a) and of the fan beam geometry (b).

Consider the plane Π corresponding to the section of the body to be imaged. Then, it is quite natural to take the intersection of this plane with the rotation axis of the machine as the origin 0 of a coordinate system. The orientation of the Cartesian axes is arbitrary but one can take, for instance, the x_1-axis along the initial direction of movement of the source–receiver pair.

Given a direction $\boldsymbol{\theta}$ in the plane, the straight lines orthogonal to $\boldsymbol{\theta}$ are characterized by their signed distance s from the origin (see figure 8.7). If ϕ is the angle between $\boldsymbol{\theta}$ and the x_1-axis, then $\boldsymbol{\theta} = \{\cos\phi, \sin\phi\}$ while $\boldsymbol{\theta}^{\perp} = \{-\sin\phi, \cos\phi\}$ is the unit vector orthogonal to $\boldsymbol{\theta}$. In terms of these unit vectors the equation of the straight line L, orthogonal to $\boldsymbol{\theta}$ with signed distance s from the origin, is given by $\boldsymbol{x} = s\boldsymbol{\theta} + t\boldsymbol{\theta}^{\perp}$, where t is the variable defined in figure 8.7. Then the *projection of* $f^{(0)}$ *in the direction* $\boldsymbol{\theta}$ is the function of s defined by

$$\left(R_{\theta} f^{(0)}\right)(s) = \int_{-\infty}^{+\infty} f^{(0)}(s\boldsymbol{\theta} + t\boldsymbol{\theta}^{\perp})dt. \tag{8.8}$$

For a given $\boldsymbol{\theta}$, the parallel beam method, described above, provides sampled and noisy values of the projection $\left(R_{\theta} f^{(0)}\right)(s)$.

As an example, in figure 8.8 we show two projections of a function $f^{(0)}$ which is the sum of the characteristic functions of two disjoint cylinders, one with radius $1/2$ and the other with radius 1. It is obvious that when the direction $\boldsymbol{\theta}$ is parallel to the straight line joining the centres of the two cylinders, their projections are separated. When the direction $\boldsymbol{\theta}$ is orthogonal to this line, the projections of the two cylinders are superimposed. For other directions one can have intermediate situations.

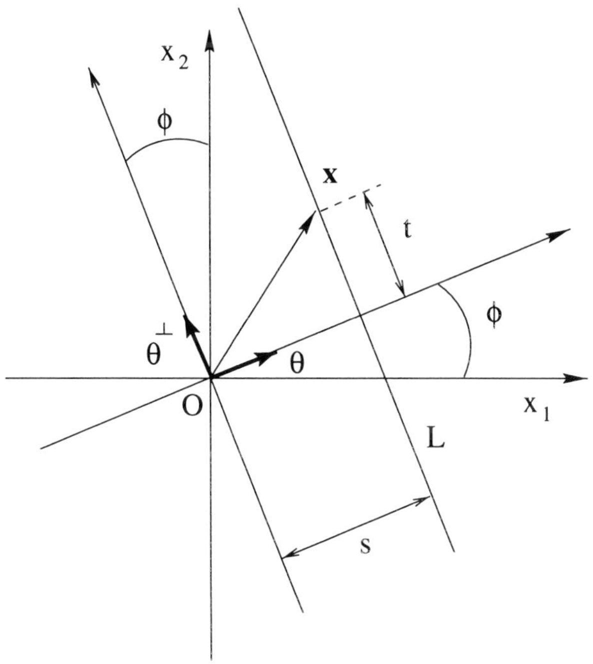

Figure 8.7. Definition of the geometrical variables for the Radon transform.

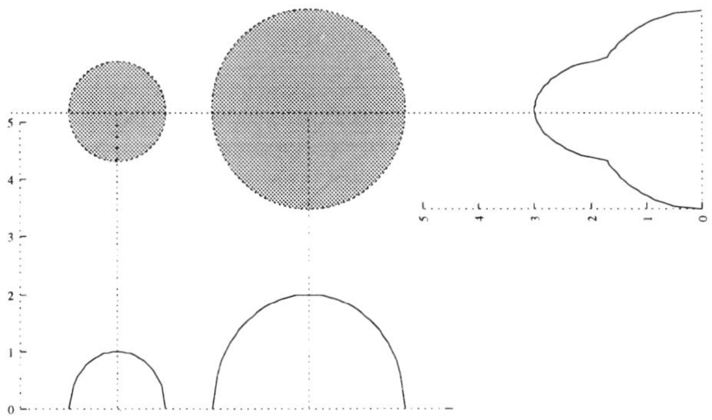

Figure 8.8. Two projections of an object which consists of a pair of cylinders.

If we consider all projections of $f^{(0)}$, corresponding to all possible directions $\boldsymbol{\theta}$, then a function of two variables, the angle ϕ and the signed distance s, is defined. This function is denoted by

$$\left(Rf^{(0)}\right)(s,\boldsymbol{\theta}) = \left(R_{\boldsymbol{\theta}} f^{(0)}\right)(s) \tag{8.9}$$

and is called the *Radon transform* of $f^{(0)}$, in honour of the mathematician Johann Radon who first investigated the problem of recovering a function of two variables from its line integrals. The result of Radon was even more general, since he proved formulas for the reconstruction of a function of q variables from knowledge of its integrals over all hyperplanes with dimension $q-1$ [5].

The Radon transform defines a mapping R which transforms a function of two space variables into a function of one space and one angular variable. The domain of definition of this function in the $\{s,\phi\}$ plane is the strip defined by: $-\infty < s < \infty$, $-\pi < \phi \le \pi$. In this strip the Radon transform has the symmetry property

$$\left(Rf^{(0)}\right)(-s,-\boldsymbol{\theta}) = \left(Rf^{(0)}\right)(s,\boldsymbol{\theta}) \tag{8.10}$$

which easily follows from the explicit representation of $Rf^{(0)}$ in terms of the variables s,ϕ

$$\left(Rf^{(0)}\right)(s,\boldsymbol{\theta}) = \int_{-\infty}^{\infty} f^{(0)}(s\cos\phi - t\sin\phi, s\sin\phi + t\cos\phi)dt. \tag{8.11}$$

The $\{s,\phi\}$-plot of the Radon transform of $f^{(0)}$, obtained by representing its values as grey levels, is called the *sinogram* of $f^{(0)}$ [6]. This picture is, in a sense, the image of $f^{(0)}$ provided by a CT scanner. Then the sinogram obtained from the measured values of the Radon transform is a discrete and noisy image of $f^{(0)}$.

If the support of the function $f^{(0)}$ is interior to the disc of radius a, then the support of the sinogram is the rectangle: $|s| \le a$, $|\phi| \le \pi$. Each point in the rectangle corresponds to a straight line crossing the disc of radius a in the $\{x_1,x_2\}$-plane. The points corresponding to straight lines through a fixed point in the $\{x_1,x_2\}$-plane describe a sinusoidal curve in the $\{s,\phi\}$-plane, given by: $s = x_1\cos\phi + x_2\sin\phi$. As a consequence, if the object consists of a bright spot over a black background, then its sinogram is precisely a bright sinusoidal curve over a black background (see figure 8.9). This property has given rise to the name sinogram. In figure 8.9 we also give an example of a sinogram of a complex object.

8.3 Emission tomography

X-ray tomography is also called *transmission computed tomography* (TCT) because the image is obtained by detecting the X-rays transmitted by the body.

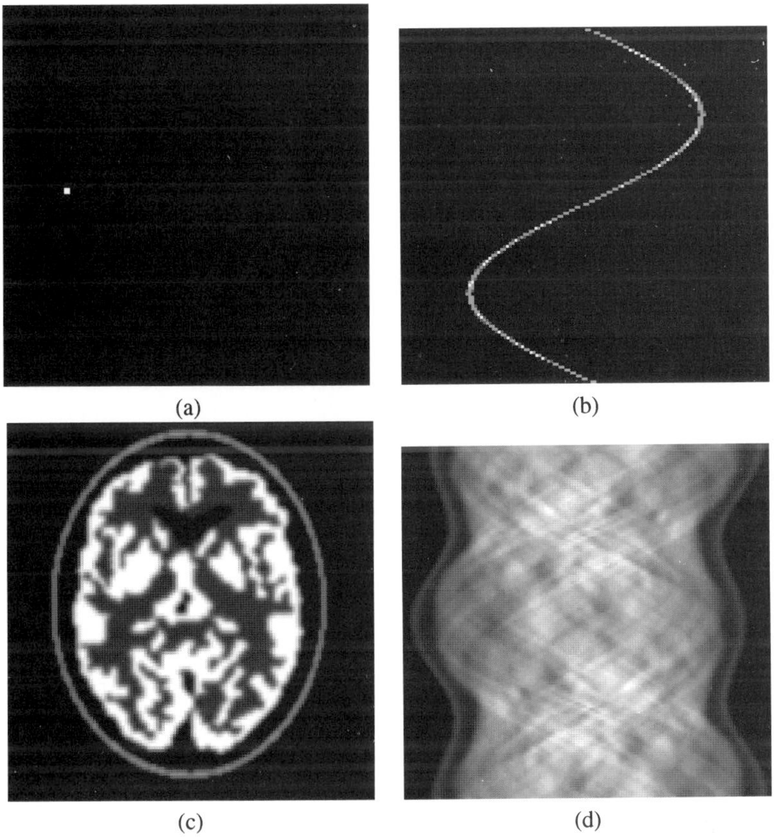

Figure 8.9. Illustrating the sinogram: the object in (a) consists of a unique bright spot and the corresponding sinogram is the white curve in (b). Each point of this curve corresponds to a straight line passing through the bright spot in (a). In (c) a more structured object is given and its sinogram is shown in (d).

This technique provides information about anatomical details of the body organs: the map of the linear attenuation function is essentially the map of the density of the tissues.

A quite different type of information is obtained by the so-called *emission computed tomography* (ECT). This technique is based on the administration, either by injection or by inhalation, of radionuclide-labelled agents known as radiopharmaceuticals. Their distribution in the body of the patient depends on factors such as blood flow, metabolic processes, etc. Then a map of this distribution is obtained by detecting the γ-rays produced by the decay of the radionuclides. Therefore ECT yields functional information, in the sense that the images produced by ECT show the function of the biological tissues of the organs.

Two different modalities of ECT are usually considered:

- *single-photon emission tomography* (SPECT), which makes use of radioisotopes such as ^{99m}Tc, where a single γ-ray is emitted per nuclear disintegration;
- *positron emission tomography* (PET), which makes use of β^+-emitters such as ^{11}C,^{15}O,^{13}N and ^{18}F, where the final result of a nuclear disintegration is a pair of γ-rays, propagating in opposite directions, produced by the annihilation of the emitted positron in the tissue.

In both cases it is necessary to detect the γ-rays coming from well-defined regions of the body and, to this purpose, different methods are used in the two cases.

In SPECT the discrimination of the γ-rays is obtained by a *collimator* which is a large slab covering the crystal detector face, consisting of holes separated by lead septa (see figure 8.10). The axes of the holes are perpendicular to

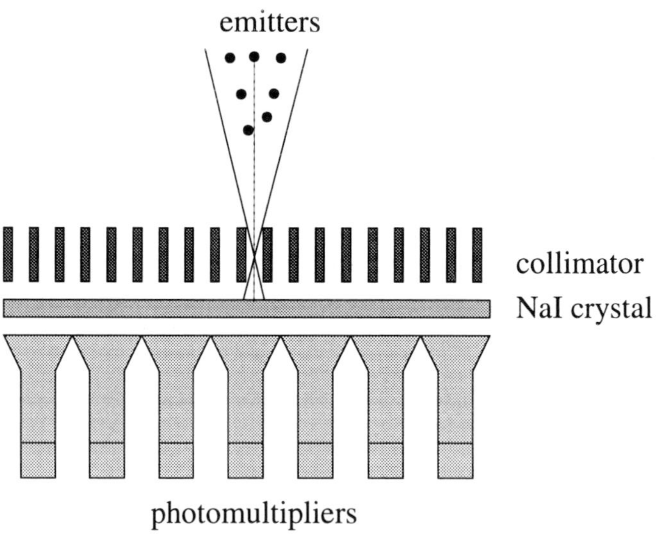

Figure 8.10. Scheme of the detection system in SPECT. The γ-rays, transmitted by the collimator, cause scintillations in the NaI crystal. The energy and position of the interaction of the γ-quantum in the crystal are determined by an array of photomultipliers. The acceptance cone of a hole of the collimator is also indicated.

the detector face. Therefore each hole mainly collects the γ-quanta which are emitted in the direction of its axis by the radionuclides located in voxels also crossed by this axis. These γ-quanta are counted by the detection system and their number (in a given time) is proportional to the number of radionuclides contained in the voxels discriminated by the hole. In this way we obtain a number which is roughly proportional to the integral of the radionuclides concentration along the axis of the hole. We can denote again the unknown concentration by $f^{(0)}(\boldsymbol{x})$ since this is the object to be imaged by the SPECT machine.

If we consider now a line of holes of the collimator and the plane Π, defined by this line and the axis of the holes, then each hole provides a sample of the projection (in the direction of the line) of the concentration function $f^{(0)}(\boldsymbol{x})$ in the plane Π. By rotating the collimator–detector system around the body in a way similar to that of the parallel-beam scanning in X-ray tomography, we obtain information which roughly consists of a set of projections of $f^{(0)}(\boldsymbol{x})$. These data can also be represented in the form of a sinogram. Even if the basic reconstruction technique of TCT, i.e. the filtered back projection which will be discussed in chapter 11, is often used in the reconstruction of SPECT images, several corrections should be necessary. The principal ones are due to the following effects.

- *Collimator blur*: Because the holes of the collimator are not infinitely narrow, photons moving in directions other than that perpendicular to the detector plane can also be detected (see figure 8.10). In other words, each hole of the collimator collects photons coming from the radionuclides interior to a small cone around its axis. This effect causes a substantial loss of resolution as shown in figure 8.11.
- *Attenuation and scatter*: The photons emitted may be absorbed or scattered by the body tissues before reaching the detectors. Because of the attenuation, the radionuclides concentration is underestimated, the effect being not uniform over the cross-section of the body (it is greater at the centre than at the boundary). Moreover, as an effect of the scattering, a photon detected in a certain direction may correspond to a radionuclide which is not located in the acceptance cone of the hole.
- *Noise*: The amount of radiopharmaceutical administrated to the patient as well as the acquisition time must be subjected to severe limitations. As a consequence, at each position in the detector plane the number of counts is relatively low, i.e. the data are affected by a rather large amount of noise which is accurately described by Poisson statistics. As a consequence a filtering of the projections before reconstruction or the use of statistical methods (see chapter 7) may be required.

In PET the collimation is obtained by the use of pairs of detectors in coincidence. This technique, which is also called electronic collimation, is more accurate than the physical collimation (as described above) and can be used for designing a detection system with a great efficiency. The procedure is illustrated in figure 8.12: annihilation photons, arising in a voxel intersected by a line joining the centres of the two detectors, are detected if their direction

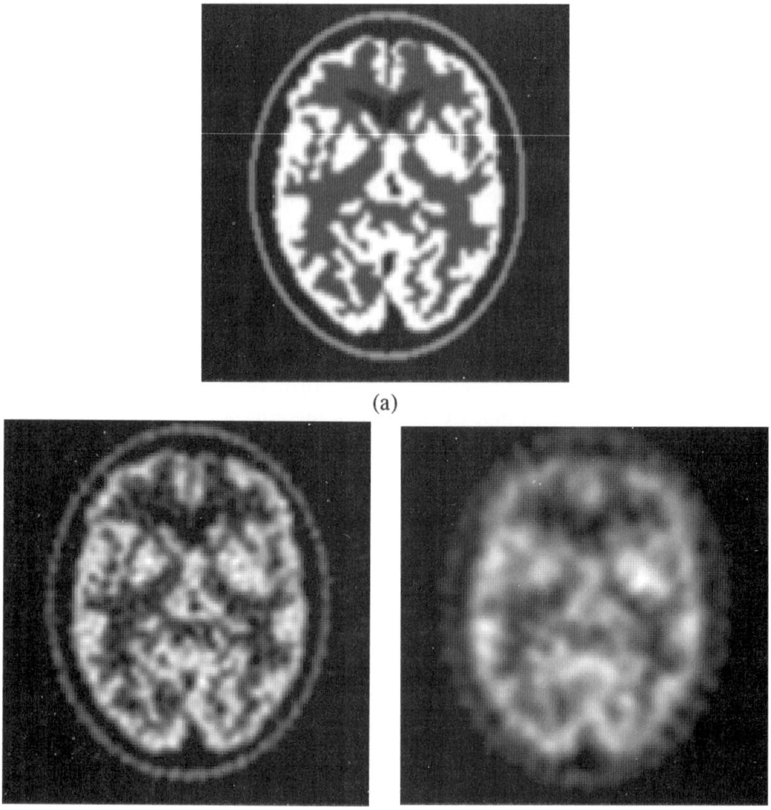

(a)

(b) (c)

Figure 8.11. The reconstruction of a digital phantom (shown in (a)) as provided by the filtered back projection in the case of TCT data (shown in (b)) and in the case of SPECT data (shown in (c)). In this example the SPECT projections have been computed by taking into account only the collimator blur and the Poisson noise.

of emission is parallel, or close, to the previous line. Since the number of positrons (and therefore of γ-pairs) emitted in a given time and in a given volume is proportional to the concentration of the radionuclides, the number of detected photons is proportional to the total number of radionuclides interior to the tube defined by the two detectors and shown in figure 8.12, hence to the integral of their concentration over this tube. If the cross-section of the tube is sufficiently narrow, this integral can be approximated by a line integral and therefore the fundamental information we obtain has the same nature as that of X-ray tomography.

The description of the various geometries of the PET scanners is outside

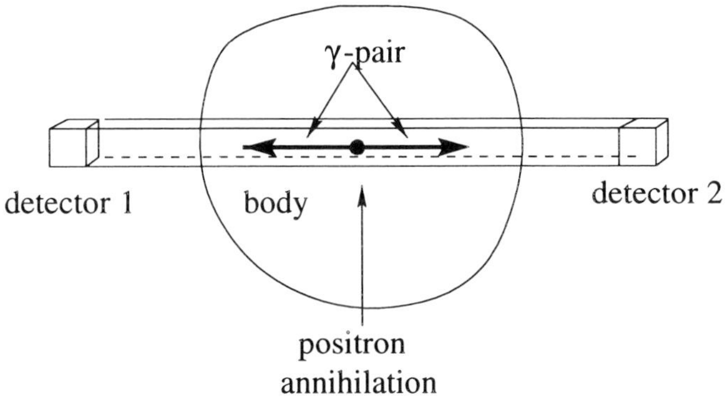

Figure 8.12. Scheme of the electronic collimation in PET.

the scope of this book. We only observe that data are, in general, organized into a sinogram so that the basic algorithm of tomography can be used also in this case.

Several corrections should be applied also to PET data, some of them being due to effects which are also present in SPECT. However, the effects of collimator blur and Poisson noise are not as important as in SPECT, thanks to the great precision and efficiency of the electronic collimation. Moreover the effect of attenuation can be corrected fairly easily. In practice, each PET scan is preceded by two separate measurements, the main purpose of which is to estimate the attenuation correction to the measured data.

The brief discussion we have given of the principles of SPECT and PET imaging should make clear that the basic approximation, which consists of considering SPECT and PET data as line integrals of the radionuclides concentration, is rather rough. In order to improve the quality of the reconstructed images, an improved physical model of the acquisition process should be used. Such a model must take into account the features of the specific scanner under consideration. Therefore we can only give here a rather general description of these models.

The first point is that, even if the geometry of the detection system is designed in order to define planar cross-sections of the body to be imaged, the reconstruction problem is always 3D both as a consequence of the scatter effect and also as a consequence of the collimation blur (the second effect is especially important in the case of SPECT). Therefore, from the beginning, the region of interest is a certain domain $\mathcal{D}$ in 3D space. This domain is replaced by a fine grid consisting of N voxels, characterized by the index n. Moreover the detection system defines M tubes (or cones) characterized by the index m.

The basic physical quantity describing the acquisition process is the probability that a decay occurring at voxel n is detected in tube m. This probability is denoted by $P_{m,n}$. If we denote by $f_n^{(0)}$ the average number of decays occurring in the voxel n, and by $g_m^{(0)}$ the average number of γ-rays detected in the tube (or cone) m, then the relationship between the quantities $f_n^{(0)}$ and $g_m^{(0)}$ is given by

$$g_m^{(0)} = \sum_{n=1}^{N} P_{m,n} f_n^{(0)}. \tag{8.12}$$

As we see, the probabilities $P_{m,n}$ are the elements of a matrix which is called the *transition matrix*. Since not all decays occurring at voxel n are detected, these matrix elements must satisfy the conditions

$$\sum_{m=1}^{M} P_{m,n} < 1 \tag{8.13}$$

for $n = 1, 2, \cdots, N$.

The quantities $g_m^{(0)}$ are the noise-free data. Since $g_m^{(0)}$ is the average number of counts in the tube (or cone) m, the measured number of counts g_m is the value of a Poisson random variable whose expectation value is $g_m^{(0)}$. The difference $g_m - g_m^{(0)}$ represents the so-called Poisson noise.

From this scheme it turns out that the key problem, in the modelling of a SPECT or PET scanner, is the determination or approximation of the matrix elements $P_{m,n}$. This computation must take into account the geometry of the detection system as well as the physical processes affecting the propagation of the photons through the tissues, namely attenuation and scattering. Since these effects depend on the density of the tissues, their estimation should require the knowledge of the linear attenuation function, i.e. the function imaged by the CT scanners. As already observed, one of the advantages of the PET scanners, with respect to the SPECT scanners, is that they allow the estimate of the corrections of the transition matrix due to attenuation by means of two separate measurements which precede the imaging scan.

8.4 Inverse diffraction and inverse source problems

Inverse diffraction and inverse source problems are sometimes considered as intermediate steps in the solution of inverse scattering problems. However, they may also have some direct application. For instance, as we mentioned in section 3.6, inverse diffraction is the basic problem of acoustic holography. In that section only the case of planar surfaces was considered. Here we briefly discuss the more general case of arbitrary surfaces. On the other hand, the inverse source problem is, for instance, the problem of determining the charge-current distribution of a radiating antenna from the knowledge of its radiation pattern. These problems have been investigated both in the scalar case (acoustic waves) and in the vector case (electromagnetic waves). Here we describe these problems only in the most simple case, i.e. the scalar one.

As a general comment we observe that, while the solution of an inverse diffraction problem is unique, the inverse source problem is always affected by a rather serious non-uniqueness. This property is due to the so-called *non-radiating sources*, i.e. to the existence of sources (for instance an oscillating body or an oscillating charge-current distribution) which do not produce a radiation field. A survey of this subject can be found in [7].

Before introducing the two problems, we describe the physical situation we are considering. We assume that a scalar field is generated by oscillating sources contained in a bounded domain $\mathcal{D}$, with a regular boundary Σ. If the sources oscillate with a frequency ν, then the field is monochromatic and can be represented by

$$U(\boldsymbol{x}, t) = u(\boldsymbol{x})e^{-i2\pi\nu t} \tag{8.14}$$

where $\boldsymbol{x} = \{x_1, x_2, x_3\}$. The free propagation of the field outside Σ is described by the Helmholtz equation

$$\Delta u + k^2 u = 0 \tag{8.15}$$

where $k = 2\pi\nu/c$ is the wave-number of the monochromatic field, c being the velocity of the propagating waves. To the wave-number k one can also associate the wave-length λ, which is given by

$$\lambda = \frac{2\pi}{k} = \frac{c}{\nu} \tag{8.16}$$

and is a characteristic length of the problem under consideration.

The field amplitude $u(\boldsymbol{x})$ must also satisfy the *Sommerfeld radiation condition* at infinity

$$\lim_{r\to\infty}\left[r\left(\frac{\partial u}{\partial r} - iku\right)\right] = 0, \quad r = |\boldsymbol{x}| \tag{8.17}$$

which expresses the fact that we are considering an outgoing wave, i.e. a wave propagating from the sources, and not an ingoing wave, i.e. a wave propagating toward the sources.

A. *Inverse diffraction*

Consider two closed surfaces, Σ_1 and Σ_2, both surrounding the domain $\mathcal{D}$ of the sources, such that Σ_1 is interior to Σ_2 (see figure 8.13). Then the problem of *inverse diffraction* associated with the surfaces Σ_1, Σ_2 is the following: given the values of the propagating field $u(\boldsymbol{x})$ on the surface Σ_2, determine its values on the surface Σ_1. The solution of this problem provides a map of the field over Σ_1; if Σ_1 is close to the domain $\mathcal{D}$ of the sources, this map is a sort of imaging of the sources.

In order to give the mathematical formulation of the inverse problem we must first solve the direct one: determine the field amplitude $u(\boldsymbol{x})$ which satisfies

Helmholtz equation in the region exterior to Σ_1, takes prescribed values, say $f^{(0)}(\boldsymbol{x})$ on Σ_1

$$u(\boldsymbol{x})|_{\Sigma_1} = f^{(0)}(\boldsymbol{x}) \tag{8.18}$$

and satisfies the Sommerfeld radiation condition at infinity.

If $G(k; \boldsymbol{x}, \boldsymbol{x}')$ is the Green function of this problem, then $u(\boldsymbol{x})$ can be expressed as follows

$$u(\boldsymbol{x}) = \int_{\Sigma_1} G(k; \boldsymbol{x}, \boldsymbol{x}') f^{(0)}(\boldsymbol{x})' d\sigma(\boldsymbol{x}') \tag{8.19}$$

where $d\sigma(\boldsymbol{x}')$ is the measure on the surface Σ_1. In simple cases, the Green function $G(k; \boldsymbol{x}, \boldsymbol{x}')$ can be determined analytically; otherwise it must be computed numerically.

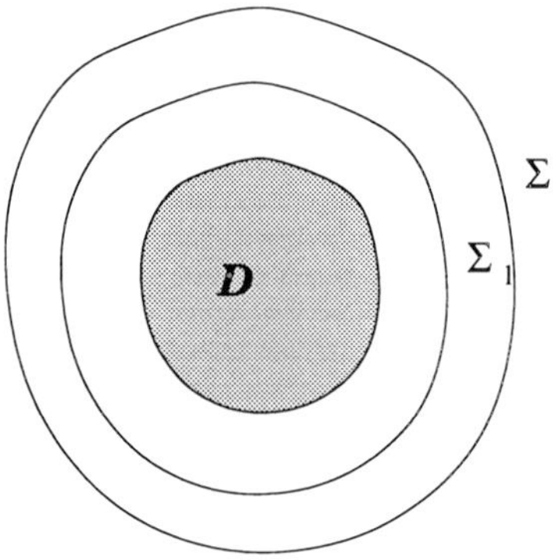

Figure 8.13. Geometry of the problem of inverse diffraction. The shaded area corresponds to the domain $\mathcal{D}$ containing the sources of the field. Σ_2 is the surface where data are recorded

If we come back now to the inverse problem, let $g^{(0)}(\boldsymbol{x})$ be the values of $u(\boldsymbol{x})$ on Σ_2

$$u(\boldsymbol{x})|_{\Sigma_2} = g^{(0)}(\boldsymbol{x}). \tag{8.20}$$

Then the problem consists of looking for boundary values $f^{(0)}(\boldsymbol{x})$ such that $u(\boldsymbol{x})$, as given by equation (8.19), satisfies condition (8.20). We obtain

$$g^{(0)}(\boldsymbol{x}) = \int_{\Sigma_1} G(k; \boldsymbol{x}, \boldsymbol{x}') f^{(0)}(\boldsymbol{x}') d\sigma(\boldsymbol{x}'), \quad \boldsymbol{x} \in \Sigma_2. \tag{8.21}$$

In such a way the problem has been reduced to the solution of a first-kind Fredholm integral equation. In a practical situation, we do not have the exact values $g^{(0)}(\boldsymbol{x})$ of the field on Σ_2 but only measured values $g(\boldsymbol{x})$, affected by noise or experimental errors: $g(\boldsymbol{x}) = g^{(0)}(\boldsymbol{x}) + w(\boldsymbol{x})$, where $w(\boldsymbol{x})$ is the term describing the noise.

The integral operator (8.21) belongs to the class of integral operators which will discussed in section 9.4 and therefore has a singular value decomposition which can be used in the investigation of the problem. The computation of this singular system, however, may be difficult if the surfaces Σ_1, Σ_2 are rather irregular.

The case of spherical surfaces is simple because spherical harmonic expansions can be used for solving the problem. Let Σ_1, Σ_2 be spheres with centre at the origin and let a_1, a_2 be their radii, $a_1 < a_2$. Moreover, let $r, \boldsymbol{\theta}$ be polar coordinates

$$r = |\boldsymbol{x}|, \quad \boldsymbol{\theta} = \frac{\boldsymbol{x}}{r}, \tag{8.22}$$

$\boldsymbol{\theta}$ being a point on the sphere S^2 of the 3D space.

The boundary values of $u(\boldsymbol{x})$ on Σ_1 define a function on S^2, denoted by $f^{(0)}(\boldsymbol{\theta})$, which can be represented by means of its expansion in terms of spherical harmonics $Y_{l,m}$

$$f^{(0)}(\boldsymbol{\theta}) = \sum_{l=0}^{\infty} \sum_{m=-l}^{l} f_{l,m}^{(0)} Y_{l,m}(\boldsymbol{\theta}) \tag{8.23}$$

where (we denote by $d\boldsymbol{\theta}$ the measure on S^2)

$$f_{l,m}^{(0)} = \int_{S^2} f^{(0)}(\boldsymbol{\theta}) Y_{l,m}^*(\boldsymbol{\theta}) d\boldsymbol{\theta}. \tag{8.24}$$

Then, in terms of the spherical Hankel functions of the first kind

$$h_l^{(1)}(r) = \sqrt{\frac{\pi}{2r}} H_{l+1/2}^{(1)}(r), \tag{8.25}$$

the solution of the direct problem is given by

$$u(r, \boldsymbol{\theta}) = \sum_{l=0}^{\infty} \sum_{m=-l}^{l} f_{l,m}^{(0)} \frac{h_l^{(1)}(kr)}{h_l^{(1)}(ka_1)} Y_{l,m}(\boldsymbol{\theta}). \tag{8.26}$$

Consider now the inverse problem. The exact data define a function on S^2, denoted by $g^{(0)}(\boldsymbol{\theta})$. The relationship between $g^{(0)}(\boldsymbol{\theta})$ and $f^{(0)}(\boldsymbol{\theta})$ can be obtained from equation (8.26) if we observe that

$$g^{(0)}(\boldsymbol{\theta}) = u(a_2, \boldsymbol{\theta}). \tag{8.27}$$

By taking into account equation (8.24), the relationship can be written as follows

$$g^{(0)}(\boldsymbol{\theta}) = \int_{S^2} G(\boldsymbol{\theta}, \boldsymbol{\theta}') f^{(0)}(\boldsymbol{\theta}') d\boldsymbol{\theta}' \tag{8.28}$$

where

$$G(\boldsymbol{\theta}, \boldsymbol{\theta}') = \sum_{l=0}^{\infty} \sum_{m=-l}^{l} \lambda_l Y_{l,m}(\boldsymbol{\theta}) Y^*_{l,m}(\boldsymbol{\theta}') \tag{8.29}$$

and

$$\lambda_l = \frac{h_l^{(1)}(ka_2)}{h_l^{(1)}(ka_1)}. \tag{8.30}$$

Equation (8.28) is just equation (8.21) in the particular case of spherical surfaces.

The integral operator defined by equations (8.28)–(8.30) has some simple and interesting properties which can be summarized as follows: *the numbers* λ_l *are the eigenvalues of the operator, associated with the eigenfunctions* $Y_{l,m}$, *so that the eigenvalue* λ_l *has multiplicity* $2l + 1$; *thanks to the completeness of the spherical harmonics (they are a basis in* $L^2(S^2)$*), no function* $f^{(0)}(\boldsymbol{\theta}) \neq 0$ *exists such that the corresponding* $g^{(0)}(\boldsymbol{\theta})$ *is exactly zero.*

The last property implies that the null space of the integral operator (8.28) does not contain functions $f^{(0)} \neq 0$ and therefore the solution of the inverse diffraction problem is unique. However, from the asymptotic behaviour of the Hankel functions for large l one has

$$\lambda_l \simeq \exp\left[-(l+1)\ln\left(\frac{a_2}{a_1}\right)\right] \tag{8.31}$$

and therefore the eigenvalues tends to zero when $l \to \infty$. This behaviour implies that the problem is ill-posed as concerns existence of the solution and continuous dependence of the solution on the data. This point will be discussed in chapter 10.

The previous analysis may be appropriate in the case of *near-field data*, i.e. when the ratio a_2/a_1 is not much larger than 1. When this ratio is much larger than 1, then we are in the case of *far-field data* and it is more convenient to consider the asymptotic limit $a_2/a_1 \to \infty$.

From the asymptotic behaviour of the spherical Hankel functions for $r \to \infty$

$$h_l^{(1)}(r) \simeq (-i)^{l+1} \frac{e^{ikr}}{r} \tag{8.32}$$

we obtain the following asymptotic behaviour of the field amplitude (8.26)

$$u(r, \boldsymbol{\theta}) \simeq \frac{e^{ikr}}{r} u_\infty(\boldsymbol{\theta}) \tag{8.33}$$

where

$$u_\infty(\boldsymbol{\theta}) = \sum_{l=0}^{\infty} \sum_{m=-l}^{l} (-i)^{l+1} \frac{f_{l,m}^{(0)}}{h_l^{(1)}(ka_1)} Y_{l,m}(\boldsymbol{\theta}). \tag{8.34}$$

The function $u_\infty(\boldsymbol{\theta})$ is usually called the *radiation pattern*. In the inverse-diffraction problem with far-field data, the exact data are precisely the values of the radiation pattern, i.e.

$$g^{(0)}(\boldsymbol{\theta}) = u_\infty(\boldsymbol{\theta}). \tag{8.35}$$

Then, from equation (8.34) we can derive an equation analogous to equation (8.28)

$$g^{(0)}(\boldsymbol{\theta}) = \int_{S^2} G_\infty(\boldsymbol{\theta}, \boldsymbol{\theta}') f^{(0)}(\boldsymbol{\theta}') d\boldsymbol{\theta} \tag{8.36}$$

where

$$G_\infty(\boldsymbol{\theta}, \boldsymbol{\theta}') = \sum_{l=0}^{\infty} \sum_{m=-l}^{l} \lambda_l^{(\infty)} Y_{l,m}(\boldsymbol{\theta}) Y_{l,m}^*(\boldsymbol{\theta}') \tag{8.37}$$

and

$$\lambda_l^{(\infty)} = \frac{(-i)^{l+1}}{h_l^{(1)}(ka_1)}. \tag{8.38}$$

Also in the case of the integral operator (8.36) the null space does not contain functions $f^{(0)} \neq 0$, so that the solution of the problem is unique. The eigenvalues $\lambda_l^{(\infty)}$, however, tend to zero much more rapidly than the eigenvalues λ_l of the problem with near-field data. Their asymptotic behaviour, indeed, is given by $|\lambda_l^{(\infty)}| \simeq \exp\left[-l \ln(2l/ka_1)\right]$. Since the components of $g^{(0)}(\boldsymbol{\theta})$ with respect to the spherical harmonics are given by $g_{m,l}^{(0)} = \lambda_l f_{m,l}^{(0)}$ in the near-field case and by $g_{m,l}^{(0)} = \lambda_l^{(\infty)} f_{m,l}^{(0)}$ in the far-field case, it follows that, for a given $f^{(0)}(\boldsymbol{\theta})$, the number of components of $g^{(0)}$ significantly different from zero is much smaller in the case of far-field data than in the case of near-field data. In other words, and in agreement with physical intuition, far-field data contain less information than near-field data because, in the presence of noise, they allow the recovery of a smaller number of components of $f^{(0)}(\boldsymbol{\theta})$. One could also say that the far-field problem has a higher degree of ill-posedness than the near-field problem.

B. *Inverse source problem*

Assume that the sources of the field, located in the domain $\mathcal{D}$, are described by a function which will be denoted again by $f^{(0)}(\boldsymbol{x})$. The precise meaning of this function is that it is an inhomogeneous term in the Helmholtz equation: in the region $\mathcal{D}$ the field amplitude $u(\boldsymbol{x})$ is a solution of the following inhomogeneous equation

$$\Delta u + k^2 u = -4\pi f^{(0)} \tag{8.39}$$

while, outside $\mathcal{D}$, is a solution of the homogeneous equation (8.15) and satisfies the Sommerfeld radiation condition, equation (8.17).

With reference to figure 8.13, the *inverse source problem* can be formulated as the problem of determining $f^{(0)}(\boldsymbol{x})$ in $\mathcal{D}$ from given values of the radiation field $u(\boldsymbol{x})$ on a closed surface Σ' (i.e. Σ_1 or Σ_2) such that $\mathcal{D}$ is interior to Σ'.

It is clear that this problem is highly underdetermined because we wish to recover a function of three variables from the knowledge of a function of two variables, i.e. the values of the field on a 2D surface. This underdetermination, however, cannot be removed by taking the values of the field in a volume because, as shown before in the case of a spherical surface (although the result is true for more general surfaces), knowledge of the field on a closed surface determines uniquely the field everywhere outside $\mathcal{D}$. This intrinsic underdetermination of the problem is due to the existence of the so-called non-radiating sources which will be discussed in the following.

As usual, in order to obtain the mathematical formulation of the inverse problem, we must first solve the direct one, i.e. the problem of determining a function u, satisfying equations (8.39),(8.15) and condition (8.17), for a given source function $f^{(0)}(\boldsymbol{x})$. This solution can be obtained by introducing the Green function

$$G_0(r) = \frac{e^{ikr}}{r} \tag{8.40}$$

and is given by

$$u(\boldsymbol{x}) = \int_{\mathcal{D}} G_0(|\boldsymbol{x} - \boldsymbol{x}'|) f^{(0)}(\boldsymbol{x}') d\boldsymbol{x}'. \tag{8.41}$$

If we assume that the values of $u(\boldsymbol{x})$ are given on the surface Σ', i.e.

$$u(\boldsymbol{x})|_{\Sigma'} = g^{(0)}(\boldsymbol{x}) \tag{8.42}$$

then the inverse source problem consists of looking for a function $f^{(0)}(\boldsymbol{x})$ such that $u(\boldsymbol{x})$, as given by equation (8.41), satisfies condition (8.42). In this way we obtain the following integral equation for $f^{(0)}(\boldsymbol{x})$

$$g^{(0)}(\boldsymbol{x}) = \int_{\mathcal{D}} G_0(|\boldsymbol{x} - \boldsymbol{x}'|) f^{(0)}(\boldsymbol{x}') d\boldsymbol{x}'. \tag{8.43}$$

In order to investigate the null space of this integral operator, which consists of the non-radiating sources, i.e. of the source functions $f^{(0)}(\boldsymbol{x})$ such that

$$\int_{\mathcal{D}} G_0(|\boldsymbol{x} - \boldsymbol{x}'|) f^{(0)}(\boldsymbol{x}') d\boldsymbol{x}' = 0, \tag{8.44}$$

we can use the previously proved result that the radiation pattern determines uniquely the field everywhere outside $\mathcal{D}$. As a consequence, if a source $f^{(0)}(\boldsymbol{x})$ produces a vanishing radiation pattern, then it produces a radiation field which is zero everywhere outside $\mathcal{D}$.

From equations (8.40) and (8.41) it follows that, for large values of $r = |\boldsymbol{x}|$, $u(\boldsymbol{x})$ has the following asymptotic behaviour

$$u(\boldsymbol{x}) \simeq \frac{e^{ikr}}{r} u_\infty(\boldsymbol{\theta}) \tag{8.45}$$

with $\boldsymbol{\theta} = \boldsymbol{x}/r$ and

$$u_\infty(\boldsymbol{\theta}) = \int_{\mathcal{D}} e^{-ik\boldsymbol{\theta}\cdot\boldsymbol{x}'} f^{(0)}(\boldsymbol{x}') d\boldsymbol{x}'. \tag{8.46}$$

Therefore, given $f^{(0)}(\boldsymbol{x})$, $u_\infty(\boldsymbol{\theta})$ can be computed by means of this equation and then the field, on an arbitrary sphere with radius a_1, can be uniquely determined by solving equation (8.36). If follows that if $f^{(0)}(\boldsymbol{x})$ is such that $u_\infty(\boldsymbol{\theta})$ vanishes then the field radiated by $f^{(0)}(\boldsymbol{x})$ vanishes everywhere outside $\mathcal{D}$.

From equation (8.46) it follows that the radiation pattern of a source function $f^{(0)}(\boldsymbol{x})$ vanishes if and only if the Fourier transform of $f^{(0)}(\boldsymbol{x})$, $\hat{f}^{(0)}(\boldsymbol{\omega})$, vanishes on the sphere with centre at the origin and radius k, i.e. $\hat{f}^{(0)}(k\boldsymbol{\theta}) = 0$. This sphere is the so-called *Ewald sphere*. Therefore *the set of the non-radiating sources with support in $\mathcal{D}$ is the set of all functions $f^{(0)}(\boldsymbol{x})$ whose Fourier transform vanishes on the Ewald sphere.*

We conclude by formulating the inverse source problem from far-field data in terms of a first-kind Fredholm integral equation. If $g^{(0)}(\boldsymbol{\theta})$ are the given values of the radiation pattern, then from equation (8.46) we see that $f^{(0)}(\boldsymbol{x})$ can be obtained by solving the integral equation

$$g^{(0)}(\boldsymbol{\theta}) = \int_{\mathcal{D}} K(\boldsymbol{\theta}; \boldsymbol{x}') f^{(0)}(\boldsymbol{x}') d\boldsymbol{x}' \tag{8.47}$$

where

$$K(\boldsymbol{\theta}; \boldsymbol{x}') = e^{-ik\boldsymbol{\theta}\cdot\boldsymbol{x}'}. \tag{8.48}$$

Let us assume that $\mathcal{D}$ is the sphere of radius a. If we use the well-known expansion of a plane wave into spherical harmonics

$$e^{-ikr'\boldsymbol{\theta}\cdot\boldsymbol{\theta}'} = 4\pi \sum_{l=0}^{\infty} \sum_{m=-l}^{l} (-i)^l j_l(kr') Y_{l,m}(\boldsymbol{\theta}) Y^*_{l,m}(\boldsymbol{\theta}'), \tag{8.49}$$

where the $j_l(x)$ are the spherical Bessel functions, and if we introduce the functions

$$v_{l,m}(\boldsymbol{x}') = \frac{4\pi}{\sigma_l} i^l j_l(kr') Y_{l,m}(\boldsymbol{\theta}'), \quad u_{l,m}(\boldsymbol{\theta}) = Y_{l,m}(\boldsymbol{\theta}) \tag{8.50}$$

where

$$\sigma_l = 4\pi \left(\int_0^a r^2 j_l^2(kr) dr \right)^{1/2}, \tag{8.51}$$

we obtain the following representation of the integral kernel (8.48)

$$K(\boldsymbol{\theta}; \boldsymbol{x}') = \sum_{l=0}^{\infty} \sum_{m=-l}^{l} \sigma_l u_{l,m}(\boldsymbol{\theta}) v^*_{l,m}(\boldsymbol{x}'). \tag{8.52}$$

This representation is just the singular value decomposition of the integral operator (8.47), as we will show in the next chapter. We conclude by observing that, from the asymptotic behaviour of the spherical Bessel functions for large l, it follows that $\sigma_l \to 0$ when $l \to \infty$.

8.5 Linearized inverse scattering problems

Inverse scattering problems are very important in many different domains, such as quantum mechanics [8], acoustics and electromagnetism [9]. These problems arise from the attempt of obtaining information on a body, the scatterer, by illuminating the body with waves of various wavelengths and directions and recording the waves scattered by the body.

We assume that the scatterer is contained in a bounded domain $\mathcal{D}$ of the 3D space. Then two different kinds of inverse scattering problems are, in general, considered: the inverse medium problem and the inverse obstacle problem.

In the inverse medium problem the scatterer is an inhomogeneous medium characterized by one or more physical quantities, varying in a continuous manner, and the inverse problem consists of estimating these parameters from scattering data. In the inverse obstacle problem, the scatterer is a homogeneous body and the problem is to estimate the shape of the body from scattering data and given boundary conditions on the surface of the body.

For the sake of simplicity we consider only the scalar case. Then, outside the bounded domain $\mathcal{D}$ containing the scatterer, the field amplitude $u(\boldsymbol{x})$ satisfies the Helmholtz equation (8.15). As concerns the boundary condition at infinity, we consider the most frequent case, where the body is illuminated by a plane wave propagating in the direction $\boldsymbol{\theta}_0$. Then $u(\boldsymbol{x})$ can be written in the following form

$$u(\boldsymbol{x}) = e^{ik\boldsymbol{\theta}_0\cdot\boldsymbol{x}} + u_s(\boldsymbol{x}). \tag{8.53}$$

The scattered wave $u_s(\boldsymbol{x})$ is also a solution of the Helmholz equation outside $\mathcal{D}$, and satisfies Sommerfeld radiation condition (8.17) at infinity. As follows from the analysis of section 8.4, $u_s(\boldsymbol{x})$ behaves as an outgoing spherical wave at infinity

$$u_s(\boldsymbol{x}) \simeq u_\infty(k; \boldsymbol{\theta}_0, \boldsymbol{\theta})\frac{e^{ikr}}{r} \tag{8.54}$$

where $\boldsymbol{\theta} = \boldsymbol{x}/r$. The function $u_\infty(k; \boldsymbol{\theta}_0, \boldsymbol{\theta})$ is called the *scattering amplitude* and is the quantity which is measured in the case of far-field scattering data. It is a function of five variables and therefore a complete determination of this function provides, in general, redundant data. Thanks to this redundancy many different experimental situations can be considered. We only mention a few of them: *back-scattering*, which corresponds to measure $u_\infty(k; \boldsymbol{\theta}_0, \boldsymbol{\theta})$ in the case $\boldsymbol{\theta} = -\boldsymbol{\theta}_0$, for various values of k and $\boldsymbol{\theta}$; *forward-scattering*, which corresponds to measure $u_\infty(k; \boldsymbol{\theta}_0, \boldsymbol{\theta})$ in the case $\boldsymbol{\theta} = \boldsymbol{\theta}_0$, for various values of k and $\boldsymbol{\theta}$; *fixed frequency scattering*, which corresponds to measure $u_\infty(k; \boldsymbol{\theta}_0, \boldsymbol{\theta})$ for a fixed value of k and various values of $\boldsymbol{\theta}_0$ and $\boldsymbol{\theta}$; and so on. In some cases, problems with near-field data are also considered.

Inverse scattering problems are nonlinear and ill-posed and therefore are difficult problems both from the mathematical and computational point of view. Under some circumstances, however, it is possible to introduce physical approximations which allow a linearization of the nonlinear problem. A well-known case is that of a weak scatterer; here the Born approximation may be used. Another kind of approximation, also leading to a linear problem, is the Rytov approximation, which is valid in the case of a slowly varying scatterer, i.e. in the case where the fluctuation length of the properties of the scatterer is large compared to the wavelength $\lambda = 2\pi/k$ of the incident radiation. These approximations apply to the inverse medium problem. An approximation leading to the linearization of the inverse obstacle problem is the so-called physical-optics approximation.

A. *Born approximation*

We consider the inverse medium problem in the case of a semitransparent body characterized by a refraction index $n(\boldsymbol{x})$. If we introduce the function $f^{(0)}(\boldsymbol{x}) = 1 - n^2(\boldsymbol{x})$, then the total field $u(\boldsymbol{x})$ (incident plus scattered) is a solution, in $\mathcal{D}$, of the wave equation

$$\Delta u + k^2 u = k^2 f^{(0)} u. \tag{8.55}$$

Since the plane wave is a solution of the Helmholtz equation, from equation (8.53) it follows that the scattered wave $u_s(\boldsymbol{x})$ is a solution in $\mathcal{D}$ of the equation

$$\Delta u_s + k^2 u_s = k^2 f^{(0)} u \tag{8.56}$$

while outside $\mathcal{D}$ satisfies the Helmholtz equation and the Sommerfeld radiation condition at infinity. Then, from the solution (8.41) of the inhomogeneous equation (8.39), we obtain the following representation of the scattered wave

$$u_s(\boldsymbol{x}) = -\frac{1}{4\pi}\int_{\mathcal{D}} G_0(|\boldsymbol{x} - \boldsymbol{x}'|) f^{(0)}(\boldsymbol{x}') u(\boldsymbol{x}') d\boldsymbol{x}', \tag{8.57}$$

the Green function $G_0(r)$ being defined in equation (8.40). Finally, from the asymptotic behaviour (8.45), by identifying the scattering amplitude (8.54) with the radiation pattern (8.46), we obtain the following representation of the scattering amplitude

$$u_\infty(k; \boldsymbol{\theta}_0, \boldsymbol{\theta}) = -\frac{k^2}{4\pi}\int_{\mathcal{D}} e^{-ik\boldsymbol{\theta}\cdot\boldsymbol{x}'} f^{(0)}(\boldsymbol{x}') u(\boldsymbol{x}') d\boldsymbol{x}' \tag{8.58}$$

where $u(\boldsymbol{x})$ is the total field.

The Born approximation, which applies to the so-called weak scatterers, consists in replacing the total field $u(\boldsymbol{x})$ by the incident field (the plane wave) in equation (8.58). The result is

$$u_\infty(k; \boldsymbol{\theta}_0, \boldsymbol{\theta}) = -\frac{k^2}{4\pi}\int_{\mathcal{D}} e^{-ik(\boldsymbol{\theta}-\boldsymbol{\theta}_0)\cdot\boldsymbol{x}'} f^{(0)}(\boldsymbol{x}') d\boldsymbol{x}'. \tag{8.59}$$

This equation shows that, if we measure the scattering amplitude for all values of $\boldsymbol{\theta}$, with k and $\boldsymbol{\theta}_0$ fixed, then we obtain the Fourier transform of $f^{(0)}(\boldsymbol{x})$ on the surface of the Ewald sphere with centre at $k\boldsymbol{\theta}_0$ and radius k. It follows that, from the mathematical point of view, this problem is analogous to the inverse source problem discussed in the previous section. However, by varying the direction of incidence $\boldsymbol{\theta}_0$, a (theoretically infinite) number of experiments would allow one to determine $\hat{f}^{(0)}(\boldsymbol{\omega})$ within the sphere with centre at the origin and radius $2k$ (limiting Ewald sphere). In this case, since $\hat{f}^{(0)}(\boldsymbol{\omega})$ is an analytic function, thanks to the boundedness of the support of $f^{(0)}(\boldsymbol{x})$, the uniqueness of the solution of the inverse problem is ensured. We also observe that this inverse problem is an example of the general problem of out-of-band extrapolation, which will be investigated in chapter 11. The scattering data, indeed, provide a bandlimited approximation of the unknown object $f^{(0)}(\boldsymbol{x})$.

The case where data are collected over planes not intersecting the domain $\mathcal{D}$ was also considered in the case of the Born approximation [10], and the same conclusions outlined before were reached. Moreover, it has been shown [11] that the Rytov approximation leads to the same mathematical problem as the Born approximation.

B. *Physical-optics approximation*

We consider the inverse obstacle problem in the case of a perfectly reflecting body (a sound-soft obstacle in acoustic or a perfectly conducting obstacle in electromagnetism). In such a case the total field u, as given by equation (8.53), must satisfy the boundary condition

$$u(\boldsymbol{x})|_\Sigma = 0 \tag{8.60}$$

where Σ is the unknown surface of the scatterer which, in general, is assumed to be a convex body. From equation (8.53) and condition (8.60) we obtain that the scattered wave satisfies the following boundary condition on Σ

$$u_s(\boldsymbol{x})|_\Sigma = -e^{ik\boldsymbol{\theta}_0\cdot\boldsymbol{x}}\,|_\Sigma\,. \tag{8.61}$$

Moreover, $u_s(\boldsymbol{x})$ is a solution of the Helmholtz equation in the exterior of the domain $\mathcal{D}$ occupied by the body, and satisfies the Sommerfeld radiation condition at infinity. Then, by means of the second Green theorem one can easily derive the following representation of the scattered wave

$$u_s(\boldsymbol{x}) = -\frac{1}{4\pi}\int_\Sigma G_0(|\boldsymbol{x}-\boldsymbol{x}'|)\frac{\partial}{\partial\boldsymbol{\nu}(\boldsymbol{x}')}u(\boldsymbol{x}')d\sigma(\boldsymbol{x}') \tag{8.62}$$

where $\boldsymbol{\nu}(\boldsymbol{x}')$ is the unit normal vector to Σ at the point $\boldsymbol{x}'$, directed towards the exterior of $\mathcal{D}$. The Green function $G_0(r)$ is defined in equation (8.40). Equation (8.62) provides the foundation of the *Huygens principle*, because it represents the scattered wave as a superposition of spherical waves emitted by a double layer located at the surface of the body.

From equation (8.62) it follows that the behaviour at infinity of $u_s(\boldsymbol{x})$ is provided by equation (8.54) with a scattering amplitude which is now given by

$$u_\infty(k; \boldsymbol{\theta}_0, \boldsymbol{\theta}) = -\frac{1}{4\pi} \int_\Sigma e^{-ik\boldsymbol{\theta}\cdot\boldsymbol{x}'} \frac{\partial}{\partial \boldsymbol{\nu}(\boldsymbol{x}')} u(\boldsymbol{x}') d\sigma(\boldsymbol{x}'). \tag{8.63}$$

The *physical-optics approximation* provides a simple expression of this scattering amplitude in the case where the wave-length $\lambda = 2\pi/k$ is much smaller than the diameter of the convex body $\mathcal{D}$.

To this purpose, for a given $\boldsymbol{\theta}_0$, we must introduce the illumination region $\Sigma_+(\boldsymbol{\theta}_0)$ and the shadow region $\Sigma_-(\boldsymbol{\theta}_0)$ of the surface Σ. The meaning of

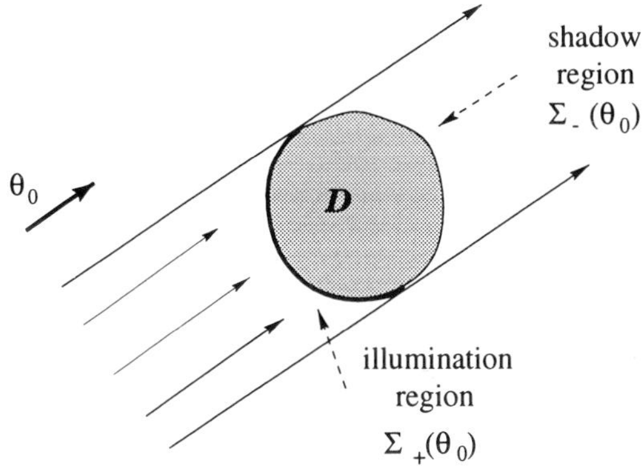

Figure 8.14. Illustrating the illumination and the shadow region of the surface Σ, for a given incidence direction $\boldsymbol{\theta}_0$.

these regions is clearly illustrated in figure 8.14. Then the physical optics approximation consists first in neglecting the contribution of the shadow region. As concerns the illumination region, one approximates locally the surface by a plane and computes the scattered wave in a point $\boldsymbol{x}'$ of the surface as the wave reflected by the tangent plane. In conclusion the following approximations are used

$$\left.\frac{\partial}{\partial \boldsymbol{\nu}(\boldsymbol{x}')} u(\boldsymbol{x}')\right|_{\Sigma_-(\boldsymbol{\theta}_0)} = 0 \tag{8.64}$$

$$\left.\frac{\partial}{\partial \boldsymbol{\nu}(\boldsymbol{x}')} u(\boldsymbol{x}')\right|_{\Sigma_+(\boldsymbol{\theta}_0)} = 2 \left.\frac{\partial}{\partial \boldsymbol{\nu}(\boldsymbol{x}')} e^{ik\boldsymbol{\theta}_0\cdot\boldsymbol{x}'}\right|_{\Sigma_+(\boldsymbol{\theta}_0)} \tag{8.65}$$

and from equation (8.63) one obtains

$$u_\infty(k;\boldsymbol{\theta}_0,\boldsymbol{\theta}) = -\frac{ik}{2\pi}\int_{\Sigma_+(\boldsymbol{\theta}_0)} (\boldsymbol{\theta}_0\cdot\boldsymbol{\nu}(\boldsymbol{x}'))\, e^{ik(\boldsymbol{\theta}_0-\boldsymbol{\theta})\cdot\boldsymbol{x}'} d\sigma(\boldsymbol{x}'). \tag{8.66}$$

This formula can be very useful in the case of back-scattering (its application to radar imaging, for instance, has been suggested), i.e. in the case $\boldsymbol{\theta}_0 = -\boldsymbol{\theta}$. From equation (8.66) we obtain

$$u_\infty(k;\boldsymbol{\theta},-\boldsymbol{\theta}) = -\frac{1}{4\pi}\int_{\Sigma_+(\boldsymbol{\theta})} \frac{\partial}{\partial\boldsymbol{\nu}(\boldsymbol{x}')} e^{2ik\boldsymbol{\theta}\cdot\boldsymbol{x}'} d\sigma(\boldsymbol{x}'). \tag{8.67}$$

If we exchange $\boldsymbol{\theta}$ and $-\boldsymbol{\theta}$, by observing that the illumination region is now $\Sigma_-(\boldsymbol{\theta})$, we also obtain

$$u_\infty(k;-\boldsymbol{\theta},\boldsymbol{\theta}) = -\frac{1}{4\pi}\int_{\Sigma_-(\boldsymbol{\theta})} \frac{\partial}{\partial\boldsymbol{\nu}(\boldsymbol{x}')} e^{-2ik\boldsymbol{\theta}\cdot\boldsymbol{x}'} d\sigma(\boldsymbol{x}'). \tag{8.68}$$

By adding the complex conjugate of equation (8.68) to equation (8.67) and using the first Green theorem, we finally obtain the *Bojarski identity* [12]

$$u_\infty(k;\boldsymbol{\theta},-\boldsymbol{\theta}) + u^*_\infty(k;-\boldsymbol{\theta},\boldsymbol{\theta}) = \frac{k^2}{\pi}\int \chi_{\mathcal{D}}(\boldsymbol{x}) e^{2ik\boldsymbol{\theta}\cdot\boldsymbol{x}} d\boldsymbol{x} \tag{8.69}$$

where $\chi_{\mathcal{D}}(\boldsymbol{x})$ is the characteristic function of the domain $\mathcal{D}$.

This formula implies that, in the physical-optics approximation the Fourier transform of the characteristic function of the scatterer (and therefore the shape of the scatterer) can be obtained from the knowledge of the back-scattering amplitude for all incident directions $\boldsymbol{\theta}$ and all wave-numbers k. In practice, back-scattering amplitudes can be measured only for a restricted region of values of k, i.e. for $k_{min} \le k \le k_{max}$. In other words one obtains an image of $\chi_{\mathcal{D}}(\boldsymbol{x})$ as provided by a band-pass filter so that a special feature of this problem is the lack of information both at low and at high frequencies. We find another example of the general problem of out-of-band extrapolation, which will be discussed in chapter 11.

References

[1] Biretta J 1994 *The restoration of HST images and Spectra II* eds Hanish R J and White R L (Baltimore, MD: Space Telescope Science Institute) 224

[2] Webb S (ed) 1988 *The Physics of Medical Imaging* (Bristol: Institute of Physics Publishing)

[3] Herman G T (ed) 1979 *Image Reconstruction from Projections, Implementation and Applications* (Berlin: Springer)

[4] Kak A C and Slaney M 1988 *Principles of Computerized Tomographic Imaging* (New York: IEEE Press)

[5] Radon J 1917 *Berichte Sächsische Akademie der Wissenschaften, Leipzig, Math. Phys. Kl* **69** 262
[6] Herman G T 1980 *Image Reconstruction from Projections, the Fundamentals of Computerized Tomography* (New York: Academic Press)
[7] Baltes H P (ed) 1978 *Inverse Source Problems in Optics* (Berlin: Springer)
[8] Chadan K and Sabatier P C 1989 *Inverse Problems in Quantum Scattering Theory* (Berlin: Springer)
[9] Colton D and Kress R 1992 *Inverse Acoustic and Electromagnetic Scattering Theory* (Berlin: Springer)
[10] Wolf E 1969 *Opt. Commun.* **1** 153
[11] Devaney A J 1981 *Opt. Lett.* **6** 374
[12] Bojarski N N 1982 *IEEE Trans. Ant. Prop.* **AP-20** 980

Chapter 9

Singular value decomposition (SVD)

One of the most fruitful tools in the theory of linear inverse problems is the singular value decomposition (SVD) of a matrix and its extension to certain classes of linear operators. Indeed, SVD is basic both for understanding the ill-posedness of linear inverse problems and for describing the effect of the regularization methods.

The SVD of a matrix was essentially introduced in 1873 by the Italian mathematician Eugenio Beltrami for solving the problem of the diagonalization of a bilinear form [1]. A more complete treatment, published independently a year later, is due to Camille Jordan [2]. A short and interesting history of SVD has been published recently [3].

Efficient and stable algorithms to compute the SVD of a matrix are available nowadays and in many problems, where the number of data is not too large, they can be used to obtain stable and approximate solutions. However, in problems of imaging, the size of the matrices is very often so large that it may be difficult to use these algorithms in practice.

In this chapter, after a description of linear imaging systems in terms of linear operators, we derive the SVD of an arbitrary matrix by assuming that the reader is familiar with the diagonalization of a symmetric (self-adjoint) matrix. Then we extend the SVD to a class of linear operators related to problems where the image is discrete but the object is assumed to belong to a class of functions. Finally we give the SVD of certain integral operators and that of the Radon transform in two dimensions.

9.1 Mathematical description of linear imaging systems

From the examples of the previous chapter it follows that a general mathematical description of a linear imaging system is the following. The object function $f^{(0)}$ is an element of a Euclidean space $\mathcal{X}$, very frequently a space of square-integrable functions, which will be called the *object space*. Analogously the image function (or data function) g is an element of another Euclidean space $\mathcal{Y}$, which will be called the *image space*.

The two spaces $\mathcal{X}$ and $\mathcal{Y}$ may coincide. This is the case, for instance, of the imaging systems considered in the first part of this book. However $\mathcal{X}$ and $\mathcal{Y}$ may be different as in the case of scattering problems or in X-ray tomography. In the last case the object, i.e. the linear attenuation function, is a function of two space variables. On the other hand the image function, i.e. the Radon transform of $f^{(0)}$, is a function of the space variable s and of the angular variable ϕ, so that the object and image spaces are different.

Once the object and image spaces have been defined, the imaging process is described by a *linear operator*, i.e. a mapping which associates functions $g^{(0)}$ of $\mathcal{Y}$ to functions $f^{(0)}$ of $\mathcal{X}$

$$g^{(0)} = Af^{(0)}. \tag{9.1}$$

As in chapter 3, $g^{(0)}$ will be called the *noise-free image* of the object $f^{(0)}$. In the case of a space-variant imaging system, the linear operator A is the integral operator defined in equation (8.1), while in the case of X-ray tomography the operator A coincides with the Radon transformation R defined in equations (8.8)–(8.9). A few basic properties of linear operators are given in appendix B.

As discussed in chapter 3, the recorded image g is affected by the noise introduced by the recording process. This effect can be described by an additional term w, which is also an element of $\mathcal{Y}$, so that

$$g = Af^{(0)} + w. \tag{9.2}$$

The function g will be called the *noisy image*. Equation (9.2) has the same structure as equation (4.2), where the linear operator has the particular form of a convolution operator.

In section 4.1 we considered not only the integral equation (4.2) but also the discretized equation (4.4), where both the object and the image are discrete. For more general inverse problems it is sometimes convenient to consider an intermediate model where the image is discrete but the object is still assumed to be a function.

It is obvious that an experimental image is always described by a finite set of numbers: one or several detectors measure variations in space (and, possibly, also in time) of the emitted or scattered radiation; the output of the detectors is digitized and the final result is precisely a set of numbers stored in the computer. We denote the elements of this set by $g_1, g_2, \cdots, g_M$. As already mentioned many times, in the case of a 2D image they can be obtained by a lexicographic ordering of the numbers corresponding to the grey levels associated with the pixels of the image. Therefore an image can also be viewed as a vector which will be called the *image vector*, or the *discrete image* and denoted by $\boldsymbol{g}$.

The components g_m of $\boldsymbol{g}$ are proportional (but, for simplicity, we omit the proportionality constant) to sampled values of the image g

$$g_m = g(\boldsymbol{x}_m); \quad m = 1, 2, \cdots, M. \tag{9.3}$$

More generally, since any detector integrates over some region around a sampling point $\boldsymbol{x}_m$, the g_m are proportional to weighted averages of g:

$$g_m = \int p_m(\boldsymbol{x}) g(\boldsymbol{x}) d\boldsymbol{x}; \quad m = 1, 2, \cdots, M. \tag{9.4}$$

The averaging function $p_m(\boldsymbol{x})$, which represents the response of the detector, has typically a peak centred at a point $\boldsymbol{x} = \boldsymbol{x}_m$ of the image plane. Then equation (9.3) is obtained from equation (9.4) when $p_m(\boldsymbol{x}) = \delta(\boldsymbol{x} - \boldsymbol{x}_m)$. It should be obvious that equation (9.4) is more realistic than equation (9.3).

The image vectors form an M-dimensional vector space $\mathcal{Y}_M$. In this vector space we can introduce the canonical scalar product of $\mathcal{E}_M$ or, more frequently, a weighted scalar product (see appendix C) characterized by a positive definite weighting matrix. This weighting matrix can be related to statistical properties of the noise (see chapter 7) or to a quadrature formula used for the discretization of the scalar product of $\mathcal{Y}$ [4].

If we assume, for simplicity, that $\mathcal{Y}$ is a space of square-integrable functions, the integral in equation (9.4) is the scalar product of the functions p_m and g, i.e. $g_m = (g, p_m)_{\mathcal{Y}}$. From equation (9.2) and from the general definition of adjoint operator (see appendix B, equation (B.6)) it follows that

$$\begin{aligned} (g, p_m)_{\mathcal{Y}} &= \left(Af^{(0)}, p_m\right)_{\mathcal{Y}} + (w, p_m)_{\mathcal{Y}} \\ &= \left(f^{(0)}, A^* p_m\right)_{\mathcal{X}} + (w, p_m)_{\mathcal{Y}}. \end{aligned} \tag{9.5}$$

Therefore, if we introduce the functions of $\mathcal{X}$ defined by

$$\varphi_m(\boldsymbol{x}') = (A^* p_m)(\boldsymbol{x}'); \quad m = 1, 2, \cdots, M \tag{9.6}$$

we obtain

$$g_m = g_m^{(0)} + w_m; \quad m = 1, 2, \cdots, M \tag{9.7}$$

where $w_m = (w, p_m)_{\mathcal{Y}}$ and

$$g_m^{(0)} = \left(f^{(0)}, \varphi_m\right)_{\mathcal{X}}; \quad m = 1, 2, \cdots, M. \tag{9.8}$$

When $\mathcal{X}$ is also a space of square-integrable functions equipped with the canonical scalar product, equation (9.8) takes the following form

$$g_m^{(0)} = \int_{\mathcal{D}} \varphi_m^*(\boldsymbol{x}') f^{(0)}(\boldsymbol{x}') d\boldsymbol{x}'; \quad m = 1, 2, \cdots, M \tag{9.9}$$

where $\mathcal{D}$ is the object domain. Therefore the components of the noise-free discrete image are suitable weighted averages of the object $f^{(0)}$, which are also called generalized moments of $f^{(0)}$.

The weighting functions φ_m must be determined for the specific problem under consideration. For instance, if we consider the space-variant imaging system described by equation (8.1) and if $p_m(\boldsymbol{x}) = \delta(\boldsymbol{x} - \boldsymbol{x}_m)$, then

$$\varphi_m(\boldsymbol{x}') = K(\boldsymbol{x}_m, \boldsymbol{x}); \quad m = 1, 2, \cdots, M. \tag{9.10}$$

In such a case the functions φ_m are obtained directly from the integral kernel of the imaging operator.

Equation (9.9) describes a linear mapping which transforms a function of the Euclidean space $\mathcal{X}$ into a vector, i.e. an element of the M-dimensional vector space $\mathcal{Y}_M$. For this reason it will be called a *semi-discrete mapping*, it will be denoted by A_M and defined by the rule

$$(A_M f)_m = (f, \varphi_m)_{\mathcal{X}}\,; \quad m = 1, 2, \cdots, M. \tag{9.11}$$

In this approach the image is discrete while the object is still considered as an element of an infinite-dimensional function space. Therefore only when an estimate of the object has been obtained one can perform a fine discretization of this estimate in order to produce a numerical or graphical result. In other words, this approach points out that in a practical inverse problem one has two distinct types of discretization. The first is the discretization of the image, which is related to the design of the experiment or of the imaging system and therefore can be affected by rather strong instrumental restrictions. The second is the discretization of the object, which depends essentially on the approximation method used by the mathematician or by the practitioner. In the approach outlined above this second kind of discretization is not performed.

In practical problems, however, the usual approach consists of a complete discretization of the problem. If we start from equation (9.8), the discretization of this equation is obtained by assuming that the object $f^{(0)}$ can be reliably approximated by a linear combination of suitable basis functions $\psi_1, \psi_2, \cdots, \psi_N$, so that we can write

$$f^{(0)}(\boldsymbol{x}') = \sum_{n=1}^{N} f_n^{(0)} \psi_n(\boldsymbol{x}'). \tag{9.12}$$

In the case of a 2D object, for instance, the functions ψ_n can be the characteristic functions of the pixels; in that case equation (9.12) is equivalent to assuming that $f^{(0)}$ can be approximated by a piecewise constant function. On the other hand, if $f^{(0)}(\boldsymbol{x})$ is bandlimited, then the ψ_n can be suitable sampling functions (see section 2.2).

By substituting equation (9.12) into equation (9.8), we get

$$g_m^{(0)} = \sum_{n=1}^{N} A_{m,n} f_n^{(0)}; \quad m = 1, 2, \cdots, M \tag{9.13}$$

where

$$A_{m,n} = (\psi_n, \varphi_m)_{\mathcal{X}}\,. \tag{9.14}$$

An expression of $A_{m,n}$ in terms of the linear operator A can be obtained by taking into account equation (9.6) and using again the relationship between A and A^*

$$A_{m,n} = \left(\psi_n, A^* p_m\right)_{\mathcal{X}} = (A\psi_n, p_m)_{\mathcal{Y}}\,. \tag{9.15}$$

The interpretation of this equation is the following: $A_{m,n}$ is the mth component of the discrete image of ψ_n.

If we denote by $\boldsymbol{f}^{(0)}$ the vector with components $f_1^{(0)}, \cdots, f_N^{(0)}$, and by $\mathbf{A}$ the matrix (in general rectangular) whose matrix elements $(\mathbf{A})_{m,n} = A_{m,n}$ are given by equation (9.14) or (9.15), from equations (9.7) and (9.13) we obtain the discrete imaging equation

$$g = \mathbf{A}\boldsymbol{f}^{(0)} + w \tag{9.16}$$

which is a generalization of equation (4.4).

In conclusion, let us say a few words about the scalar product in the space $\mathcal{X}_N$ of the vectors $\boldsymbol{f}$. Equation (9.12) defines a linear subspace of $\mathcal{X}$ which is finite dimensional. If f and h are two functions in this subspace their scalar product is given by

$$(f, h)_{\mathcal{X}} = \sum_{n,n'=1}^{N} (\psi_n, \psi_{n'})_{\mathcal{X}} f_n h_{n'}^*. \tag{9.17}$$

Therefore, if we require that the scalar product of two vectors of $\mathcal{X}_N$, $\boldsymbol{f}, \boldsymbol{h}$, coincides with the scalar product of the functions f, h they represent, i.e. $(\boldsymbol{f}, \boldsymbol{h})_N = (f, h)_{\mathcal{X}}$, we see that we must introduce in $\mathcal{X}_N$ a weighted scalar product, with a weighting matrix given by

$$C_{n,n'} = (\psi_n, \psi_{n'})_{\mathcal{X}} . \tag{9.18}$$

The matrix $C_{n,n'}$ is the *Gram matrix* of the basis functions ψ_n. If the functions ψ_n form an orthonormal set, then $C_{n,n'}^{(N)} = \delta_{n,n'}$ and $\mathcal{X}_N$ can be equipped with the canonical scalar product of $\mathcal{E}_N$. Analogously, we obtain the scalar product defined in appendix C, equation (C.18), if the functions ψ_n are orthogonal but not normalized. Otherwise we obtain a weighted scalar product having the general form given in equation (C.16). We point out that the matrix (9.18) is positive definite because if $\boldsymbol{f}$ is an element of $\mathcal{X}_N$ and $f(\boldsymbol{x})$ the corresponding function of $\mathcal{X}$ as given by equation (9.12), then equation (9.17) implies that

$$\sum_{n,n'=1}^{N} C_{n,n'} f_n f_{n'}^* = \|f\|_{\mathcal{X}}^2 > 0. \tag{9.19}$$

In conclusion the matrix $\mathbf{A}$ defines a linear mapping from a vector space $\mathcal{X}_N$ into a vector space $\mathcal{Y}_M$, both vector spaces being equipped, in general, with weighted scalar products. A change of variables, based on the Choleski factorization of the weighting matrices as discussed in appendix C, allows one to transform a problem formulated in weighted spaces into a problem formulated in canonical vector spaces.

9.2 SVD of a matrix

We first consider the case of a completely discretized imaging problem. This is characterized by an image space $\mathcal{Y}_M$ of dimension M, an object space $\mathcal{X}_N$ of dimension N and a matrix $\mathbf{A}$, $M \times N$, transforming a vector of $\mathcal{X}_N$ into a vector of $\mathcal{Y}_M$. Thanks to the remark at the end of the previous section, we assume, without loss of generality, that both $\mathcal{X}_N$ and $\mathcal{Y}_M$ are equipped with the canonical scalar product, i.e. $\mathcal{X}_N = \mathcal{E}_N$, and $\mathcal{Y}_M = \mathcal{E}_M$. This point is important because the theory and the algorithms for the singular value decomposition of a matrix are based on this assumption, even if the usual way of formulating SVD in numerical analysis does not mention at all the structure of the vector spaces.

The standard formulation is as follows: *let* $\mathbf{A}$ *be a rectangular matrix* $M \times N$, *with rank* p; *then there exists a* $p \times p$ *diagonal matrix* Σ, *with positive diagonal elements, and two isometric matrices* $\mathbf{U}$ *and* $\mathbf{V}$, *respectively* $M \times p$ *and* $N \times p$, *such that*

$$\mathbf{A} = \mathbf{U}\Sigma\mathbf{V}^*. \tag{9.20}$$

Here $\mathbf{V}^*$ denotes the adjoint of the matrix $\mathbf{V}$. We also recall that a rectangular matrix $\mathbf{V}$ is isometric if it satisfies the condition $\mathbf{V}^*\mathbf{V} = \mathbf{I}$ ($\mathbf{I}$ = unity matrix $p \times p$). A square isometric matrix is unitary (orthogonal). We must also mention that the standard algorithms of SVD apply to the case of a real matrix and, in such a case, the matrix $\mathbf{V}$ is also real, so that the adjoint matrix $\mathbf{V}^*$ coincides with the transposed matrix $\mathbf{V}^T$.

For our applications to inverse problems, however, it will be more convenient to write the decomposition (9.20) in a different way which will be derived in the following. Our starting point is the diagonalization of a self-adjoint matrix: *let* $\mathbf{A}$ *be a self-adjoint matrix* $N \times N$, *i.e. a matrix such that* $\mathbf{A}^* = \mathbf{A}$; *then there exists an* $N \times N$ *diagonal matrix* Λ *and a unitary (orthogonal) matrix* $\mathbf{V}$ *such that*

$$\mathbf{A} = \mathbf{V}\Lambda\mathbf{V}^*. \tag{9.21}$$

As is well known, this representation is a synthetic formulation of the basic results on the eigenvalue problem for the matrix $\mathbf{A}$. A self-adjoint matrix $N \times N$ has always N real eigenvalues, $\lambda_1, \lambda_2, \cdots, \lambda_N$, if each eigenvalue is counted as many times as its multiplicity. They can be ordered in such a way that: $|\lambda_1| \geq |\lambda_2| \geq \cdots \geq |\lambda_N|$. Moreover, eigenvectors associated with different eigenvalues are automatically orthogonal while m orthogonal eigenvectors can always be associated to each eigenvalue with multiplicity m. If all eigenvectors are normalized, then one can conclude that the solution of the eigenvalue problem for a self-adjoint matrix $\mathbf{A}$ provides a set of eigenvectors $\boldsymbol{v}_1, \boldsymbol{v}_2, \cdots, \boldsymbol{v}_N$ which constitute an orthonormal basis in $\mathcal{E}_N$. Here the eigenvector $\boldsymbol{v}_k$ is associated with the eigenvalue $\lambda_k : \mathbf{A}\boldsymbol{v}_k = \lambda_k \boldsymbol{v}_k$. Finally the representation (9.21) is related to the solution of the eigenvalue problem as follows: the eigenvalues λ_k are the diagonal elements of the diagonal matrix Λ while the eigenvectors $\boldsymbol{v}_k$ are the columns of the unitary matrix $\mathbf{V}$.

If $\boldsymbol{f}$ is an arbitrary vector of $\mathcal{E}_N$, then $\mathbf{V}^*\boldsymbol{f}$ is the vector whose components are the scalar products of $\boldsymbol{f}$ with the eigenvectors $\boldsymbol{v}_k$, i.e. $(\boldsymbol{f} \cdot \boldsymbol{v}_k)_N$. It follows that the representation (9.21) implies the following equation

$$\mathbf{A}\boldsymbol{f} = \sum_{k=1}^{N} \lambda_k (\boldsymbol{f} \cdot \boldsymbol{v}_k)_N \boldsymbol{v}_k, \tag{9.22}$$

where the subscript of the scalar product indicates the dimension of the vector space.

If the matrix $\mathbf{A}$ has rank $p < N$, then $\mathbf{A}$ has the eigenvalue $\lambda = 0$ with multiplicity $N - p$ (see appendix C) so that, by taking into account the ordering of the eigenvalues, we conclude that $\lambda_{p+1} = \lambda_{p+2} = \cdots = \lambda_N = 0$. It follows that in equation (9.22) the summation extends only up to p

$$\mathbf{A}\boldsymbol{f} = \sum_{k=1}^{p} \lambda_k (\boldsymbol{f} \cdot \boldsymbol{v}_k)_N \boldsymbol{v}_k. \tag{9.23}$$

This expansion will be called the *spectral representation* of the self-adjoint matrix $\mathbf{A}$. A similar representation holds true for any cyclic matrix, (see equation (2.65)), even when the cyclic matrix is not self-adjoint.

The representation (9.23) is equivalent to the representation (9.21). Indeed, in equation (9.21) we can assume that Λ is a $p \times p$ diagonal matrix formed with the non-zero eigenvalues of $\mathbf{A}$ and that $\mathbf{V}$ is an $N \times p$ isometric matrix whose columns are the p eigenvectors associated with the non-zero eigenvalues.

We consider now the case of an arbitrary matrix $\mathbf{A}$ with M rows and N columns. If the matrix is rectangular, i.e. $M \neq N$, then the eigenvalue problem is meaningless. If the matrix is square, i.e. $M = N$, but not self-adjoint, it may have eigenvalues and eigenvectors but they do not have, in general, the nice properties which hold true for self-adjoint matrices. The cyclic matrices considered in section 2.5 represent a very particular case because they always have N orthonormal eigenvectors. This is not true in general and, therefore, for an arbitrary matrix, the eigenvalue problem may not be very interesting because it does not provide a representation of the matrix similar to the spectral respresentation (9.23). The singular value representation of the matrix provides the required generalization of equation (9.23).

By means of an arbitrary $M \times N$ matrix $\mathbf{A}$, with rank $p \leq \min\{M, N\}$, it is always possible to form two self-adjoint matrices and precisely

$$\bar{\mathbf{A}} = \mathbf{A}^*\mathbf{A}, \quad \tilde{\mathbf{A}} = \mathbf{A}\mathbf{A}^*. \tag{9.24}$$

The first is $N \times N$ while the second is $M \times M$. They have the following properties, proved in appendix C:

- both matrices are self-adjoint and positive semi-definite;
- both matrices have rank p.

Since both matrices have exactly p positive eigenvalues, the zero eigenvalue has multiplicity $N-p$ for the matrix $\bar{\mathbf{A}}$ and $M-p$ for the matrix $\tilde{\mathbf{A}}$. If $M \neq N$, at least one of the two matrices has the zero eigenvalue and precisely the matrix which has the largest dimension.

The first basic result is the following: *the matrices* $\bar{\mathbf{A}}$ *and* $\tilde{\mathbf{A}}$ *have exactly the same positive eigenvalues with the same multiplicity*. The proof of this result is easy. Let us consider the matrix $\bar{\mathbf{A}}$. Since it is symmetric and positive semi-definite, from the properties mentioned above it follows that it has p positive eigenvalues $\sigma_1^2 \geq \sigma_2^2 \geq \cdots \geq \sigma_p^2$; we will denote by $\boldsymbol{v}_1, \boldsymbol{v}_2, \cdots, \boldsymbol{v}_p$ the eigenvectors associated with these eigenvalues. These eigenvectors form an orthonormal basis in the orthogonal complement of the null space of the matrix $\mathbf{A}$, $\mathcal{N}(\mathbf{A})$ (see appendix C, where it is proved that $\mathcal{N}(\mathbf{A}) = \mathcal{N}(\bar{\mathbf{A}})$). To each eigenvector $\boldsymbol{v}_k$, which is a vector of $\mathcal{E}_N$, we can associate a vector of $\mathcal{E}_M$ as follows

$$\boldsymbol{u}_k = \frac{1}{\sigma_k}\mathbf{A}\boldsymbol{v}_k. \tag{9.25}$$

All vectors $\boldsymbol{u}_k$ are different from zero and they are eigenvectors of $\tilde{\mathbf{A}}$ associated with the eigenvalue σ_k^2. Indeed

$$\begin{aligned} \tilde{\mathbf{A}}\boldsymbol{u}_k &= \mathbf{A}\mathbf{A}^*\boldsymbol{u}_k = \frac{1}{\sigma_k}(\mathbf{A}\mathbf{A}^*)\mathbf{A}\boldsymbol{v}_k \\ &= \frac{1}{\sigma_k}\mathbf{A}(\mathbf{A}\mathbf{A}^*)\boldsymbol{v}_k = \frac{1}{\sigma_k}\mathbf{A}(\sigma_k^2\boldsymbol{v}_k) = \sigma_k^2\boldsymbol{u}_k. \end{aligned} \tag{9.26}$$

This simple computation implies that all positive eigenvalues of $\bar{\mathbf{A}}$ are also positive eigenvalues of $\tilde{\mathbf{A}}$. If all σ_k^2 have multiplicity one, the two matrices have precisely the same eigenvalues because also $\tilde{\mathbf{A}}$ can have only p positive eigenvalues. The proof is not complete in the case of eigenvalues with multiplicity > 1. However, if we compute the scalar products of the vectors $\boldsymbol{u}_k$, we obtain

$$\begin{aligned} \left(\boldsymbol{u}_k \cdot \boldsymbol{u}_j\right)_M &= \frac{1}{\sigma_k\sigma_j}\left(\mathbf{A}\boldsymbol{v}_k \cdot \mathbf{A}\boldsymbol{v}_j\right)_M \\ &= \frac{1}{\sigma_k\sigma_j}\left(\mathbf{A}^*\mathbf{A}\boldsymbol{v}_k \cdot \boldsymbol{v}_j\right)_N = \frac{\sigma_k}{\sigma_j}\left(\boldsymbol{v}_k \cdot \boldsymbol{v}_j\right)_N = \delta_{k,j}, \end{aligned} \tag{9.27}$$

where we have indicated with a suffix the dimension of the space. This result implies that the eigenvectors $\boldsymbol{u}_k$ are orthonormal, so that they are linearly independent even when they are associated with the same positive eigenvalue. Since we have obtained p linearly independent eigenvectors and since $\tilde{\mathbf{A}}$ has precisely p linearly independent eigenvectors associated with positive eigenvalues, we have proved that equation (9.25) provides all these eigenvectors of $\tilde{\mathbf{A}}$. This remark completes the proof of the result stated above.

If we multiply by $\mathbf{A}^*$ both sides of equation (9.25) we obtain

$$\frac{1}{\sigma_k}\mathbf{A}^* u_k = \frac{1}{\sigma_k^2}\mathbf{A}\mathbf{A}^* v_k = v_k. \tag{9.28}$$

Therefore equations (9.25) and (9.28) show that the eigenvectors u_k, v_k are solution of the following *shifted eigenvalue problem*, a term introduced by Lanczos [5]

$$\mathbf{A} v_k = \sigma_k u_k, \quad \mathbf{A}^* u_k = \sigma_k v_k. \tag{9.29}$$

These equations can be written in the form of a standard eigenvalue problem if we define the following vectors with dimension $M + N$

$$\begin{pmatrix} u_k \\ v_k \end{pmatrix} \tag{9.30}$$

and the following symmetric matrix with dimension $(M + N) \times (M + N)$:

$$\begin{pmatrix} 0 & \mathbf{A} \\ \mathbf{A}^* & 0 \end{pmatrix}. \tag{9.31}$$

Indeed, it is easy to verify that the equations (9.29) imply that

$$\begin{pmatrix} 0 & \mathbf{A} \\ \mathbf{A}^* & 0 \end{pmatrix}\begin{pmatrix} u_k \\ v_k \end{pmatrix} = \sigma_k \begin{pmatrix} u_k \\ v_k \end{pmatrix} \tag{9.32}$$

i.e. the vectors (9.30) are eigenvectors of the matrix (9.31) associated with the eigenvalues σ_k. Since the matrix (9.31) has rank $2p$, it must have $2p$ non-zero eigenvalues. If we consider the following vectors

$$\begin{pmatrix} u_k \\ -v_k \end{pmatrix} \tag{9.33}$$

it is easy to verify that they are eigenvectors of the matrix (9.31) associated with the eigenvalues $-\sigma_k$.

The positive numbers σ_k are called the *singular values* of the matrix $\mathbf{A}$. This term comes from the theory of integral equations and it seems that it was used for the first time by Smithies [3]. The vectors u_k, v_k are called the *singular vectors* of the matrix $\mathbf{A}$ and the set of the triples $\{\sigma_k; u_k, v_k\}$ is called the *singular system* of the matrix $\mathbf{A}$. As in the case of the diagonalization of a self-adjoint matrix, each singular value is counted as many times as required by its multiplicity (which coincides with the multiplicity of σ_k^2 as an eigenvalue of $\bar{\mathbf{A}}$ and $\tilde{\mathbf{A}}$) and the following ordering is used: $\sigma_1 \geq \sigma_2 \geq \cdots \geq \sigma_p$.

The singular vectors v_k are orthogonal to the null space of $\mathbf{A}$, as one can verify directly in the following way. Let v be an element of $\mathcal{N}(\mathbf{A})$; then from the second of the equations (9.29) we get

$$(v_k \cdot v)_N = \frac{1}{\sigma_k}\left(\mathbf{A}^* u_k \cdot v\right)_N = \frac{1}{\sigma_k}(u_k \cdot \mathbf{A} v)_M = 0. \tag{9.34}$$

In a similar way, we can prove that the u_k are orthogonal to the null space of $\mathbf{A}^*$. Therefore the v_k form an orthonormal basis in $\mathcal{N}(\mathbf{A})^\perp = \mathcal{R}(\mathbf{A}^*)$ while the u_k form an orthonormal basis in $\mathcal{N}(\mathbf{A}^*)^\perp = \mathcal{R}(\mathbf{A})$ (see equation (B.8)). The matrix $\mathbf{A}$ transforms the basis v_k into the basis u_k, except for the scaling factors σ_k, and analogously the matrix $\mathbf{A}^*$ transforms the basis u_k into the basis v_k. We have represented this situation in figure 9.1.

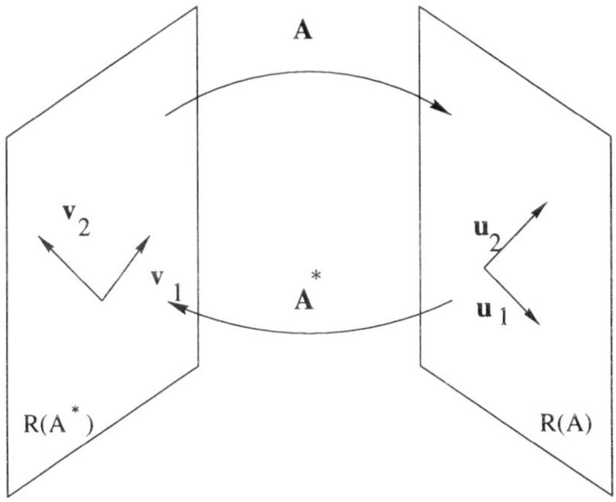

Figure 9.1. Schematic representation of the relationship between the two bases of singular vectors.

We can now obtain the singular value decomposition of the matrix $\mathbf{A}$ in a form similar to the spectral representation (9.23) of a self-adjoint matrix. Since the singular vectors u_k form a basis in $\mathcal{R}(\mathbf{A})$, for any f in $\mathcal{E}_N$ we can write $\mathbf{A}f$ as a linear combination of the u_k

$$\mathbf{A}f = \sum_{k=1}^{p} (\mathbf{A}f \cdot u_k)_M \, u_k. \tag{9.35}$$

Then, by the use of the second of the equations (9.29) we have

$$(\mathbf{A}f \cdot u_k)_M = \left(f \cdot \mathbf{A}^* u_k\right)_N = \sigma_k \,(f \cdot v_k)_N \tag{9.36}$$

and by substituting in equation (9.35) we obtain

$$\mathbf{A}f = \sum_{k=1}^{p} \sigma_k \,(f \cdot v_k)_N \, u_k. \tag{9.37}$$

This is precisely the *singular value decomposition* of the matrix $\mathbf{A}$; in a similar way we obtain the SVD of $\mathbf{A}^*$

$$\mathbf{A}^* g = \sum_{k=1}^{p} \sigma_k \,(g \cdot u_k)_M \, v_k. \tag{9.38}$$

In order to show that the representation (9.37) is equivalent to the representation (9.20) it is sufficient to remark that the isometric matrix $\mathbf{U}$ is the matrix whose columns are the p orthonormal vectors u_k, while the isometric matrix $\mathbf{V}$ is the matrix whose columns are the p orthonormal vectors v_k. Then equation (9.37) follows from equation (9.20) precisely in the same way as equation (9.23) follows from equation (9.21).

By means of the representation (9.37) it is easy to show that the maximum singular value σ_1 is just the norm of the matrix $\mathbf{A}$. Indeed, from the Parseval equality (see appendix A, equation (A.18)) we obtain

$$\|\mathbf{A}f\|_M^2 = \sum_{k=1}^{p} \sigma_k^2 \left|(f \cdot v_k)_N\right|^2 . \tag{9.39}$$

Since $\sigma_k \leq \sigma_1$ it follows that

$$\|\mathbf{A}f\|_M^2 \leq \sigma_1^2 \sum_{k=1}^{p} \left|(f \cdot v_k)_N\right|^2 \leq \sigma_1^2 \|f\|_N^2 \tag{9.40}$$

and therefore $\|\mathbf{A}\| \leq \sigma_1$. If we observe that equality holds true in equation (9.40) when $f = v_1$, we conclude that

$$\|\mathbf{A}\| = \sigma_1 . \tag{9.41}$$

In a similar way we can prove that $\|\mathbf{A}^*\| = \sigma_1$.

Let us comment now about the analogies and differences between the spectral representation (9.23) of a self-adjoint matrix and the SVD (9.37) of an arbitrary matrix. In both cases there exists an orthonormal basis in the orthogonal complement of $\mathcal{N}(\mathbf{A})$, i.e. that formed by the vectors v_k, such that the action of the matrix on a vector f is simply to multiply the components of f (with respect to this basis) by scaling factors (the eigenvalues in the case of a self-adjoint matrix, the singular values in the case of an arbitrary matrix). In this way one obtains the components of $\mathbf{A}f$. The difference is that in the self-adjoint case these are the components of $\mathbf{A}f$ with respect to the same basis, i.e. that formed by the vectors v_k, while in the general case they are the components of $\mathbf{A}f$ with respect to a different basis, i.e. that formed by the vectors u_k. In other words only one basis is needed in the self-adjoint case while two bases are needed in the general case.

We conclude this comparison by observing that also the spectral representation (9.23) can be written in the form (9.37) if we put $\sigma_k = |\lambda_k|$ and $u_k = sgn(\lambda_k)v_k$, without modifying the definition of the v_k. In particular we find that $u_k = v_k$, for any k, only in the case of a symmetric and positive semi-definite matrix.

Efficient algorithms for the computation of the SVD of a matrix are available. They are based on a method developed by Golub and Reinsch [6] (see also [7]). The input is a matrix $M \times N$ with $M \geq N$ (if this condition is not satisfied one takes the transposed matrix). The output is a vector of length N (the singular values) and two matrices **U** and **V**, respectively $M \times N$ and $N \times N$, i.e. the matrices of equation (9.20). If $p < N$, then $N - p$ singular values will be zero or of the order of the machine precision. It is also important to point out that the usual routines apply to the case of a real-valued matrix and cannot be trivially extended to the case of a complex-valued one. Routines specially designed for complex-valued matrices are also available (for instance in the LAPACK package).

9.3 SVD of a semi-discrete mapping

In section 9.1 we have considered the case where the image is discrete while the object is a function of continuous variables. This imaging problem can be formulated in terms of a semi-discrete mapping, transforming functions into vectors. In this section we show that such an operator has a singular value decomposition and we provide a method which can be used, in principle, for computing its singular system. To this purpose we need results given in section 9.2 because, as we will show, it is always possible to reduce the problem to the diagonalization of a self-adjoint matrix.

We assume, for simplicity, that $\mathcal{Y}_M$ is the canonical Euclidean vector space $\mathcal{E}_M$, while $\mathcal{X}$ can be any Euclidean (Hilbert) function space. It is not difficult to extend the analysis to the case where $\mathcal{Y}_M$ is a weighted Euclidean vector space [4].

The semi-discrete mappings introduced in section 9.1 have the following general structure

$$(A_M f)_m = (f, \varphi_m)_{\mathcal{X}}\,; \quad m = 1, \cdots, M \tag{9.42}$$

where $\varphi_1, \varphi_2, \cdots, \varphi_M$ are given functions of $\mathcal{X}$ which describe the imaging process. Then, for investigating the singular system of A_M we need the adjoint operator A_M^* which is defined by the usual rule (we denote again with a subscript M the scalar product of $\mathcal{E}_M$)

$$(A_M f \cdot g)_M = \left(f, A_M^* g\right)_{\mathcal{X}}\,. \tag{9.43}$$

From this definition it is evident that A_M^* is a mapping which transforms a vector into a function.

By taking into account the definition (9.42) of A_M we get

$$\begin{aligned} (A_M f \cdot g)_M &= \sum_{m=1}^{M} (A_M f)_m\, g_m^* = \sum_{m=1}^{M} (f, \varphi_m)_{\mathcal{X}}\, g_m^* \\ &= \left(f, \sum_{m=1}^{M} g_m \varphi_m\right)_{\mathcal{X}} = \left(f, A_M^* g\right)_{\mathcal{X}} \end{aligned} \tag{9.44}$$

and therefore

$$(A_M^* g)(x) = \sum_{m=1}^{M} g_m \varphi_m(x). \tag{9.45}$$

We conclude that the range of A_M^* is the finite-dimensional subspace spanned by the functions $\varphi_1, \varphi_2, \cdots, \varphi_M$. If these functions are linearly independent then this subspace has dimension M. In the following we assume that this condition is satisfied.

From the general rule $\mathcal{R}(A_M^*)^{\perp} = \mathcal{N}(A_M)$ (see appendix B), we see that the null space of A_M, i.e. the subspace of the invisible objects, is the infinite-dimensional subspace of all functions which are orthogonal to the subspace spanned by the functions $\varphi_1, \varphi_2, \cdots, \varphi_M$. This result can also be obtained directly from the definition (9.42). On the other hand, if the φ_m are linearly independent, the null space of A_M^* contains only the zero vector and therefore the relationship $\mathcal{N}(A_M^*) = \mathcal{R}(A_M)^{\perp}$ implies that the range of A_M coincides with $\mathcal{E}_M$.

As in the case of a matrix, we investigate the operators $\bar{A}_M = A_M^* A_M$ and $\tilde{A}_M = A_M A_M^*$. Since A_M maps functions into vectors while A_M^* maps vectors into functions, the operator $\bar{A}_M$ maps functions into functions, i.e. it is an operator in $\mathcal{X}$. In the particular case where $\mathcal{X}$ is a space of square-integrable functions, we have

$$(f, \varphi_m)_{\mathcal{X}} = \int_{\mathcal{D}} f(x) \varphi_m^*(x) dx \tag{9.46}$$

and from equation (9.45) with $g_m = (A_M f)_m = (f, \varphi_m)_{\mathcal{X}}$ we derive that $\bar{A}_M$ is the integral operator

$$(\bar{A}_M f)(x) = \int \bar{K}_M(x, x') f(x') dx' \tag{9.47}$$

with

$$\bar{K}_M(x, x') = \sum_{m=1}^{M} \varphi_m(x) \varphi_m^*(x'). \tag{9.48}$$

Such an operator is called a *finite-rank integral operator* because its range is finite dimensional (and has dimension M). It is a particular case of the integral operators considered in the next section.

As concerns the operator $\tilde{A}_M$, it maps vectors into vectors and therefore is an operator in $\mathcal{E}_M$, which can be represented by a matrix. From equation (9.45) we get

$$\left(\tilde{A}_M g\right)_m = (A_M^* g, \varphi_m)_{\mathcal{X}} = \sum_{m'=1}^{M} (\varphi_{m'}, \varphi_m)_{\mathcal{X}}\, g_m \tag{9.49}$$

so that the matrix associated with $\tilde{A}_M$ is given by

$$\left(\tilde{\mathbf{A}}_M\right)_{m,m'} = (\varphi_{m'}, \varphi_m)_{\mathcal{X}}\,. \tag{9.50}$$

This is the *Gram matrix* of the functions φ_m. Its rank is M since the φ_m are linearly independent. Moreover it is a positive-definite matrix, as follows from the relationship

$$\left(\tilde{\mathbf{A}}_M g \cdot g\right)_M = \sum_{m,m'}^{M} (\varphi_m, \varphi_{m'})_{\mathcal{X}}\, g_m g_{m'}^* = \left\| \sum_{m=1}^{M} g_m \varphi_m \right\|_{\mathcal{X}}^2 \geq 0. \tag{9.51}$$

Since $\tilde{\mathbf{A}}_M$ is a self-adjoint and positive-definite matrix, it has exactly M positive eigenvalues.

By repeating the arguments used in section 9.2 in the case of a matrix we conclude that the operators $\bar{A}_M$ and $\tilde{A}_M$ have the same non-zero eigenvalues which are, precisely, the M non-zero eigenvalues of the Gram matrix of the functions φ_m.

Let us denote by $\sigma_1^2 \geq \sigma_2^2 \geq \cdots \geq \sigma_M^2$ the eigenvalues of this Gram matrix and by $\boldsymbol{u}_1, \boldsymbol{u}_2, \cdots, \boldsymbol{u}_M$ the corresponding eigenvectors. Eigenfunctions $v_1(\boldsymbol{x}), v_2(\boldsymbol{x}) \cdots, v_M(\boldsymbol{x})$ of the operator $\bar{A}_M$ (i.e. of the integral operator (9.47) in the case where $\mathcal{X}$ is a space of square-integrable functions) can be associated to the eigenvectors $\boldsymbol{u}_k$ by means of the same procedure used in the case of matrices: $v_k = \sigma_k^{-1} A_M^* \boldsymbol{u}_k$. In conclusion we find that the singular system of the operator A_M consists of the solutions of the shifted eigenvalue problem

$$A_M v_k = \sigma_k \boldsymbol{u}_k, \quad A_M^* \boldsymbol{u}_k = \sigma_k v_k. \tag{9.52}$$

A possible procedure for the computation of this singular system is the following:

- compute the Gram matrix of the functions φ_m;
- compute the eigenvalues σ_k^2 and the eigenvectors $\boldsymbol{u}_k$ of the Gram matrix; the square roots of the eigenvalues are the singular values while the the $\boldsymbol{u}_k$ are the singular vectors of A_M in the image space;
- compute the singular functions $v_k(\boldsymbol{x})$ by means of the second of the equations (9.52) and equation (9.45); the result is

$$v_k(\boldsymbol{x}) = \frac{1}{\sigma_k} \sum_{m=1}^{M} (\boldsymbol{u}_k)_m\, \varphi_m(\boldsymbol{x}). \tag{9.53}$$

In the case of a large image, the Gram matrix is also large and the computation of the singular system may be difficult in practice. The method, however, can be very useful in the solution of inverse problems with few data, as those where functions of only one variable are involved. Examples of applications to instrumental physics can be found in [8].

We conclude by deriving the singular value decomposition of the operators A_M and A_M^* in a form analogous to that obtained for the matrices $\mathbf{A}$ and $\mathbf{A}^*$, equations (9.37) and (9.38) respectively. Indeed, since $A_M f$ is a vector of $\mathcal{E}_M = \mathcal{Y}_M$ and since the singular vectors $\boldsymbol{u}_k$ form a basis in $\mathcal{E}_M$, we can write

$$A_M f = \sum_{k=1}^{M} (A_M f \cdot \boldsymbol{u}_k)_M \, \boldsymbol{u}_k. \tag{9.54}$$

From the second of the equations (9.52) we get

$$(A_M f \cdot \boldsymbol{u}_k)_M = \left(f, A_M^* \boldsymbol{u}_k\right)_{\mathcal{X}} = \sigma_k \, (f, v_k)_{\mathcal{X}} \tag{9.55}$$

and the substitution of this relation in equation (9.54) provides the SVD of A_M

$$A_M f = \sum_{k=1}^{M} \sigma_k \, (f, v_k)_{\mathcal{X}} \, \boldsymbol{u}_k. \tag{9.56}$$

Moreover, equation (9.45) implies that the range of A_M^* is the subspace spanned by the (linearly independent) functions $\varphi_m(\boldsymbol{x})$. The singular functions $v_k(\boldsymbol{x})$, which are just linear combinations of the $\varphi_m(\boldsymbol{x})$, form an orthonormal basis in this subspace so that

$$A_M^* \boldsymbol{g} = \sum_{k=1}^{M} \left(A_M^* \boldsymbol{g}, v_k\right)_{\mathcal{X}} v_k. \tag{9.57}$$

From the first of the equations (9.52), we obtain

$$\left(A_M^* \boldsymbol{g}, v_k\right)_{\mathcal{X}} = (\boldsymbol{g} \cdot A_M v_k)_M = \sigma_k \, (\boldsymbol{g} \cdot \boldsymbol{u}_k)_M \, . \tag{9.58}$$

It follows that

$$A_M^* \boldsymbol{g} = \sum_{k=1}^{M} \sigma_k \, (\boldsymbol{g} \cdot \boldsymbol{u}_k)_M \, v_k \tag{9.59}$$

and this is the SVD of the adjoint operator A_M^*.

9.4 SVD of an integral operator with square-integrable kernel

In section 8.1 a space variant imaging system is described in terms of an integral operator of the following form

$$(Af)(\boldsymbol{x}) = \int_{\mathcal{D}} K(\boldsymbol{x}, \boldsymbol{x}') f(\boldsymbol{x}') d\boldsymbol{x}', \quad \boldsymbol{x} \in \mathcal{D}' \tag{9.60}$$

where $\mathcal{D}$ and $\mathcal{D}'$ are respectively the object and the image domain. Other examples of such operators are provided in section 8.4 and section 8.5 (see, for instance, equations (8.28) and (8.43)) while an example of finite-rank integral operator is given in section 9.3 (see equations (9.47) and (9.48)). This operator is self-adjoint. A more general type of finite-rank integral operators is obtained if the integral kernel has the following structure

$$K(\boldsymbol{x}, \boldsymbol{x}') = \sum_{m=1}^{M} \varphi_m(\boldsymbol{x})\psi_m^*(\boldsymbol{x}') \tag{9.61}$$

the functions φ_m and ψ_m being, in general, different.

The analysis of the present section also applies to the case of space-invariant imaging systems, i.e. systems described by convolution operators, when the symmetry of the system with respect to translations is destroyed by the fact that the objects are localized in a bounded domain $\mathcal{D}$. In such a case, which will be investigated in detail in chapter 11, the integral operator takes the following form

$$(Af)(\boldsymbol{x}) = \int_{\mathcal{D}} K(\boldsymbol{x} - \boldsymbol{x}')f(\boldsymbol{x}')d\boldsymbol{x}', \quad \boldsymbol{x} \in \mathcal{D}' \tag{9.62}$$

where $\mathcal{D}'$ can possibly be the complete image plane.

We assume, for simplicity, that both the object and the image are square-integrable functions of the space variables, so that we have $\mathcal{X} = L^2(\mathcal{D})$ and $\mathcal{Y} = L^2(\mathcal{D}')$. Then equation (9.60) defines an operator from $L^2(\mathcal{D})$ into $L^2(\mathcal{D}')$. This operator is continuous if the integral kernel is square-integrable, i.e.

$$\|K\|^2 = \int_{\mathcal{D}'} d\boldsymbol{x} \int_{\mathcal{D}} d\boldsymbol{x}' |K(\boldsymbol{x}, \boldsymbol{x}')|^2 < \infty. \tag{9.63}$$

An integral operator satisfying this condition is usually called an integral operator of the *Hilbert–Schmidt class*.

In order to show that this operator is continuous we apply the Schwarz inequality (see appendix A) to the r.h.s. of equation (9.60) which, for $\boldsymbol{x}$ fixed, is the scalar product of two square integrable functions. It follows

$$|(Af)(\boldsymbol{x})|^2 \leq \left(\int_{\mathcal{D}} |K(\boldsymbol{x}, \boldsymbol{x}')|^2 d\boldsymbol{x}'\right)\left(\int_{\mathcal{D}} |f(\boldsymbol{x}')|^2 d\boldsymbol{x}'\right). \tag{9.64}$$

If we integrate both sides of this inequality with respect to $\boldsymbol{x}$, and we take the square root of the result, we get

$$\|Af\|_{\mathcal{Y}} \leq \|K\| \;\; \|f\|_{\mathcal{X}}, \tag{9.65}$$

i.e. the operator is bounded. This property implies the continuity of the operator (see appendix B).

Remark 9.1. Condition (9.63) is satisfied in the case of the integral operators of section 8.4 and section 8.5 as well as in the case of the finite rank operator (9.61) *(if the functions $\varphi_m(\boldsymbol{x})$ and $\psi_m(\boldsymbol{x})$ are square-integrable). Moreover it is also satisfied in the case of the operator (9.62) if the domain $\mathcal{D}$ or the domain $\mathcal{D}'$ is bounded and the PSF $K(\boldsymbol{x})$ is square-integrable. Indeed when $\mathcal{D}$ is bounded, by means of a change of variables, we get*

$$\|K\|^2 \leq m(\mathcal{D}) \int |K(\boldsymbol{x})|^2 d\boldsymbol{x} \tag{9.66}$$

where $m(\mathcal{D})$ is the measure (area or volume) of $\mathcal{D}$. Equality holds true when $\mathcal{D}'$ coincides with the complete image plane. A similar inequality holds true when $m(\mathcal{D}') < \infty$, with $m(\mathcal{D})$ replaced by $m(\mathcal{D}')$.

The adjoint A^* of the operator A is given by

$$(A^*g)(\boldsymbol{x}') = \int_{\mathcal{D}'} K^*(\boldsymbol{x}, \boldsymbol{x}')g(\boldsymbol{x})d\boldsymbol{x}, \quad \boldsymbol{x}' \in \mathcal{D} \tag{9.67}$$

as we can easily verify by checking that equation (B.6) is satisfied. Then, as in the case of a matrix, section 9.2, we can introduce the operators $\bar{A} = A^*A$ and $\tilde{A} = AA^*$. Both are integral operators with integral kernels given by

$$\bar{K}(\boldsymbol{x}, \boldsymbol{x}') = \int_{\mathcal{D}'} K^*(\boldsymbol{x}'', \boldsymbol{x})K(\boldsymbol{x}'', \boldsymbol{x}')d\boldsymbol{x}''; \quad \boldsymbol{x}, \boldsymbol{x}' \in \mathcal{D} \tag{9.68}$$

in the case of $\bar{A}$, and by

$$\tilde{K}(\boldsymbol{x}, \boldsymbol{x}') = \int_{\mathcal{D}} K(\boldsymbol{x}, \boldsymbol{x}'')K^*(\boldsymbol{x}', \boldsymbol{x}'')d\boldsymbol{x}''; \quad \boldsymbol{x}, \boldsymbol{x}' \in \mathcal{D}' \tag{9.69}$$

in the case of $\tilde{A}$.

The integral operators $\bar{A}$, $\tilde{A}$ have the following properties:

- both operators are self-adjoint, i.e. for any pair of functions f, h in $\mathcal{X}$ and any pair of functions g, w in $\mathcal{Y}$

$$\left(\bar{A}f, h\right)_{\mathcal{X}} = \left(f, \bar{A}h\right)_{\mathcal{X}}, \quad \left(\tilde{A}g, w\right)_{\mathcal{Y}} = \left(g, \tilde{A}w\right)_{\mathcal{Y}}; \tag{9.70}$$

  this property is a consequence of the following relations

$$\bar{K}^*(\boldsymbol{x}, \boldsymbol{x}') = \bar{K}(\boldsymbol{x}', \boldsymbol{x}), \quad \tilde{K}^*(\boldsymbol{x}, \boldsymbol{x}') = \tilde{K}(\boldsymbol{x}', \boldsymbol{x}) \tag{9.71}$$

  which can be easily checked by means of equations (9.68) and (9.69);
- both operators are of the Hilbert–Schmidt class because their integral kernels are square-integrable (the proof of this result is similar to the proof of the inequality (9.65); the starting point is the application of the Schwarz inequality to the r.h.s. of equation (9.68) and equation (9.69));

- both operators are positive semi-definite

$$(\bar{A}f, f)_{\mathcal{X}} \geq 0, \quad \left(\tilde{A}g, g\right)_{\mathcal{Y}} \geq 0. \tag{9.72}$$

Remark 9.2. We sketch the proof of this property in the case of $\bar{A}$. Indeed, from equation (9.68), by means of an exchange of the integration order we have

$$\begin{aligned}(\bar{A}f, f)_{\mathcal{X}} &= \int_{\mathcal{D}} \left(\int_{\mathcal{D}} \bar{K}(\boldsymbol{x}, \boldsymbol{x}') f(\boldsymbol{x}') d\boldsymbol{x}' \right) f^*(\boldsymbol{x}) d\boldsymbol{x} \\ &= \int_{\mathcal{D}'} \left| \int_{\mathcal{D}} K(\boldsymbol{x}'', \boldsymbol{x}) f(\boldsymbol{x}) d\boldsymbol{x} \right|^2 d\boldsymbol{x}'' \geq 0. \end{aligned} \tag{9.73}$$

A similar proof applies to the case of $\tilde{A}$.

According to the Hilbert–Schmidt theory [9], an integral operator with a symmetric and square-integrable kernel has real eigenvalues with finite multiplicity. Moreover the eigenfunctions associated with different eigenvalues are orthogonal. The eigenvalues form, in general, a countable set and accumulate to zero. However a finite rank integral operator, with rank M, has exactly M eigenvalues if each eigenvalue is counted as many times as its multiplicity. In such a case the zero eigenvalue has infinite multiplicity.

The integral operator of equation (8.28) is an example of a self-adjoint integral operator with an infinite set of eigenvalues. Indeed, its kernel is symmetric (because $G^*(\boldsymbol{\theta}, \boldsymbol{\theta}') = G(\boldsymbol{\theta}', \boldsymbol{\theta})$, as follows from equation (8.29)), and is square-integrable (because $G(\boldsymbol{\theta}, \boldsymbol{\theta}')$ is a bounded function defined over a bounded domain). In agreement with the general result stated above, its eigenvalues, given in equation (8.30), have finite multiplicity (the multiplicity of λ_l is $2l+1$) and accumulate to zero, as follows from equation (8.31). Moreover the representation (8.29) is a particular case of the general result which we give now.

If $K(\boldsymbol{x}, \boldsymbol{x}')$ is a symmetric and square-integrable kernel, let $\lambda_1, \lambda_2, \lambda_3, \cdots$ be the sequence of the eigenvalues of the corresponding integral operator, ordered in such a way that $|\lambda_1| \geq |\lambda_2| \geq |\lambda_3| \geq \cdots$, each eigenvalue being counted as many times as its multiplicity. Moreover, let $v_1(\boldsymbol{x}), v_2(\boldsymbol{x}), v_3(\boldsymbol{x}), \cdots$ be the sequence of the eigenfuctions associated with these eigenvalues. They constitute an orthonormal set of square-integrable functions. Then the basic results of the Hilbert–Schmidt theory is the following *spectral representation* of the kernel $K(\boldsymbol{x}, \boldsymbol{x}')$

$$K(\boldsymbol{x}, \boldsymbol{x}') = \sum_{k=1}^{\infty} \lambda_k v_k(\boldsymbol{x}) v_k^*(\boldsymbol{x}') \tag{9.74}$$

the series being convergent in the sense of the L^2-norm. Accordingly, the eigenvalues λ_k satisfy the following condition

$$\sum_{k=1}^{\infty} \lambda_k^2 < \infty. \tag{9.75}$$

It is not difficult to understand that equation (9.74) provides an extension, to the case of an integral operator, of the spectral representation (9.23) of a symmetric matrix.

The results stated above apply to the operators $\bar{A}$ and $\tilde{A}$, whose non-zero eigenvalues are positive because they are positive semi-definite operators. By generalizing to the present case the method used in the case of a matrix, we can show that the operators $\bar{A}$ and $\tilde{A}$ have the same positive eigenvalues with the same multiplicity.

Indeed, let us denote by σ_k^2 the positive eigenvalues of $\bar{A}$, with the ordering $\sigma_1^2 \geq \sigma_2^2 \geq \cdots \geq \sigma_k^2 \geq \cdots$ and $\sigma_k^2 \to 0$ for $k \to \infty$. Again each eigenvalue has been counted as many times as its multiplicity, which is certainly finite as follows from the Hilbert–Schmidt theory.

A normalized eigenfunction v_k is associated to each eigenvalue σ_k^2 and these eigenfunctions constitute an orthonormal system, i.e. $\left(v_k, v_j\right)_{\mathcal{X}} = \delta_{kj}$. Then, to each eigenfunction v_k of $\bar{A}$ we can associate a function u_k in $\mathcal{Y}$ defined by

$$u_k = \frac{1}{\sigma_k} A v_k. \tag{9.76}$$

Using the relation $\tilde{A}A = A\bar{A}$, which can be easily proved by an exchange of integration order in the definition of the integral kernels, we obtain

$$\tilde{A}u_k = \frac{1}{\sigma_k} A\bar{A}v_k = \sigma_k^2 \left(\frac{1}{\sigma_k} A v_k\right) = \sigma_k^2 u_k \tag{9.77}$$

and also

$$\begin{aligned} \left(u_k, u_j\right)_{\mathcal{Y}} &= \frac{1}{\sigma_k \sigma_j} \left(A v_k, A v_j\right)_{\mathcal{Y}} \\ &= \frac{1}{\sigma_k \sigma_j} \left(v_k, \bar{A} v_j\right)_{\mathcal{X}} = \frac{\sigma_j}{\sigma_k} \left(v_k, v_j\right)_{\mathcal{X}} = \delta_{kj}. \end{aligned} \tag{9.78}$$

Therefore all eigenvalues σ_k^2 of $\bar{A}$ are also eigenvalues of $\tilde{A}$ and the u_k are the corresponding eigenfunctions, which constitute an orthonormal system in $\mathcal{Y}$. In order to show that in this way we have obtained all eigenvalues and eigenfunctions of $\tilde{A}$ it is sufficient to repeat the same argument starting from $\tilde{A}$. If its eigenvalues and eigenfunctions are denoted by σ_k and u_k respectively, then we can show, by means of the relation $\bar{A}A^* = A^*\tilde{A}$, that the functions of $\mathcal{X}$ defined by

$$v_k = \frac{1}{\sigma_k} A^* u_k \tag{9.79}$$

are orthonormal eigenfunctions of $\bar{A}$ associated with the eigenvalues σ_k^2.

The equations (9.76) and (9.79) define the usual shifted eigenvalue problem, which we write now explicitly in terms of the integral operators

$$\int_{\mathcal{D}} K(\boldsymbol{x}, \boldsymbol{x}')v_k(\boldsymbol{x}')d\boldsymbol{x}' = \sigma_k u_k(\boldsymbol{x}) \tag{9.80}$$
$$\int_{\mathcal{D}'} K^*(\boldsymbol{x}, \boldsymbol{x}')u_k(\boldsymbol{x})d\boldsymbol{x} = \sigma_k v_k(\boldsymbol{x}').$$

The proof of the SVD of the operators A and A^* requires the completeness of the Hilbert space $\mathcal{X}$ and $\mathcal{Y}$ and therefore we do not report this proof here (see, for instance, [10]). We only give the result, which takes the usual form except for the fact that now the sums are replaced by series. We have

$$Af = \sum_{k=1}^{\infty} \sigma_k \, (f, v_k)_{\mathcal{X}} \, u_k \tag{9.81}$$

and also

$$A^* f = \sum_{k=1}^{\infty} \sigma_k \, (g, u_k)_{\mathcal{Y}} \, v_k, \tag{9.82}$$

the series being convergent with respect to the norms of $\mathcal{Y}$ and $\mathcal{X}$ respectively.

We also remark that the SVD of A is equivalent to the following series expansion of the kernel $K(\boldsymbol{x}, \boldsymbol{x}')$

$$K(\boldsymbol{x}, \boldsymbol{x}') = \sum_{k=1}^{\infty} \sigma_k u_k(\boldsymbol{x}) v_k^*(\boldsymbol{x}') \tag{9.83}$$

the series being convergent with respect to the L^2-norm. From this expansion (which is a generalization of equation (9.74)), from the orthonormality of the singular functions and from the definition (9.63) of $\|K\|^2$ it follows that

$$\|K\|^2 = \sum_{k=1}^{\infty} \sigma_k^2 \tag{9.84}$$

and therefore *the sum of squares of the singular values of a Hilbert–Schmidt integral operator is convergent.*

Using the SVD (9.81) it is also possible to show that the norm of the integral operator A (see appendix B for the definition) is given by

$$\|A\| = \sigma_1. \tag{9.85}$$

From this equation and equation (9.84), it follows that the *the norm of the integral operator is never greater than the L^2-norm of the integral kernel; the two norms coincide if and only if the integral operator has rank one.*

9.5 SVD of the Radon transform

The SVD of the Radon transform, which was defined in section 8.2, has been established by Davison [11]. In his paper the general case of functions of n variables is considered and it is assumed that both the object and the data space are suitable weighted spaces of square-integrable functions (weighted spaces are defined in appendix A). Here we consider the case of functions of two variables, which is the most important one for the applications, and, in this case, we consider the most natural choice for the weight in the image space.

In practical applications it is quite natural to assume that the objects are described by functions with support interior to a disc of radius a. It is not restrictive to take $a = 1$ because it is always possible to satisfy this condition by rescaling the space variables. Next we assume that the object functions f are square-integrable, so that $\mathcal{X} = L^2(\mathcal{D})$, $\mathcal{D}$ being the disc of radius 1, and

$$\|f\|_{\mathcal{X}}^2 = \int_{\mathcal{D}} |f(\boldsymbol{x})|^2 d\boldsymbol{x}. \tag{9.86}$$

We denote by $w(s)$ the half length of the chord obtained by intersecting the disc with a straight line having signed distance s from the origin (see figure 9.2). We have

$$w(s) = (1 - s^2)^{1/2}. \tag{9.87}$$

Then the Radon transform of a function which is zero outside $\mathcal{D}$ is given by

$$(Rf)(s, \boldsymbol{\theta}) = \int_{-w(s)}^{w(s)} f(s\boldsymbol{\theta} + t\boldsymbol{\theta}^{\perp})dt, \quad |s| \leq 1. \tag{9.88}$$

A simple property of the functions in the range of the operator R is obtained by applying the Schwarz inequality (see appendix A) to the r.h.s. of equation (9.88) if we consider the integrand as the scalar product of the function f and of the function equal to one over the interval $[-w(s), w(s)]$. We get

$$|(Rf)(s, \boldsymbol{\theta})|^2 \leq 2w(s) \int_{-w(s)}^{w(s)} |f(s\boldsymbol{\theta} + t\boldsymbol{\theta}^{\perp})|^2 dt \tag{9.89}$$

so that

$$\begin{aligned} \int_{-1}^{1} w^{-1}(s)\, |(Rf)(s, \boldsymbol{\theta})|^2 ds &\leq 2 \int_{-1}^{1} ds \int_{-w(s)}^{w(s)} |f(s\boldsymbol{\theta} + t\boldsymbol{\theta}^{\perp})|^2 dt \\ &= 2 \int_{\mathcal{D}} |f(\boldsymbol{x})|^2 d\boldsymbol{x}. \end{aligned} \tag{9.90}$$

This remark suggests that a quite natural norm for the functions $g = g(s, \boldsymbol{\theta})$ of the data space $\mathcal{Y}$, is the following one

$$\|g\|_{\mathcal{Y}}^2 = \int_0^{2\pi} d\phi \int_{-1}^{1} \frac{ds}{w(s)} |g(s, \boldsymbol{\theta})|^2. \tag{9.91}$$

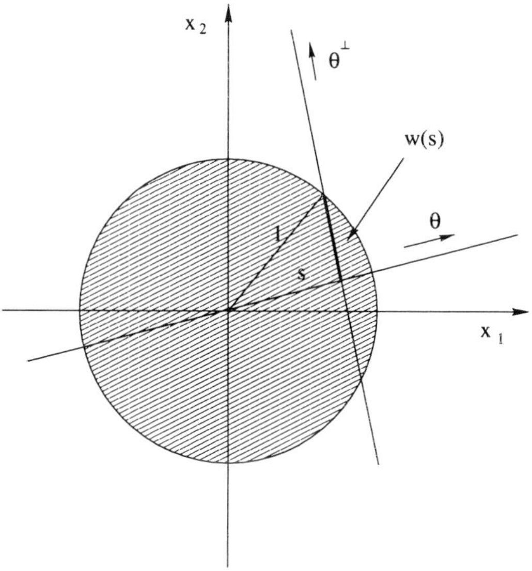

Figure 9.2. Geometry of the Radon transform in the case of functions with support interior to the disc of radius 1.

Then, from the inequality (9.90), by integrating with respect to the angle ϕ, it follows that R is a bounded operator from $\mathcal{X}$ into $\mathcal{Y}$

$$\|Rf\|_{\mathcal{Y}} \leq \sqrt{4\pi}\,\|f\|_{\mathcal{X}}. \tag{9.92}$$

From the results reported in the following it turns out that the largest singular value of R is just $\sqrt{4\pi}$ so that $\|R\| = \sqrt{4\pi}$. In other words inequality (9.92) is precise and cannot be improved.

The singular system of R can be obtained by representing a function $g(s, \boldsymbol{\theta})$ of $\mathcal{Y}$, for fixed $\boldsymbol{\theta}$, as a series of *Chebyshev polynomials of the second kind* [12], which are defined by

$$U_m(s) = \frac{\sin\left[(m+1)\arccos s\right]}{\sin\left(\arccos s\right)}; \quad m = 0, 1, 2, \cdots \tag{9.93}$$

These polynomials are orthogonal with respect to the weight function $w(s)$; more precisely they satisfy the following orthogonality and normalization conditions

$$\int_{-1}^{1} w(s) U_m(s) U_{m'}(s) ds = \frac{\pi}{2}\delta_{m,m'}. \tag{9.94}$$

Remark 9.3. The Chebyshev polynomials are related to trigonometric functions and equation (9.94) follows from the orthogonality of these functions. Indeed, by means of the change of variable $s = \cos\xi$, $0 \le \xi \le \pi$, *we get*

$$U_m(\cos\xi) = \frac{\sin(m+1)\xi}{\sin\xi} \tag{9.95}$$

and therefore

$$\begin{aligned} \int_{-1}^{1} w(s)U_m(s)U_{m'}(s)ds &= \int_0^{\pi} \sin[(m+1)\xi]\sin\left[(m'+1)\xi\right]d\xi \\ &= \frac{\pi}{2}\delta_{m,m'}. \end{aligned} \tag{9.96}$$

If we observe that a function $g(s)$, *which is square-integrable with respect to the weight* $w^{-1}(s)$, *is also a square-integrable function of* ξ

$$\int_{-1}^{1} |g(s)|^2 \frac{ds}{w(s)} = \int_0^{\pi} |g(\cos\xi)|^2 d\xi \tag{9.97}$$

and that $g(\cos\xi)$ *can be represented by the trigonometric series*

$$g(\cos\xi) = \sum_{m=0}^{\infty} c_m \sin[(m+1)\xi] \tag{9.98}$$

with

$$c_m = \frac{2}{\pi}\int_0^{\pi} g(\cos\xi)\sin[(m+1)\xi]\,d\xi, \tag{9.99}$$

we conclude that $g(s)$ *can be represented by a series expansion in terms of the functions* $u_m(s) = w(s)U_m(s)$.

The previous remark implies that any function $g = g(s,\boldsymbol{\theta})$ in $\mathcal{Y}$ can be represented, for fixed $\boldsymbol{\theta}$, in terms of the functions $w(s)U_m(s)$ and this fact suggests to consider the subspaces $\mathcal{Y}_m$ of $\mathcal{Y}$, which are defined as the subspaces containing the functions

$$g_m(s,\boldsymbol{\theta}) = \sqrt{\frac{2}{\pi}}\,w(s)U_m(s)u(\boldsymbol{\theta}); \quad m = 0, 1, \cdots \tag{9.100}$$

where $u(\boldsymbol{\theta})$ is an arbitrary square-integrable function of $\boldsymbol{\theta}$. From equations (9.91) and (9.94) it follows that

$$\|g_m\|_{\mathcal{Y}}^2 = \int_0^{2\pi} |u(\boldsymbol{\theta})|^2 d\phi. \tag{9.101}$$

We also remark that the subspaces $\mathcal{Y}_m$ are mutually orthogonal.

The next step is to show that the operator RR^* transforms a function of $\mathcal{Y}_m$ into another function of $\mathcal{Y}_m$. To this purpose we need the expression of the adjoint operator R^*.

From the definition of the adjoint operator and from the definition of the scalar products in $\mathcal{X}$ and $\mathcal{Y}$ we easily derive that R^* is given by

$$\left(R^*g\right)(\boldsymbol{x}) = \int_0^{2\pi} g(\boldsymbol{\theta}\cdot\boldsymbol{x},\boldsymbol{\theta})w^{-1}(\boldsymbol{\theta}\cdot\boldsymbol{x})d\phi. \tag{9.102}$$

Indeed, by means of some exchange of integration order and of a change of variables, we have

$$\begin{aligned}(Rf,g)_{\mathcal{Y}} &= \int_{-1}^{1}\frac{ds}{w(s)}\int_0^{2\pi}d\phi\left(\int_{-w(s)}^{w(s)}dtf(s\boldsymbol{\theta}+t\boldsymbol{\theta}^{\perp})\right)g^*(s,\boldsymbol{\theta})\\ &= \int_0^{2\pi}\left(\int_{-1}^{1}ds\int_{-w(s)}^{w(s)}dtf(s\boldsymbol{\theta}+t\boldsymbol{\theta}^{\perp})\frac{g^*(s,\boldsymbol{\theta})}{w(s)}\right)d\phi\\ &= \int_{\mathcal{D}}d\boldsymbol{x}\,f(\boldsymbol{x})\left(\int_0^{2\pi}\frac{g^*(\boldsymbol{\theta}\cdot\boldsymbol{x},\boldsymbol{\theta})}{w(\boldsymbol{\theta}\cdot\boldsymbol{x})}d\phi\right) = \left(f,R^*g\right)_{\mathcal{X}}.\end{aligned} \tag{9.103}$$

The variables $\boldsymbol{x} = s\boldsymbol{\theta}+t\boldsymbol{\theta}^{\perp}$ have been introduced (for fixed $\boldsymbol{\theta}$) at the place of the variables $\{s,t\}$ and the relation $s = \boldsymbol{\theta}\cdot\boldsymbol{x}$ has been used. Finally the integrals with respect to ϕ and $\boldsymbol{x}$ have been exchanged.

If g_m is a function of $\mathcal{Y}_m$ as defined in equation (9.100), then, from equations (9.88) and (9.102), we obtain

$$\left(RR^*g_m\right)(s,\boldsymbol{\theta}) = \sqrt{\frac{2}{\pi}}\int_{-w(s)}^{w(s)}dt\int_0^{2\pi}d\phi' U_m\left[\boldsymbol{\theta}'\cdot(s\boldsymbol{\theta}+t\boldsymbol{\theta}^{\perp})\right]u(\boldsymbol{\theta}') \tag{9.104}$$

and by means of some straightforward computations we get

$$\left(RR^*g_m\right)(s,\boldsymbol{\theta}) = \frac{4\pi}{m+1}\sqrt{\frac{2}{\pi}}w(s)U_m(s)\bar{u}(\boldsymbol{\theta}) \tag{9.105}$$

where

$$\bar{u}(\boldsymbol{\theta}) = \frac{1}{2\pi}\int_0^{2\pi}\frac{\sin\left[(m+1)(\phi-\phi')\right]}{\sin\left(\phi-\phi'\right)}u(\boldsymbol{\theta}')d\phi'. \tag{9.106}$$

Remark 9.4. Equations (9.105)–(9.106) can be proved as follows. We exchange the integration order in equation (9.104) and we consider the integral

$$I(s;\boldsymbol{\theta},\boldsymbol{\theta}') = \int_{-w(s)}^{w(s)}U_m\left[\boldsymbol{\theta}'\cdot(s\boldsymbol{\theta}+t\boldsymbol{\theta}^{\perp})\right]dt. \tag{9.107}$$

If $\boldsymbol{\theta} = \{\cos\phi, \sin\phi\}$ *and* $\boldsymbol{\theta}' = \{\cos\phi', \sin\phi'\}$, *we have* $\boldsymbol{\theta}\cdot\boldsymbol{\theta}' = \cos(\phi - \phi')$ *and* $\boldsymbol{\theta}^{\perp}\cdot\boldsymbol{\theta}' = -\sin(\phi-\phi')$. *Then, if we write* $s = \cos\xi$ *and* $\psi = \phi - \phi'$, *we obtain*

$$I(\cos\xi; \boldsymbol{\theta}, \boldsymbol{\theta}') = \int_{-\sin\xi}^{\sin\xi} U_m(\cos\xi\cos\psi - t\sin\psi)\,dt. \tag{9.108}$$

By the change of variable $u = \cos\xi\cos\psi - t\sin\psi$ *this integral becomes*

$$I(\cos\xi; \boldsymbol{\theta}, \boldsymbol{\theta}') = \int_{\cos(\xi+\psi)}^{\cos(\xi-\psi)} \frac{U_m(u)}{\sin\psi}\,du \tag{9.109}$$

and by introducing the new variable $u = \cos\eta$ *we finally obtain*

$$\begin{aligned} I(\cos\xi; \boldsymbol{\theta}, \boldsymbol{\theta}') &= \frac{1}{\sin\psi}\int_{\xi-\psi}^{\xi+\psi} \sin[(m+1)\eta]\,d\eta \\ &= \frac{2}{m+1}\,\frac{\sin[(m+1)\psi]}{\sin\psi}\,\sin[(m+1)\xi]. \end{aligned} \tag{9.110}$$

By substituting this expression in equation (9.104), with $\psi = \phi - \phi'$ *and* $\xi = \arccos s$, *we get equations (9.105) and (9.106).*

Equation (9.105) implies that the operator RR^* transforms a function of $\mathcal{Y}_m$ into another function of $\mathcal{Y}_m$, i.e. each subspace $\mathcal{Y}_m$ reduces the operator RR^*. Moreover, the restriction of the operator RR^* to the subspace $\mathcal{Y}_m$ is equivalent to the integral operator defined in equation (9.106). It follows that it is possible to diagonalize RR^* by diagonalizing its restrictions to the subspaces $\mathcal{Y}_m$. Thanks to the completeness of Chebyshev polynomials we find in such a way all eigenvalues and eigenvectors of RR^*.

The integral operator given in equation (9.106) is a finite-rank integral operator, as follows from the identity

$$\frac{1}{2\pi}\,\frac{\sin\left[(m+1)(\phi-\phi')\right]}{\sin(\phi-\phi')} = \sum_{k=0}^{m} Y_{m-2k}(\boldsymbol{\theta})Y^*_{m-2k}(\boldsymbol{\theta}') \tag{9.111}$$

where

$$Y_l(\boldsymbol{\theta}) = \frac{1}{\sqrt{2\pi}}e^{-il\phi}. \tag{9.112}$$

Since these functions are orthonormal in $L^2(0, 2\pi)$, we find that the functions $Y_{m-2k}(\boldsymbol{\theta})$ are the eigenfunctions of the integral operator (9.106) associated with the eigenvalue 1. In conclusion, this operator has the eigenvalue 1 with multiplicity $m+1$ and the eigenvalue 0 with infinite multiplicity.

If we introduce the functions

$$u_{m,k}(s, \boldsymbol{\theta}) = \sqrt{\frac{2}{\pi}}w(s)U_m(s)Y_{m-k}(\boldsymbol{\theta}) \tag{9.113}$$

with $k = 0, 1, \cdots, m$, these functions are orthonormal in $\mathcal{Y}$. Moreover, from equations (9.105), (9.106) and (9.111) it follows that

$$RR^* u_{m,k} = \sigma_m^2 u_{m,k}; \quad k = 0, 1, \cdots, m, \tag{9.114}$$

where

$$\sigma_m = \left(\frac{4\pi}{m+1} \right)^{1/2}. \tag{9.115}$$

Since we have obtained all eigenvalues and eigenfunctions of RR^*, by means of the arguments already used in the previous sections we can prove that the σ_m^2 are also the eigenvalues of the operator R^*R, associated with the eigenfunctions

$$v_{m,k} = \frac{1}{\sigma_m} R^* u_{m,k}. \tag{9.116}$$

As follows from the Fourier slice theorem, which will be proved in section 11.1, these eigenfunctions constitute a complete set of orthonormal functions in $L^2(\mathcal{D})$. They are given by [13]

$$v_{m,k}(\boldsymbol{x}) = (2m+2)^{1/2} Q_{m,|m-2k|}(|\boldsymbol{x}|) Y_{m-2k}\left(\frac{\boldsymbol{x}}{|\boldsymbol{x}|} \right) \tag{9.117}$$

where

$$Q_{m,l}(r) = r^l P^{(0,l)}_{\frac{1}{2}(m+l)}(2r^2 - 1) \tag{9.118}$$

$P_n^{(\alpha,\beta)}(t)$ being the Jacobi polynomial of degree n [12].

In conclusion, the singular values of the Radon transform are given by equation (9.115) while the corresponding singular functions are given by equations (9.113) and (9.117). Since we have $m+1$ singular functions associated with σ_m, corresponding to $k = 0, 1, \cdots, m$ in equation (9.113), it follows that σ_m has multiplicity $m + 1$. Moreover $\sigma_m \to 0$, even if the decay is rather slow. If each singular value is counted as many times as its multiplicity we see that the number of singular values with $m \leq M$ is given by $(M + 1)(M + 2)/2$.

References

[1] Beltrami E 1873 *Giornale di Matematiche ad Uso degli Studenti delle Università* **11** 98

[2] Jordan C 1874 *J. Math. Pures Appl., Deuxième Sèries* **19** 35

[3] Stewart G W 1993 *SIAM Review* **35** 551

[4] Bertero M, De Mol C and Pike E R 1985 *Inverse Problems* **1** 301

[5] Lanczos C 1961 *Linear Differential Operators* (London: Van Nostrand)

[6] Golub G H and Reinsch C 1970 *Numer. Math.* **14** 403

[7] Press W H, Teukolsky S A, Vetterling W T and Flannery B P 1992 *Numerical Recipes* (Cambridge: Cambridge University Press)

[8] Bertero M and Pike E R 1993 *Handbook of Statistics* vol 10, ed N K Bose and C R Rao (Amsterdam: Elsevier) p 1

[9] Mikhlin S G 1957 *Integral Equations* (London: Pergamon Press)

[10] Groetsch C W 1993 *Inverse Problems in the Mathematical Sciences* (Wiesbaden: Vieweg)

[11] Davison M E 1981 *Numer. Funct. Anal. Optimiz.* **3** 321

[12] Abramovitz M and Stegun I A 1965 *Handbook of Mathematical Functions* (New York: Dover)

[13] Louis A K 1984 *SIAM J. Math. Anal.* **15 621**

Chapter 10

Inversion methods revisited

In chapters 5, 6 and 7 some of the most important inversion methods have been presented and discussed in the case of image deconvolution. Here we extend these methods to the more general linear inverse problems introduced in chapter 8 and the singular value decomposition presented in chapter 9 is used for describing and understanding their main features. This chapter can also be used as a short summary of these methods.

10.1 The generalized solution

According to the mathematical description of a linear imaging system introduced in section 9.1, the noisy image of an object $f^{(0)}$ is given by

$$g = Af^{(0)} + w \tag{10.1}$$

where A is a linear operator from the object space $\mathcal{X}$ into the image space $\mathcal{Y}$ and w is an element of $\mathcal{Y}$ describing the noise contribution to g.

As in the case of image restoration, we neglect the noise term in equation (10.1) and we formulate the inverse problem as follows: given the noisy image g and the linear operator A describing the imaging system, solve the linear equation

$$Af = g. \tag{10.2}$$

In this chapter we assume that the operator A has a singular value decomposition so that we can write

$$Af = \sum_{j=1}^{p} \sigma_j \left(f, v_j\right)_{\mathcal{X}} u_j \tag{10.3}$$

where $p < \infty$ for a discrete or a semi-discrete problem (section 9.2 and section 9.3, respectively), while $p = \infty$ for the operators considered in section 9.4 and section 9.5. Therefore, as far as possible, we treat simultaneously finite and infinite-dimensional problems. Only at the end of this section will we separately consider the two cases.

The problem (10.2) is, in general, ill-posed in the sense that the solution is not unique, does not exist, or else does not depend continuously on the data.

Uniqueness does not hold when the null-space of the operator A, $\mathcal{N}(A)$, i.e. the set of the invisible objects f such that $Af = 0$, is not trivial. For instance uniqueness never holds true in the case of the semi-discrete problems of section 9.3, while it does hold true in the case of a discrete problem if the rank of the matrix coincides with the number of unknowns. Uniqueness holds true also in the case of the inversion of the Radon transform, section 9.5.

The procedure most frequently used for restoring uniqueness is the following one. Any element f of the object space $\mathcal{X}$ can be represented by

$$f = \sum_{j=1}^{p} (f, v_j)_{\mathcal{X}}\, v_j + v \tag{10.4}$$

where v is the projection of f onto $\mathcal{N}(A)$, while the first term is the component of f orthogonal to $\mathcal{N}(A)$. The term v can be called the invisible component of the object f because it does not contribute to the image of f, Af. Since this invisible component cannot be determined from equation (10.2), it may be natural to look for a solution of this equation whose invisible component is zero. Such a solution is unique because from equation (10.3) and equation (10.4), with $v = 0$, we easily deduce that $Af = 0$ implies $f = 0$. If this solution exists, it is denoted by $f^{\dagger}$ and called the *minimal norm solution*. Indeed, any solution of equation (10.2) is given by

$$f = f^{\dagger} + v \tag{10.5}$$

where v is an arbitrary element of $\mathcal{N}(A)$. Since v is orthogonal to $f^{\dagger}$, we have

$$\|f\|_{\mathcal{X}}^2 = \left\|f^{\dagger}\right\|_{\mathcal{X}}^2 + \|v\|_{\mathcal{X}}^2 \tag{10.6}$$

and therefore the solution with $v = 0$, i.e. $f^{\dagger}$, is the solution of minimal norm.

As concerns the existence of a solution of equation (10.2) and, in particular, of $f^{\dagger}$, we first have to distinguish between the following two cases.

- The null space of A^*, $\mathcal{N}(A^*)$, contains only the zero element, $g = 0$. Then the singular functions (vectors) u_j constitute an orthonormal basis in $\mathcal{Y}$ and the noisy image g can be represented by

$$g = \sum_{j=1}^{p} (g, u_j)_{\mathcal{Y}}\, u_j. \tag{10.7}$$

By comparing this representation with the SVD of A, equation (10.3), we see that a solution of equation (10.2) may exist. We reconsider this point in a moment.

- The null space of A^*, $\mathcal{N}(A^*)$, contains non-zero elements. In such a case the singular functions (vectors) u_j do not constitute an orthonormal basis in $\mathcal{Y}$ and the noisy image g can be represented as follows

$$g = \sum_{j=1}^{p} (g, u_j)_{\mathcal{Y}} u_j + u \tag{10.8}$$

where u is the component of g in $\mathcal{N}(A^*)$, i.e. the component of g orthogonal to the range of A (see equation (B.8)). This term is analogous to the out-of-band noise which affects the noisy images formed by a bandlimited imaging system (see section 4.5). Notice that, if the mathematical model of the imaging system is physically correct, the presence of this term is an effect due to the noise. If $u \neq 0$, by comparing the representation (10.3) of Af with the representation (10.8) of g, we see that there does not exists any object f such that Af coincides with g. Then we can look for objects f such that Af is as close as possible to f, i.e. for objects which minimize the discrepancy functional

$$\|Af - g\|_{\mathcal{Y}} = minimum. \tag{10.9}$$

Any solution of this variational problem is called a *least-squares solution.*

The concept of least-squares solution is more general than the concept of solution because a solution of equation (10.2) is also a least-squares solution. More precisely, the set of the least-squares solutions coincide with the set of the solutions if and only if the minimum of the discrepancy functional (10.9) is zero. This remark shows that, without loss of generality, we can investigate the problem of existence in the case of the least-squares solutions.

As shown in appendix E, solving problem (10.9) is equivalent to solving its Euler equation, which is given by

$$A^*Af = A^*g. \tag{10.10}$$

From the SVD of the operator A, equation (10.3), and of the operator A^*

$$A^*g = \sum_{j=1}^{p} \sigma_j (g, u_j)_{\mathcal{Y}} v_j \tag{10.11}$$

we obtain

$$A^*Af = \sum_{j=1}^{p} \sigma_j^2 (f, v_j)_{\mathcal{X}} v_j. \tag{10.12}$$

If we insert these representations in equation (10.10) and we compare the coefficients of v_k, we find that the components of any solution f of equation (10.10) are given by

$$\sigma_j^2 (f, v_j)_{\mathcal{X}} = \sigma_j (g, u_j)_{\mathcal{Y}}, \tag{10.13}$$

and therefore

$$\left(f, v_j\right)_{\mathcal{X}} = \frac{1}{\sigma_j}\left(g, u_j\right)_{\mathcal{Y}}. \tag{10.14}$$

In such a way the problem of the existence of least-squares solutions has been reduced to the problem of the existence of elements of the object space $\mathcal{X}$ whose components with respect to the singular functions v_j are given by equation (10.14). To this purpose we consider separately the case $p < \infty$ and the case $p = +\infty$. For the sake of simplicity, we consider first the case $p < \infty$ even if this is not the natural order. Discrete or semi-discrete problems come, in general from the discretization of infinite-dimensional problems but, for these problems, we must also consider convergence questions which can reduce the clarity of the treatment.

(A) *The case* $p < +\infty$

According to equation (10.14) we introduce the following function (vector)

$$f^{\dagger} = \sum_{j=1}^{p} \frac{1}{\sigma_j}\left(g, u_j\right)_{\mathcal{Y}} v_j \tag{10.15}$$

which is a solution of equation (10.10) and therefore a least-squares solution.

As concerns the uniqueness, let us first observe that the null space of the operator A and the null space of the operator A^*A coincide, $\mathcal{N}(A) = \mathcal{N}(A^*A)$. The proof of this result is given in appendix C in the case of a matrix but it can be easily generalized to any linear operator. Thanks to this property, $f^{\dagger}$ is the unique least-squares solution if and only if $\mathcal{N}(A)$ contains only the zero element.

If uniqueness does not hold true as a consequence of the existence of invisible objects, then any least-squares solution is given by equation (10.5), with $f^{\dagger}$ defined by equation (10.15) and therefore we can conclude that $f^{\dagger}$ is the *least-squares solution of minimal norm*. It is also called the *generalized solution* of the inverse problem and the results of the analysis outlined above can be summarized as follows: *in the case of a discrete or semidiscrete problem, for any image g (noise-free or noisy) there exists a unique generalized solution* $f^{\dagger}$, *whose singular function (vector) expansion is given by equation (10.15).*

Equation (10.15) defines an operator from $\mathcal{Y}$ to $\mathcal{X}$, which is called the *generalized inverse operator* of A and denoted by $A^{\dagger}$

$$f^{\dagger} = A^{\dagger}g. \tag{10.16}$$

We observe that $A^{\dagger}g = 0$ if and only if $A^*g = 0$, so that $\mathcal{N}(A^{\dagger}) = \mathcal{N}(A^*)$.

In the case of a matrix $\mathbf{A}$, the generalized inverse matrix $\mathbf{A}^{\dagger}$ is also called the *Moore–Penrose inverse* of $\mathbf{A}$. If the SVD of $\mathbf{A}$ is given by equation (9.20), then the SVD of $\mathbf{A}^{\dagger}$ is given by

$$\mathbf{A}^{\dagger} = \mathbf{V}\Sigma^{-1}\mathbf{U}^*, \tag{10.17}$$

as one can easily deduce from equation (10.15).

The generalized solution depends continuously on the image g and therefore $A^\dagger$ is a linear and continuous operator. This property can be proved rather easily and, to this purpose as well as to the purpose of investigating the numerical stability of the solution, we first observe that only the component of g in the range of A, $\mathcal{R}(A)$, contributes to $f^\dagger$. This component is the first term at the r.h.s. of equation (10.8) and will be denoted by $g^\dagger$, since it can be considered as the image of $f^\dagger$, i.e. $g^\dagger = Af^\dagger$. We can also write $f^\dagger = A^\dagger g^\dagger$. We give explicitly these relationships in terms of the singular systems of the operator A

$$f^\dagger = \sum_{j=1}^{p} \frac{1}{\sigma_j} \left(g^\dagger, u_j\right)_{\mathcal{Y}} v_j \tag{10.18}$$

$$g^\dagger = \sum_{j=1}^{p} \sigma_j \left(f^\dagger, v_j\right)_{\mathcal{X}} u_j. \tag{10.19}$$

We also observe that the procedure which transforms g into $g^\dagger$ is analogous to the filtering operation considered in section 4.5 for suppressing the out-of-band noise.

If $\delta g^\dagger$ is a variation of the image $g^\dagger$ and $\delta f^\dagger$ is the corresponding variation of $f^\dagger$, from equation (10.18), taking into account the linearity of the relationship as well as the fact that $\sigma_1 \geq \sigma_2 \geq \cdots \geq \sigma_p$, we obtain

$$\|\delta f^\dagger\|_{\mathcal{X}}^2 = \sum_{j=1}^{p} \frac{1}{\sigma_j^2} \left|\left(\delta g^\dagger, u_j\right)_{\mathcal{Y}}\right|^2 \leq \frac{1}{\sigma_p^2} \|\delta g^\dagger\|_{\mathcal{Y}}^2. \tag{10.20}$$

This inequality proves, as we did already state, that $f^\dagger$ depends continuously on $g^\dagger$ because, when $\delta g^\dagger$ tends to zero, also $\delta f^\dagger$ tends to zero.

Continuity, however, is not sufficient to assure numerical stability which can be investigated by estimating the relative error induced on $f^\dagger$ by the error on $g^\dagger$. To this purpose we can use the relation (10.19) for obtaining the following inequality

$$\|g^\dagger\|_{\mathcal{Y}}^2 = \sum_{j=1}^{p} \sigma_j^2 \left|\left(f^\dagger, v_j\right)_{\mathcal{X}}\right|^2 \leq \sigma_1^2 \|f^\dagger\|_{\mathcal{X}}^2, \tag{10.21}$$

which, combined with the inequality (10.20), provides the following estimate of the relative error

$$\frac{\|\delta f^\dagger\|_{\mathcal{X}}}{\|f^\dagger\|_{\mathcal{X}}} \leq \frac{\sigma_1}{\sigma_p} \frac{\|\delta g^\dagger\|_{\mathcal{Y}}}{\|g^\dagger\|_{\mathcal{X}}}. \tag{10.22}$$

The quantity

$$\alpha = \frac{\sigma_1}{\sigma_p} \tag{10.23}$$

is the condition number of the operator A, controlling the numerical stability of the inverse problem. It is just the extension of the condition number introduced in section 4.4, equation (4.25), and, as in that case, it is not difficult to prove that the inequality (10.22) cannot be improved even if it may be pessimistic. According to the analysis of Twomey, already mentioned in section 4.4, a better estimate of error propagation can be provided by the AREMF β which is now given by

$$\beta = \frac{1}{p}\left(\sum_{j=1}^{p}\sigma_j^2\right)\left(\sum_{j=1}^{p}\sigma_j^{-2}\right). \tag{10.24}$$

As in the case of the problems considered in section 4.4, the condition number α or the AREMF β of a discrete or semidiscrete inverse problem can be quite large. Indeed, if the problem we are considering comes from the discretization of a problem with singular values tending to zero and if the discretization is fine enough, we expect some of its singular values to be small (even very small) so that its condition number is large. In such a case the generalized solution $f^\dagger$ is deprived of any physical meaning and cannot be accepted as a reasonable estimate of the unknown object $f^{(0)}$. We find the same difficulty discussed in the case of the solution provided by the inverse filtering method for deconvolution problems.

(B) *The case* $p = +\infty$

If we proceed as in the case $p < +\infty$, we can introduce the following formal solution

$$f^\dagger = \sum_{j=1}^{\infty}\frac{1}{\sigma_j}\left(g, u_j\right)_{\mathcal{Y}} v_j. \tag{10.25}$$

We say that this solution is formal because it is given by a series expansion so that the solution exists if and only if the series is convergent.

If we consider convergence in the sense of the norm of $\mathcal{X}$, then this convergence is assured if and only if the sum of the squares of the coefficients of the singular functions v_j is convergent (see appendix A). We obtain the following condition

$$\sum_{j=1}^{\infty}\frac{1}{\sigma_j^2}\left|\left(g, u_j\right)_{\mathcal{Y}}\right|^2 < \infty \tag{10.26}$$

which is also called the *Picard criterion* for the existence of solutions or least-squares solutions of the linear inverse problem we are considering [1].

It is important to point out that, if the singular values σ_j accumulate to zero, as shown in several examples of chapter 8 and chapter 9, then condition (10.26) may not be satisfied by an arbitrary noisy image g. If it is not satisfied, then *no solution or least-squares solution of the inverse problem exists*.

The functions g satisfying the Picard criterion are the images in the range of A, $\mathcal{R}(A)$. For anyone of these functions, the series (10.25) defining $f^\dagger$ is convergent. Then by repeating the same arguments used in the case $p < \infty$ we can conclude that *for any image g satisfying the Picard criterion there exists a unique generalized solution $f^\dagger$, whose singular function expansion is given by equation (10.25).*

The generalized solution defines a generalized inverse operator $A^\dagger$ as in the case $p < \infty$. The main difference is that now this operator is not defined everywhere on $\mathcal{Y}$ but only on the set of the functions g satisfying the Picard criterion. This set is the domain of the operator $A^\dagger$, $\mathcal{D}(A^\dagger)$. Moreover the operator $A^\dagger$ is not continuous or, in other words, the generalized solution $f^\dagger$ does not depend continuously on the image g.

In order to prove this statement, let us assume that g is an image satisfying the Picard criterion and let us consider a sequence of images given by

$$g_j = g + \sqrt{\sigma_j} u_j. \tag{10.27}$$

It is obvious that

$$\|g_j - g\|_{\mathcal{Y}} = \sqrt{\sigma_j} \to 0 \tag{10.28}$$

because we have assumed that the singular values tend to zero. On the other hand, if we denote by $f^\dagger$ the generalized solution associated with g, and by $f_j^\dagger$ the generalized solution associated with g_j, we have

$$f_j^\dagger = f^\dagger + \frac{1}{\sqrt{\sigma_j}} v_j \tag{10.29}$$

so that

$$\left\| f_j^\dagger - f^\dagger \right\| = \frac{1}{\sqrt{\sigma_j}} \to \infty. \tag{10.30}$$

In such a way we have found a sequence of images, converging to g, such that the sequence of the corresponding generalized solutions does not converge to $f^\dagger$. This pathology, related to the fact that the singular values tend to zero, generates the numerical instability, already discussed, of the discrete versions of the problem.

10.2 The Tikhonov regularization method

The generalized solution of an ill-posed or ill-conditioned problem is not physically meaningful because it is completely corrupted by the noise propagation from data to solution. This point has been discussed in detail in chapter 4. For this reason we must look for approximate solutions satisfying additional constraints suggested by the physics of the problem and regularization is a way for obtaining such solutions. We reconsider here the Tikhonov regularization method which has been introduced in chapter 5 in the case of deconvolution problems.

The starting point is to define a family of *regularized solutions* f_μ, depending on the *regularization parameter* $\mu > 0$, as the family of the functions minimizing the functionals

$$\Phi_\mu(f; g) = \|Af - g\|_{\mathcal{Y}}^2 + \mu \, \|f\|_{\mathcal{X}}^2 \tag{10.31}$$

where g is the given image. The main formal difference with respect to the functional (5.6) is that the objects and the images may now belong to different spaces.

The Euler equation associated with the minimization of this functional is derived in appendix E and is given by

$$(A^*A + \mu I)f = A^*g. \tag{10.32}$$

An object is a minimum point f_μ of the functional (10.31) if and only if it is a solution of equation (10.32).

In order to solve this equation, let us represent an arbitrary element f of $\mathcal{X}$ in terms of the singular functions v_j of the operator A and of the elements v (orthogonal to all the v_j) of the null space of A, as in equation (10.4). Here p can be finite or infinite. If we insert this representation in equation (10.32) and we take into account equations (10.11) and (10.12), we obtain

$$\sum_{j=1}^{p} (\sigma_j^2 + \mu) \left(f, v_j\right)_{\mathcal{X}} v_j + \mu v = \sum_{j=1}^{p} \sigma_j \left(g, u_j\right)_{\mathcal{Y}} v_j. \tag{10.33}$$

It follows that there exists a unique solution f_μ of equation (10.32), which can be obtained from equation (10.4) with $v = 0$ and with coefficients $\left(f, v_j\right)_{\mathcal{X}}$ given by

$$(\sigma_j^2 + \mu) \left(f, v_j\right)_{\mathcal{X}} = \sigma_j \left(g, u_j\right)_{\mathcal{Y}}. \tag{10.34}$$

In conclusion we get

$$f_\mu = \sum_{j=1}^{p} \frac{\sigma_j}{\sigma_j^2 + \mu} \left(g, u_j\right)_{\mathcal{Y}} v_j. \tag{10.35}$$

If $p = \infty$, the series at the r.h.s. of this equation does always converge (thanks to the factors σ_j, the coefficients tend to zero more rapidly than the components $(g, u_j)_{\mathcal{Y}}$ of g) and therefore the regularized solution f_μ exists for any noisy image g.

The regularized solution f_μ can be rewritten in the form given by equations (5.25), (5.26), i.e.

$$f_\mu = R_\mu g \tag{10.36}$$

$$R_\mu = \left(A^*A + \mu I\right)^{-1} A^*. \tag{10.37}$$

The operator R_μ is an approximation of the generalized inverse $A^\dagger$, equation (10.16), in the sense explained in section 5.3. Indeed it is not difficult to prove that:

- for any $\mu > 0$, R_μ is a linear and continuous operator from $\mathcal{Y}$ into $\mathcal{X}$
- for any image g, in the case $p < \infty$, and for any image g satisfying the Picard criterion in the case $p = \infty$

$$\lim_{\mu \to 0} R_\mu g = A^\dagger g. \tag{10.38}$$

Remark 10.1. In the case of a discrete problem it may be convenient to compute the generalized solution by computing R_μ, i.e. by inverting the matrix $\mathbf{A}_\mu = \mathbf{A}^*\mathbf{A} + \mu\mathbf{I}$. When this matrix must be inverted for several values of μ, a considerable saving of operations can be obtained by a method due to Elden [2].

By means of equation (10.35), and the Parseval equality given in appendix A, equation (A.13), we can compute all quantities which are relevant for analysing the method.

- *Energy functional*

$$E^2(f_\mu) = \|f_\mu\|_{\mathcal{X}}^2 = \sum_{j=1}^{p} \frac{\sigma_j^2}{(\sigma_j^2 + \mu)^2} \left|(g, u_j)_{\mathcal{Y}}\right|^2 \tag{10.39}$$

- *Discrepancy functional*

$$\varepsilon^2(f_\mu; g) = \|Af_\mu - g\|_{\mathcal{Y}}^2 = \sum_{j=1}^{p} \frac{\mu^2}{(\sigma_j^2 + \mu)^2} \left|(g, u_j)_{\mathcal{Y}}\right|^2 + \|u\|_{\mathcal{Y}}^2. \tag{10.40}$$

In equation (10.40) the representation (10.8) of the image g has been used. We observe that, as in the case of image deconvolution, the energy is a decreasing function of the regularization parameter μ, while the discrepancy is an increasing function of μ.

If we assume that the noisy image g is given by equation (10.1), the difference between the regularized solution f_μ and the true object $f^{(0)}$ is given by

$$f_\mu - f^{(0)} = \left(R_\mu A f^{(0)} - f^{(0)}\right) + R_\mu w. \tag{10.41}$$

We have:

- *Approximation error*

$$\|R_\mu A f^{(0)} - f^{(0)}\|_{\mathcal{X}}^2 = \sum_{j=1}^{p} \frac{\mu^2}{(\sigma_j^2 + \mu)^2} \left|(f^{(0)}, v_j)_{\mathcal{X}}\right|^2 + \|v^{(0)}\|_{\mathcal{X}}^2 \tag{10.42}$$

where $v^{(0)}$ is the component of $f^{(0)}$ in the null space of A (i.e. the invisible component of $f^{(0)}$).

- *Noise-propagation error*

$$\| R_\mu w \|_{\mathcal{X}}^2 = \sum_{j=1}^{p} \frac{\sigma_j^2}{(\sigma_j^2 + \mu)^2} \left| (w, u_j)_{\mathcal{Y}} \right|^2 . \tag{10.43}$$

We observe that, as in the case of image deconvolution, the approximation error is an increasing function of μ while the noise-propagation error is a decreasing function of μ. As a consequence the total restoration error, given by

$$\varrho(\mu) = \| f_\mu - f^{(0)} \|_{\mathcal{X}} \tag{10.44}$$

has a minimum, i.e. there exists an optimum value of the regularization parameter. We do not repeat the discussion of this property, called semiconvergence, which was widely discussed in section 5.3.

As concerns the choice of the regularization parameter, the methods introduced in section 5.6 can also be applied to the present case.

10.3 Truncated SVD

The representation (10.35) of the regularized solution can be recast in the following form

$$f_\mu = \sum_{j=1}^{p} \frac{W_{\mu,j}}{\sigma_j} (g, u_j)_{\mathcal{Y}} \, v_j \tag{10.45}$$

where

$$W_{\mu,j} = \frac{\sigma_j^2}{\sigma_j^2 + \mu} . \tag{10.46}$$

This expression shows that the regularized solution f_μ can be obtained by a filtering of the singular value decomposition of the generalized solution: the components of $f^\dagger$ corresponding to singular values much larger than μ are taken without any significant modification, whereas the components corresponding to singular values much smaller than μ are essentially removed.

This remark suggests that we consider other kinds of filters in the vein of the filtering methods discussed in section 5.4 for image deconvolution. Here we only discuss a method which has been widely used in the solution of linear inverse problem, the so-called *truncated SVD*.

The idea is to replace the smooth filter given in equation (10.46) by a sharp one, i.e. to take in the singular function expansion of the generalized solution only the terms corresponding to singular values greater than a certain threshold value. Since the singular values are ordered to form a non-decreasing sequence, those greater than the threshold value are those corresponding to values of the index less than a certain maximum integer.

Let us denote by J the number of singular values satisfying the condition

$$\sigma_j^2 \geq \mu; \tag{10.47}$$

then the approximate solution provided by the truncated SVD is as follows

$$f_J = \sum_{j=1}^{J} \frac{1}{\sigma_j} \left(g, u_j\right)_{\mathcal{Y}} v_j. \tag{10.48}$$

This equation can be obtained from equation (10.45) by taking $W_{\mu,j} = 1$ when $\sigma_j^2 \geq \mu$ and $W_{\mu,j} = 0$ when $\sigma_j^2 < \mu$. Moreover it can be conveniently written in a form similar to that of equation (10.36) by introducing the operator $R^{(J)}$ from $\mathcal{Y}$ into $\mathcal{X}$ such that

$$f_J = R^{(J)} g. \tag{10.49}$$

If we observe that $J = J(\mu)$ increases when μ decreases and that $J(\mu) \to p$ when $\mu \to 0$, then we can prove that $R^{(J)}$ has properties similar to R_μ: for any J, $R^{(J)}$ is a bounded operator from $\mathcal{Y}$ into $\mathcal{X}$; moreover $R^{(J)}$ converges to $A^\dagger$, in the sense of equation (10.38), when $J \to p$.

The computation of the quantities which are relevant for the analysis of the method is again very easy.

- *Energy functional*

$$E^2(f_J) = \|f_J\|_{\mathcal{X}}^2 = \sum_{j=1}^{J} \frac{1}{\sigma_j^2} \left|\left(g, u_j\right)_{\mathcal{Y}}\right|^2 \tag{10.50}$$

- *Discrepancy functional*

$$\varepsilon^2(f_J; g) = \|Af_J - g\|_{\mathcal{Y}}^2 = \sum_{j=J+1}^{p} \left|\left(g, u_j\right)_{\mathcal{Y}}\right|^2 + \|u\|_{\mathcal{Y}}^2. \tag{10.51}$$

These expressions imply that the energy is an increasing function of J (and therefore a decreasing function of μ) while the discrepancy is a decreasing function of J (and therefore an increasing function of μ). These properties correspond to similar properties of the Tikhonov regularized solutions.

Let us consider now the difference between the truncated solution and the true object $f^{(0)}$. Using equation (10.49) and the model (10.1) for the noisy image g, we obtain

$$R^{(J)} g - f^{(0)} = \left(R^{(J)} A f^{(0)} - f^{(0)}\right) + R^{(J)} w. \tag{10.52}$$

We have:

- *Approximation error*

$$\| R^{(J)} A f^{(0)} - f^{(0)} \|_{\mathcal{X}}^2 = \sum_{j=J+1}^{p} \left|(f^{(0)}, v_j)_{\mathcal{X}}\right|^2 + \| v^{(0)} \|_{\mathcal{X}}^2 \tag{10.53}$$

where $v^{(0)}$ is the component of $f^{(0)}$ in the null space of A.

- *Noise-propagation error*

$$\|R^{(J)}w\|_{\mathcal{X}}^2 = \sum_{j=1}^{J} \frac{1}{\sigma_j^2} \left|(w, u_j)_{\mathcal{Y}}\right|^2. \tag{10.54}$$

The approximation error is a decreasing function of J while the noise-propagation error is an increasing function of J. Therefore the total restoration error, $\rho_J = \|f_J - f^{(0)}\|_{\mathcal{X}}$, has a minimum for a certain value of J: when J increases the truncated SVD solution first approaches the true object $f^{(0)}$ and then goes away. It follows that also this method has the semiconvergence property which is typical of the regularization methods.

The previous remark implies that there exists an optimum number of terms in the truncated SVD. As usual the problem is to estimate this optimum number in the case of real data.

It has been shown [3] that the discrepancy principle can provide such an estimate. Let ε^2 be an estimate of the noise norm

$$\|w\|_{\mathcal{Y}}^2 \leq \varepsilon^2 \tag{10.55}$$

and let $b > 1$ be a fixed but arbitrary positive parameter. Then an estimate of the optimum number of terms is given by the integer $\tilde{J}$ such that

$$\|Af_{\tilde{J}} - g\|^2 \geq b\varepsilon^2, \quad \|Af_{\tilde{J}+1} - g\|^2 < b\varepsilon^2. \tag{10.56}$$

The parameter b is introduced for theoretical reasons because one needs the condition $b > 1$ to prove the convergence of the method (in the case $p = \infty$) for $\epsilon \to 0$. In practice, we can take $b = 1$. The method provides a unique value $\tilde{J}$ because, as shown before, the discrepancy functional is a decreasing function of the number of terms in the truncated SVD.

If an estimate E^2 of the norm of the true object is also known

$$\|f^{(0)}\|_{\mathcal{X}}^2 \leq E^2 \tag{10.57}$$

then another method proposed by Miller [4] consists of taking as an estimate of the optimum number of terms the number of singular values satisfying the condition

$$\sigma_k^2 \geq \left(\frac{\varepsilon}{E}\right)^2. \tag{10.58}$$

In conclusion let us observe that in 2D problems the number of terms to be used in the truncated SVD can be rather large. For instance, in the case of tomography, the singular values are given by equation (9.115). They depend only on the index m and, for a given m, the multiplicity of the singular value is $m + 1$. Therefore, the total number of singular values with $m \leq m_0$ is $J_0 = (m_0 + 1)(m_0 + 2)/2$. If we consider, for instance, all the singular values greater than 10^{-1}, the maximum value of m is $m_0 = 1255$ and their total number is $J_0 = 788140$. If the threshold is 10^{-2}, then $m_0 = 125662$ and $J_0 \simeq 9 \times 10^9$. These large numbers occur because tomography is a mildly ill-posed problem. For other more severely ill-posed problems the number of terms can be much smaller.

10.4 Iterative regularization methods

In this section we discuss the extension of the iterative methods introduced in chapter 6 to more general linear inverse problems. We consider the *Landweber*, *projected Landweber*, *steepest descent* and *conjugate gradient* methods. In the case of discrete or semi-discrete problems, the common feature of these methods is that the basic operation required for their implementation is a matrix-vector multiplication. Therefore these methods are particularly useful in practice when the matrix **A** is large and sparse.

(A) *The Landweber method*

As in the case of image deconvolution, the iterative scheme can be written as follows

$$f_{k+1} = f_k + \tau \left(A^* g - A^* A f_k\right), \tag{10.59}$$

where τ is the relaxation parameter. For simplicity we consider only the most usual choice for the initial guess, i.e. $f_0 = 0$.

The first problem is to establish conditions on τ ensuring the convergence of the method. To this purpose we can insert in equation (10.59) the representations (10.11) and (10.12) of the operators A^* and A^*A. It follows that the components of the iterates with respect to the orthonormal system formed by the singular functions v_j satisfy the following recursive relation

$$\left(f_{k+1}, v_j\right)_{\mathcal{X}} = \tau\sigma_j \left(g, u_j\right)_{\mathcal{Y}} + (1 - \tau\sigma_j^2) \left(f_k, v_j\right)_{\mathcal{X}} . \tag{10.60}$$

On the other hand, the component of f_k in the null space of A is not modified by the iterations so that it must coincide with the corresponding component of f_0. Since we assume $f_0 = 0$, this component is also zero.

From equation (10.60) we can easily derive by induction that

$$(f_k, v_j)_{\mathcal{X}} = \tau\{1 + (1 - \tau\sigma_j^2) + \cdots + (1 - \tau\sigma_j^2)^{k-1}\}\sigma_j \left(g, u_j\right)_{\mathcal{Y}} . \tag{10.61}$$

This formula is analogous to that of equation (6.10), with σ_j^2 replacing $\hat{H}(\omega)$ and $\sigma_j \left(g, u_j\right)_{\mathcal{Y}}$ replacing $\hat{\hat{g}}(\omega)$. Therefore, if we use equation (6.12), we get

$$\left(f, v_j\right)_{\mathcal{X}} = \left\{1 - (1 - \tau\sigma_j^2)^k\right\} \frac{1}{\sigma_j} \left(g, u_j\right)_{\mathcal{Y}} . \tag{10.62}$$

Convergence is assured for any j, in the limit $k \to \infty$, if and only if τ satisfies the condition

$$|1 - \tau\sigma_j^2| < 1 \tag{10.63}$$

or equivalently

$$0 < \tau < \frac{2}{\sigma_j^2}. \tag{10.64}$$

Since σ_1 is the largest singular value of the operator A, the conditions (10.64) are satisfied for any j if and only if

$$0 < \tau < \frac{2}{\sigma_1^2}. \tag{10.65}$$

This condition is analogous to condition (6.25) for the problem of image deconvolution. When it is satisfied, we get from equation (10.62) that

$$\lim_{k\to\infty} \left(f_k, v_j\right)_{\mathcal{X}} = \frac{1}{\sigma_j} \left(g, u_j\right)_{\mathcal{Y}}. \tag{10.66}$$

For a finite value of k, equation (10.62) implies the following representation of f_k in terms of the singular system of A

$$f_k = \sum_{j=1}^{p} \left\{1 - (1 - \tau\sigma_j^2)^k\right\} \frac{\left(g, u_j\right)_{\mathcal{Y}}}{\sigma_j} v_j. \tag{10.67}$$

Hence we find that f_k is a filtered version of the generalized solution $f^\dagger$. Moreover we can deduce from equation (10.66) that, in the case $p = \infty$, the iterates f_k converge to $f^\dagger$ if the data satisfy the Picard criterion (10.26). In the discrete or in the semi-discrete case, the iterates f_k always converge to $f^\dagger$.

Since the limit is not physically meaningful in the case of an ill-conditioned or ill-posed problem, we must investigate the regularization properties of the algorithm. We give again the expression of the most relevant quantities.

- *Energy functional*

$$E^2(f_k) = \|f_k\|_{\mathcal{X}}^2 = \sum_{j=1}^{p} \frac{1}{\sigma_j^2} |1 - (1 - \tau\sigma_j^2)^k|^2 \left|\left(g, u_j\right)_{\mathcal{Y}}\right|^2 \tag{10.68}$$

- *Discrepancy functional*

$$\varepsilon^2(f_k; g) = \|Af_k - g\|_{\mathcal{Y}}^2 = \sum_{j=1}^{p} (1 - \tau\sigma_j^2)^{2k} \left|\left(g, u_j\right)_{\mathcal{Y}}\right|^2 + \|u\|_{\mathcal{Y}}^2. \tag{10.69}$$

We observe once more that the energy is an increasing function of k while the discrepancy is a decreasing function of k.

If we introduce the operator $R^{(k)}$ from $\mathcal{Y}$ into $\mathcal{X}$ associated with the mapping from g into f_k defined by equation (10.67)

$$f_k = R^{(k)} g, \tag{10.70}$$

then, by means of the usual model for the process of image formation, we can write the difference between f_k and the true object $f^{(0)}$ as follows

$$f_k - f^{(0)} = \left(R^{(k)} A f^{(0)} - f^{(0)}\right) + R^{(k)} w. \tag{10.71}$$

We have:

- *Approximation error*

$$\| R^{(k)} A f^{(0)} - f^{(0)} \|_{\mathcal{X}}^2 = \sum_{j=1}^{p} (1 - \tau\sigma_j^2)^{2k} \left|(f^{(0)}, v_j)_{\mathcal{X}}\right|^2 \tag{10.72}$$

- *Noise-propagation error*

$$\| R^{(k)} w \|_{\mathcal{X}}^2 = \sum_{j=1}^{p} \frac{1}{\sigma_j^2} |1 - (1 - \tau\sigma_j^2)^k|^2 \left|(g, u_j)_{\mathcal{Y}}\right|^2. \tag{10.73}$$

It follows that the approximation error is a decreasing function of k while the noise-propagation error is an increasing function of k. As a result the restoration error, $\rho_k = \| f_k - f^{(0)} \|_{\mathcal{X}}$, has a minimum for a certain value of k, k_{opt}, so that the semiconvergence property holds true also in this case.

In the case of real data, an estimate $\tilde{k}$ of k_{opt} can be obtained by means of the discrepancy principle already discussed in section 6.1. This method provides a unique value of $\tilde{k}$ because, as follows from equation (10.69), the discrepancy is a decreasing function of k.

Let us comment now on the implementation of the method for discrete or semi-discrete problems. If the singular system of the operator has been computed, then one can compute directly f_k by means of equation (10.67), i.e. one can consider the method as a filtering method and not as an iterative method. However, as we have already observed, the computation of the singular system may not be practical for imaging problems. Then one has to implement the iterative scheme (10.59).

In the case of a discrete problem, the basic operation in the implementation of equation (10.59) is a matrix-vector multiplication. This is not so obvious in the case of a semi-discrete problem but it is not difficult to modify equation (10.59) in such a way that the same property holds also in this case.

Indeed, let us consider a semi-discrete mapping A_M, section 9.3, defined by M linearly independent functions $\varphi_1, \varphi_2, \cdots, \varphi_M$. Then, as follows from equation (10.59), in the case $f_0 = 0$, f_k belongs to this subspace because it is in the range of the operator A_M^*. As a consequence, for any k, there exists a unique vector $\boldsymbol{f}_k$ such that

$$f_k = A^* \boldsymbol{f}_k. \tag{10.74}$$

If we insert this representation in equation (10.59) and if we take into account that the unique vector $\boldsymbol{h}$ such that $A^*\boldsymbol{h} = 0$ is $\boldsymbol{h} = 0$, we obtain

$$f_{k+1} = f_k + \tau(g - A_M A_M^* f_k). \tag{10.75}$$

The operator $A_M A_M^*$ is the Gram matrix of the functions φ_m, so that we find that also in this case the implementation of equation (10.75) only requires a matrix-vector multiplication.

(B) *The projected Landweber method*

In the case of constrained least-squares solutions which belong to a closed and convex set $\mathcal{C}$, the projected Landweber method can be used for obtaining approximations for these solutions. It is defined by the iterative scheme

$$f_{k+1}^{(\mathcal{C})} = P_{\mathcal{C}}\left[f_k^{(\mathcal{C})} + \tau(A^*g - A^*Af_k^{(\mathcal{C})})\right]. \tag{10.76}$$

where $P_{\mathcal{C}}$ is again the projection operator onto the set $\mathcal{C}$. As in the case of the Landweber method one usually takes $f_0^{(\mathcal{C})} = 0$. Moreover the relaxation parameter τ must satisfy condition (10.65).

In the case of a discrete or semi-discrete problem the iterates $f_k^{(\mathcal{C})}$ converge, for $k \to \infty$, to a solution of the constrained least-squares problem

$$\|Af - g\|_{\mathcal{Y}} = minimum, \quad f \in \mathcal{C}. \tag{10.77}$$

When $p = \infty$, then it has only been proved [4] that the iterates f_k converge weakly to a solution of this problem if g is the image of a function of $\mathcal{C}$.

When the solutions of problem (10.77) are unphysical, the method usually shows the semiconvergence property which is typical of the regularization methods. Examples are given in section 6.2.

From the computational point of view, the basic operations required at each step for the implementation of the method are a matrix-vector multiplication and a projection onto the set $\mathcal{C}$. In order to choose correctly the value of the relaxation parameter, an estimate of σ_1 is also required.

(C) *The steepest descent method*

The steepest descent method, shortly discussed at the beginning of section 6.4 is defined by the following iterative scheme

$$f_{k+1} = f_k + \tau_k \bar{r}_k \tag{10.78}$$

where

$$\bar{r}_k = A^*g - A^*Af_k, \quad \tau_k = \frac{\|\bar{r}_k\|_{\mathcal{X}}^2}{\|A\bar{r}_k\|_{\mathcal{Y}}^2}. \tag{10.79}$$

It can be viewed as a Landweber method with a choice of the relaxation parameter depending on the iteration step.

The iterates f_k converge, for $k \to \infty$, to a least-squares solution in the case of a discrete or semi-discrete problem. If $p = \infty$ then it has been proved [5] that the same result holds true if g is in the range of the operator A, or, in other words, if g satisfies the Picard criterion (10.26). Also the steepest descent method has the semiconvergence property and therefore can be used as a regularization method.

(D) *The conjugate gradient method*

The method, discussed in section 6.4, is defined by the following iterative scheme

$$\bar{r}_0 = \bar{p}_0 = A^* g, \quad f_0 = 0$$

$$\bar{\alpha}_k = \frac{\|\bar{r}_k\|^2_{\mathcal{X}}}{\|A\bar{p}_k)\|^2_{\mathcal{Y}}}, \quad \bar{r}_{k+1} = \bar{r}_k - \bar{\alpha}_k A^* A \bar{p}_k \tag{10.80}$$

$$\bar{\beta}_k = \frac{\|\bar{r}_{k+1}\|^2_{\mathcal{X}}}{\|\bar{r}_k\|^2_{\mathcal{X}}}, \quad \bar{p}_{k+1} = \bar{r}_{k+1} + \bar{\beta}_k \bar{p}_k$$

$$f_{k+1} = f_k + \bar{\alpha}_k \bar{p}_k.$$

In the case of discrete or semi-discrete problems the method converges to the generalized solution in M steps (M is the dimension of the data space), if $p = M$ and if the Krylov subspace $\mathcal{K}^{(M)}(\bar{A}; \bar{g})$ has dimension M so that it coincides with the data space $\mathcal{Y}$. In the case $p = \infty$, it has been proved [6], that the iterates f_k converge, for $k \to \infty$, to the generalized solution $f^\dagger$, if g is in the range of the operator A and therefore satisfies the Picard criterion. It is also known that the method has the semiconvergence property, so that it is a powerful regularization method for the solution of linear inverse problems.

A glance to the iterative scheme (10.80) shows that also here the basic operation required at each step is a matrix-vector multiplication.

10.5 Statistical methods

The statistical methods for the solution of linear inverse problems are presented in chapter 7 in a form which is already suitable for any discrete problem and not only for deconvolution problems. Therefore we briefly summarize here the main results and we emphasize only the points which are relevant for the more general problems considered in this part of the book.

We have divided the statistical methods into two classes:

- Maximum likelihood methods

- Bayesian methods.

In ML methods only the image is treated in a probabilistic way as a consequence of the presence of noise, while the object is treated in a deterministic way, more precisely as a set of parameters characterizing the probability distribution of the image. In Bayesian methods both the image and the object are treated in a probabilistic way and this approach allows the use of *a priori* information on the object expressed in the form of its probability distribution.

The ML methods are essentially variational methods because the estimates they provide are obtained by the maximization of the likelihood function (see section 7.1). Only two cases were discussed in chapter 7, corresponding to two different kinds of noise, and precisely

- additive Gaussian noise
- Poisson noise.

If the noise is Gaussian, then the ML method is equivalent to the least-squares method because it is equivalent to the following minimization problem

$$\left(\mathbf{S}_\nu^{-1}(\mathbf{A}f - g) \cdot (\mathbf{A}f - g)\right)_M = minimum \tag{10.81}$$

where $\mathbf{S}_\nu$ is the covariance matrix of the noise and the scalar product is that of $\mathcal{E}_M$, the canonical Euclidean space with dimension M. Indeed, the problem (10.81) is just the problem (10.9) if $\mathcal{Y}$ is a weighted Euclidean space (as defined in appendix C) with a weighting matrix given by $\mathbf{C} = \mathbf{S}_\nu^{-1}$.

The equivalence between maximum likelihood and least-squares implies that the ML method leads to an ill-posed problem, the basic reason being that no *a priori* information on the object is used. If such an information is introduced then we can re-obtain the regularization methods investigated in part I as well as in this chapter.

Let us briefly comment about the ill-conditioning of the problem (10.81). If the noise is white, i.e. $\mathbf{S}_\nu = s^2\mathbf{I}$, then the condition number of the problem (10.81) is given by the ratio between the maximum and minimum singular value of the matrix $\mathbf{A}$ and therefore depends only on the imaging system. When the noise is not white, then the condition number depends, in a sense, both on the imaging system and on the noise.

The most simple case is that of a diagonal covariance matrix, i.e. the case where the components of the image are independent random variables with variance s_m^2, so that equation (10.81) becomes

$$\sum_{m=1}^{M} \frac{1}{s_m^2} \left|(\mathbf{A}f)_m - g_m\right|^2 = minimum. \tag{10.82}$$

This problem can be reduced to a least-squares problem in the canonical space if we introduce the matrix

$$(\mathbf{A}_\nu)_{m,n} = \frac{1}{s_m} A_{m,n} \tag{10.83}$$

which depends on the imaging system through the matrix $\mathbf{A}$ and on the noise ν through the variances s_m^2. Then the condition number of the problem (10.82) is the condition number of the matrix $\mathbf{A}_\nu$, which can be greater or smaller than the condition number of the matrix $\mathbf{A}$, according to the properties of the noise.

In the case of correlated noise, the previous elementary analysis can be generalized using Choleski factorization of the covariance matrix $\mathbf{S}_\nu$ (see appendix C), which we write now in the following form

$$\mathbf{S}_\nu = \mathbf{H}_\nu \mathbf{H}_\nu^T \tag{10.84}$$

where $\mathbf{H}_\nu$ is a lower triangular matrix. Since $\mathbf{S}_\nu$ is, by assumption, invertible, the matrix $\mathbf{H}_\nu$ is also invertible and we can write

$$\mathbf{S}_\nu^{-1} = \left(\mathbf{H}_\nu^T\right)^{-1} \mathbf{H}_\nu^{-1}. \tag{10.85}$$

If we insert this representation in equation (10.81) we re-obtain a least-squares problem in the canonical space with the following modified image and matrix

$$g_\nu = \mathbf{H}_\nu^{-1} g, \quad \mathbf{A}_\nu = \mathbf{H}_\nu^{-1} \mathbf{A}. \tag{10.86}$$

Again, the condition number of the problem (10.81) is just the condition number of the matrix $\mathbf{A}_\nu$.

In the case of Poisson noise, the existence of ML estimates, i.e. of objects which maximize the likelihood function for a given image is proved, but it is not proved that they provide reasonable approximations of the unknown object. Numerical experiments suggest that, in general, they do not. As in the case of Gaussian noise, the ML problem is ill-posed, even if the positivity constraint is now implicit in the method. Clearly, this constraint is not sufficient for obtaining stable and sound solutions.

The EM method is an iterative method which provides approximations of the ML estimates, converging to these estimates when the number of iterations tends to infinity. Numerical experiments indicate that this method may have the semiconvergence property which is typical of the iterative regularization methods (Landweber, CG, etc). Therefore the iterative process should be stopped after a suitable number of iterations, even if, as far as we know, stopping rules have not been established.

As concerns the implementation of the algorithm, it takes the most simple form if the matrix $\mathbf{A}$ satisfies the conditions

$$\alpha_n = \sum_{m=1}^{M} A_{m,n} = 1; \quad n = 1, 2, \cdots, N. \tag{10.87}$$

If these conditions are not satisfied, then it is possible to modify the matrix and the object as follows

$$B_{m,n} = \frac{1}{\alpha_n} A_{m,n}; \quad h_n = \alpha_n f_n \tag{10.88}$$

so that

$$\mathbf{A}f = \mathbf{B}h; \quad \sum_{m=1}^{M} B_{m,n} = 1. \tag{10.89}$$

In such a way the iterative algorithm takes the very simple form

$$\tilde{h}_{k+1} = \tilde{h}_k \left(\mathbf{B}^* \frac{g}{\mathbf{B}\tilde{h}_k} \right) \tag{10.90}$$

where the division and the product of two vectors are defined, respectively, as the division and the product component by component (see section 7.3). The iterate $\tilde{h}_k$ provides an approximation $\tilde{f}_k$ of the unknown object which is obtained by inverting the second of the relations (10.88), i.e. $\left(\tilde{f}_k\right)_n = \alpha_n^{-1} \left(\tilde{h}_k\right)_n$.

As a final remark we point out that also in this case, as in the other iterative methods, the basic operation in the implementation of the method is a matrix-vector multiplication, so that also this method can be very useful in cases where the matrix $\mathbf{A}$ is large and sparse.

Bayesian methods can be considered as a form of statistical regularization of the maximum likelihood methods. Indeed, the basic feature of these methods (where both the images and the objects are viewed as realizations of random processes) is the use of *a priori* information on the object in the form of its probability distribution. Then the joint probability distribution of the image and of the object is given by the product of the likelihood function and of the object probability distribution. Bayesian estimates can be obtained, for instance, by maximizing the joint probability distribution with respect to the object components, once the image is given. These points are discussed in section 7.4.

The determination of the Bayesian estimates is, in general, rather heavy from the computational point of view. Only in the case of Gaussian processes the treatment is rather elementary and was presented in section 7.5. In this case the analogy between Bayesian methods and regularization is especially clear. Here we only discuss the formal relationship with the Tikhonov regularization method.

As shown in section 7.5, the determination of the Bayesian estimate is equivalent to the minimization of the following functional

$$\Phi(f; g) = \left(\mathbf{S}_\nu^{-1}(\mathbf{A}f - g) \cdot (\mathbf{A}f - g)\right)_M + \left(\mathbf{S}_\phi^{-1} f \cdot f\right)_N \tag{10.91}$$

where $\mathbf{S}_\nu$ is the covariance matrix of the noise while $\mathbf{S}_\phi$ is the covariance matrix of the object. This functional is just the Tikhonov functional, defined in equation (10.31), if $\mathcal{X}$ and $\mathcal{Y}$ are Euclidean spaces equipped with weighted scalar products, the weighting matrices being respectively

$$\mathbf{B} = \mathbf{S}_\phi^{-1}, \quad \mathbf{C} = \mathbf{S}_\nu^{-1}. \tag{10.92}$$

As proved in appendix C, the adjoint of the matrix $\mathbf{A}$ in these weighted spaces is given by

$$\mathbf{A}_w^* = \mathbf{B}^{-1}\mathbf{A}^*\mathbf{C} = \mathbf{S}_\phi \mathbf{A}^* \mathbf{S}_\nu^{-1} \tag{10.93}$$

and therefore the standard expression (10.36)–(10.37) of the Tikhonov regularized solution (with $\mu = 1$) provides the following expression for the Bayesian estimate $\tilde{f}$

$$\begin{aligned} \tilde{f} &= \left(\mathbf{A}_w^*\mathbf{A} + \mathbf{I}\right)^{-1} \mathbf{A}_w^* g \\ &= \left(\mathbf{S}_\phi \mathbf{A}^* \mathbf{S}_\nu^{-1}\mathbf{A} + \mathbf{I}\right)^{-1} \mathbf{S}_\phi \mathbf{A}^* \mathbf{S}_\nu^{-1} g \\ &= \left(\mathbf{A}^* \mathbf{S}_\nu^{-1}\mathbf{A} + \mathbf{S}_\phi^{-1}\right)^{-1} \mathbf{A}^* \mathbf{S}_\nu^{-1} g \end{aligned} \tag{10.94}$$

which is precisely the expression derived in section 7.5.

References

[1] Groetsch C W 1993 *Inverse Problems in the Mathematical Sciences* (Stuttgart: Vieweg)

[2] Elden L 1997 *BIT* **17** 134

[3] Defrise M and De Mol C 1887 *Inverse Problems: An Interdisciplinary Study* ed P C Sabatier *Advances Electron. Elect. Physics Supplement 19* (London: Academic Press) 261

[4] Eicke B 1992 *Num. Funct. Anal. Optimiz.* **13** 413

[5] McCormick F and Rodriguez G H 1975 *J. Math. Anal. Appl.* **49** 275

[6] Gilyazov S F 1977 *Moscow Univ. Comput. Math. Cybernet.* **3** 78

Chapter 11

Fourier-based methods for specific problems

In this chapter we consider two problems where Fourier-based methods can be used in place of the singular value decomposition: tomographic imaging and super-resolution in image deconvolution. These methods are equivalent to the use of the SVD but, of course, are much faster from the computational point of view. The SVD can be used as a theoretical tool for investigating the resolution provided by these methods.

11.1 The Fourier slice theorem in tomography

In section 8.2 it has been shown that the basic problem in X-ray tomography is the inversion of the so-called Radon transform R, defined by

$$(Rf)(s, \boldsymbol{\theta}) = \int_{-\infty}^{+\infty} f(s\boldsymbol{\theta} + t\boldsymbol{\theta}^{\perp})dt. \tag{11.1}$$

If we denote by $g(s, \boldsymbol{\theta})$ the measured values of the Radon transform of an unknown function f, then the problem is to solve the linear equation

$$Rf = g. \tag{11.2}$$

In section 9.5 the singular system of R has been computed, assuming that the object f is a square-integrable function with support interior to a disc of radius 1, while g belongs to a suitable weighted space of square-integrable functions. From this result we deduce that the problem (11.2) is ill-posed because the singular values of R accumulate to zero. The singular system, however, is not very convenient in practice for computing the regularized solutions because, as observed in section 10.3, for a given noise level the number of singular values which must be used is very large. This fact, which is negative from the computational point of view, is positive from the point of view of the accuracy of the regularized solution, because it implies that the problem is mildly ill-posed. Therefore, it is very important that methods based on the Fourier transform can be used for obtaining approximate and stable solutions of equation (11.2). These methods lead to fast algorithms for Radon transform inversion.

The basic result is the so-called *Fourier slice theorem*, which can be formulated as follows: *let* $(R_{\boldsymbol{\theta}} f)(s) = (Rf)(s, \boldsymbol{\theta})$ *be the projection of* f *in the direction* $\boldsymbol{\theta}$*, then the Fourier transform of* $R_{\boldsymbol{\theta}} f$ *is the Fourier transform of* f *on the straight line passing through the origin and having direction* $\boldsymbol{\theta}$*, i.e. the following relation holds true*

$$(R_{\boldsymbol{\theta}} f)\hat{}(\omega) = \hat{f}(\omega\boldsymbol{\theta}). \tag{11.3}$$

We point out that the Fourier transform at the l.h.s. of equation (11.3) is one-dimensional while the Fourier transform at the r.h.s. is two-dimensional. Figure 11.1 provides an illustration of this result.

In order to prove equation (11.3) we insert the definition of $R_{\boldsymbol{\theta}} f$ in the expression of its Fourier transform. We get

$$\begin{aligned}(R_{\boldsymbol{\theta}} f)\hat{}(\omega) &= \int_{-\infty}^{+\infty} e^{-i\omega s}\,(R_{\boldsymbol{\theta}} f)(s)ds \\ &= \int_{-\infty}^{+\infty} e^{-i\omega s}\left(\int_{-\infty}^{+\infty} f(s\boldsymbol{\theta} + t\boldsymbol{\theta}^{\perp})dt\right)ds.\end{aligned} \tag{11.4}$$

The variables $\{s, t\}$ are the coordinates of a point $\boldsymbol{x} = \{x_1, x_2\}$ with respect to the Cartesian system defined by the unit vectors $\boldsymbol{\theta}, \boldsymbol{\theta}^{\perp}$, so that the relation between $\{s, t\}$ and $\{x_1, x_2\}$ is given by $\boldsymbol{x} = s\boldsymbol{\theta} + t\boldsymbol{\theta}^{\perp}$. Since this change of variables is a rotation, its Jacobian is one. If we further remark that $s = \boldsymbol{\theta} \cdot \boldsymbol{x}$, we see that equation (11.4) can be written as follows

$$(R_{\boldsymbol{\theta}} f)\hat{}(\omega) = \int f(\boldsymbol{x}) \exp(-i\omega\boldsymbol{\theta} \cdot \boldsymbol{x})d\boldsymbol{x} \tag{11.5}$$

and this is precisely the content of equation (11.3).

This very simple result has several important consequences:

- *The solution of equation (11.2), when it exists, is unique.* Indeed, if a function f is such that $Rf = 0$, this means that all projections of f are zero as well as their Fourier transforms. Fourier slice theorem implies that $\hat{f} = 0$, so that $f = 0$.
- *The singular functions* $v_{m,k}$ *of* R*, computed in section 9.5, form an orthonormal basis in the space of all square-integrable functions with support in the disc of radius 1.* Indeed, the singular functions in object space form, in general, an orthonormal basis in the orthogonal complement of the null space of R. Since this null space contains only the null element they are a basis in the object space.

- *The knowledge of the Radon transform of f implies the knowledge of the Fourier transform of f.* This point will be used in the next section. We only observe that, if the projections g are affected by noise, so that the usual model holds true

$$g = Rf^{(0)} + w, \tag{11.6}$$

then, by computing the Fourier transforms of $g(s, \boldsymbol{\theta})$, for any $\boldsymbol{\theta}$, we obtain a noisy version of the Fourier transform of $f^{(0)}$.

Before investigating the inversion method based on the Fourier slice theorem, we point out a few consequences of this result in the concrete case where one has only a finite set of projections.

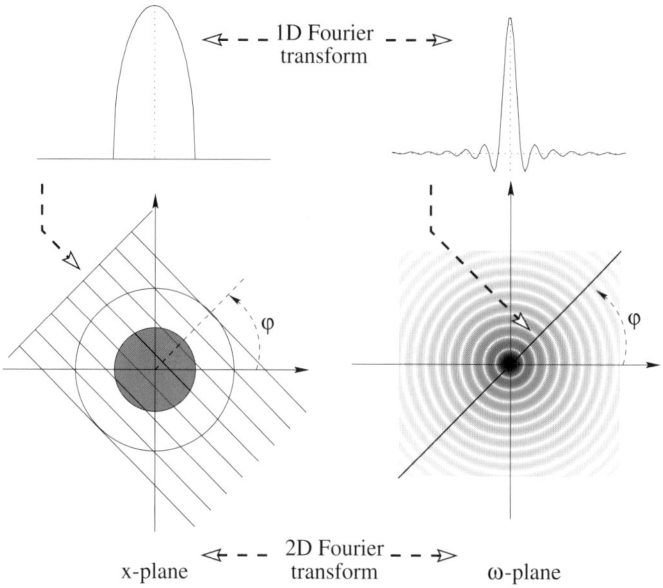

Figure 11.1. Illustrating the Fourier slice theorem in the case where the function f is the characteristic function of a disc.

First of all, in the applications of tomography the object (for instance the density distribution of a section of a human body) is typically spacelimited, i.e. is zero outside a bounded domain. As a consequence it is always possible to find a disc, with centre at the origin and radius a, such that the support of $f^{(0)}$ is interior to this disc. As we know, its Fourier transform $\hat{f}^{(0)}$ can be represented by means of the sampling expansions discussed in section 2.2. We can use, for instance, the square sampling lattice with cells of size π/a or the (more efficient) 120° rhombic sampling lattice with cells of size $2\pi/3a = 1.15\pi/a$

(see figure 2.4). The Fourier slice theorem, however, indicates that the data provide a different sampling of the Fourier transform of the unknown object.

Indeed, let us assume that we have p equispaced projections corresponding to angles in $[0, \pi)$, i.e. $\boldsymbol{\theta}_j = \{\cos(\phi_j), \sin(\phi_j\}$ with $\phi_j = \pi j/p$, $(j = 0, 1, \cdots, p-1)$. Since $f^{(0)}$ has support in the disc of radius a, each one of these projections has support in the interval $[-a, a]$. Therefore, if we denote by $g(s, \boldsymbol{\theta}_j)$ these projections, their Fourier transforms $\hat{g}(\omega, \theta_j)$ are analytic and can be represented by the usual Whittaker–Shannon theorem with sampling distance given by π/a. The uniform sampling of $\hat{g}(\omega, \boldsymbol{\theta}_j)$ provides a uniform sampling of $\hat{f}^{(0)}(\omega\boldsymbol{\theta}_j)$ (more precisely of its noisy version). If we do this sampling for each projection we obtain a sampling of $\hat{f}^{(0)}$ as that shown in figure 11.2.

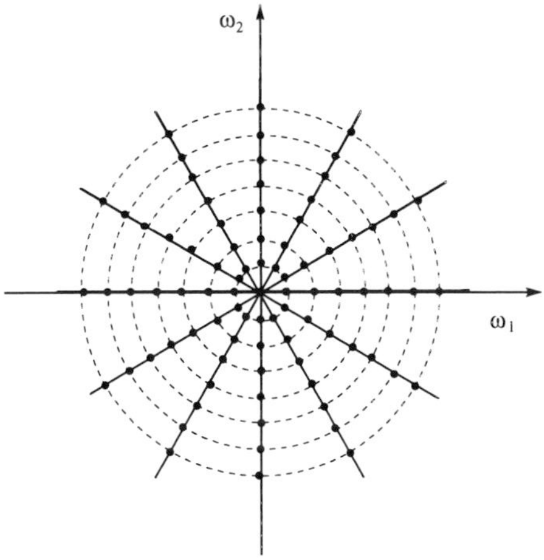

Figure 11.2. Sampling of the Fourier transform of the object obtained, through the Fourier slice theorem, from the sampling of six equispaced projections.

This sampling is not uniform, in the sense that the density of the sampling points per unit area is not constant. The sampling points are equispaced over circles with centres at the origin, the number of sampling points being constant over the circles. As a consequence the density of the sampling points is large in the vicinity of the origin, while it becomes smaller and smaller at higher frequencies. In other words we have an oversampling at low frequencies and an undersampling at high frequencies.

If we take the quantity π/a as the appropriate sampling distance for $\hat{f}^{(0)}(\boldsymbol{\omega})$, it is possible to estimate by simple arguments ([1], chapter 5) the radius of the disc in the ω-plane where $\hat{f}^{(0)}(\boldsymbol{\omega})$ is sufficiently sampled when p equispaced projections are given. To this purpose, let us remark that if we consider a circle of radius Ω in the ω-plane, this circle intersects the straight lines with directions

θ_j at points whose distance is given approximately by $\Delta = \Omega\pi/p$. Therefore $\hat{f}^{(0)}$ is correctly sampled within this disc if this distance is of the order of π/a, i.e. if

$$\Omega = \frac{p}{a}. \tag{11.7}$$

If it is correctly recovered within this disk, we obtain an Ω-bandlimited approximation of $f^{(0)}$. This also means (this point will be further discussed in section 11.3) that is is possible to restore $f^{(0)}$ with a resolution

$$\delta \cong \frac{\pi}{\Omega} = \frac{\pi a}{p}. \tag{11.8}$$

This formula illustrates a basic fact in tomography: *the resolution distance is inversely proportional to the number of projections.*

11.2 The filtered backprojection (FBP) method in tomography

The Fourier slice theorem implies that, if we have all projections of f, then we have also the Fourier transform of f and therefore the solution of equation (11.2) can be obtained by means of Fourier transform inversion. This inversion, however, is not completely trivial because the Fourier transform of f, as given by the Fourier transforms of the projections, is represented in polar coordinates. In the case of a finite number of projections this fact has the consequences discussed in the previous sections. In addition we observe that, since $\hat{f}(\omega)$ is not sampled over a Cartesian lattice, the DFT cannot be used straightforwardly for computing f.

In order to prepare the tools required for the inversion formula, we introduce the *backprojection operator* $R^{\#}$, defined as follows: let $g(s, \theta)$ be a function defined for $-\infty < s < \infty$, $0 \leq \phi < 2\pi$, then

$$\left(R^{\#}g\right)(\boldsymbol{x}) = \int_0^{2\pi} g(\boldsymbol{x} \cdot \boldsymbol{\theta}, \boldsymbol{\theta}) d\phi. \tag{11.9}$$

Therefore $R^{\#}$ transforms the data function (represented by the sinogram) into a function of the space variables $\boldsymbol{x}$, i.e. an element of the object space.

Remark 11.1. The notation $R^{\#}$ derives from the fact that this operator can be formally viewed as the adjoint of the operator R when we consider the usual L^2 scalar product both for the data functions and the objects. Indeed, we have

$$\begin{aligned}(Rf, g)_{\mathcal{Y}} &= \int_0^{2\pi} \left(\int_{-\infty}^{+\infty} (Rf)(s, \boldsymbol{\theta}) g(s, \boldsymbol{\theta}) ds\right) d\phi \qquad (11.10)\\ &= \int_0^{2\pi} \left\{\int_{-\infty}^{+\infty} \left(\int_{-\infty}^{+\infty} f(s\boldsymbol{\theta} + t\boldsymbol{\theta}^{\perp}) dt\right) g(s, \boldsymbol{\theta}) ds\right\} d\phi,\end{aligned}$$

and if we replace, for a given $\boldsymbol{\theta}$, the variables $\{s, t\}$ with the variables $\{x_1, x_2\}$ defined by $\boldsymbol{x} = s\boldsymbol{\theta} + t\boldsymbol{\theta}^{\perp}$, we obtain

$$
\begin{aligned}
(Rf, g)_{\mathcal{Y}} &= \int_0^{2\pi} \left(\int f(\boldsymbol{x}) g(\boldsymbol{x} \cdot \boldsymbol{\theta}, \boldsymbol{\theta}) d\boldsymbol{x} \right) d\phi \qquad (11.11) \\
&= \int f(\boldsymbol{x}) \left(\int_0^{2\pi} g(\boldsymbol{x} \cdot \boldsymbol{\theta}, \boldsymbol{\theta}) d\phi \right) d\boldsymbol{x} = \left(f, R^{\#} g\right)_{\mathcal{X}} .
\end{aligned}
$$

In the last step an exchange of the integration order has been performed.

It is interesting to point out that, while the Radon transform integrates over all points in a straight line, the backprojection operator integrates over all straight lines through a point. Indeed, the integral of equation (11.9) is the integral of $g(s, \boldsymbol{\theta})$ along the curve $s = \boldsymbol{x} \cdot \boldsymbol{\theta} = x_1 \cos\phi + x_2 \sin\phi$. For a given $\boldsymbol{x}$, the points of this curve correspond to the straight lines through the point $\boldsymbol{x}$.

Let $g(s, \boldsymbol{\theta})$ be the Radon transform of a function $f(\boldsymbol{x})$, $g = Rf$. If $f(\boldsymbol{x}_1)$ is greater (smaller) than $f(\boldsymbol{x}_2)$, with $\boldsymbol{x}_1$ close to $\boldsymbol{x}_2$, then the integrals over the straight lines through $\boldsymbol{x}_1$ are, in general, greater (smaller) than the integrals over the straight lines through $\boldsymbol{x}_2$. It follows that $\left(R^{\#} g\right)(\boldsymbol{x}_1)$ is also greater (smaller) than $\left(R^{\#} g\right)(\boldsymbol{x}_2)$. This remark suggest that $R^{\#} g = R^{\#} R f$ can provide a blurred image of f. More precisely, it is possible to prove ([2], chapter 2) that

$$
\left(R^{\#} R f\right)(\boldsymbol{x}) = 2 \int \frac{f(\boldsymbol{x}')}{|\boldsymbol{x} - \boldsymbol{x}'|} d\boldsymbol{x}'. \qquad (11.12)
$$

This result implies the possibility of using the methods of image deconvolution for estimating the unknown object $f^{(0)}$. Filtered backprojection, however provides a more elegant and practical solution of this problem.

In order to derive this algorithm let us first derive a suitable expression for the inverse Fourier transform in polar coordinates, $\omega = |\boldsymbol{\omega}|$, $\phi = \arctan(\omega_y / \omega_x)$, $\boldsymbol{\theta} = \{\cos\phi, \sin\phi\}$, which is given by

$$
f(\boldsymbol{x}) = \frac{1}{(2\pi)^2} \int_0^{2\pi} \left(\int_0^{+\infty} \omega \hat{f}(\omega\boldsymbol{\theta}) \exp(i\omega\boldsymbol{\theta} \cdot \boldsymbol{x}) d\omega \right) d\phi. \qquad (11.13)
$$

If we observe that $\hat{f}(\omega\boldsymbol{\theta})$ is a periodic function of ϕ, with period 2π, by replacing ϕ with $\phi + \pi$, i.e. $\boldsymbol{\theta}$ with $-\boldsymbol{\theta}$, we obtain

$$
f(\boldsymbol{x}) = \frac{1}{(2\pi)^2} \int_0^{2\pi} \left(\int_0^{+\infty} \omega \hat{f}(-\omega\boldsymbol{\theta}) \exp(-i\omega\boldsymbol{\theta} \cdot \boldsymbol{x}) d\omega \right) d\phi. \qquad (11.14)
$$

From this expression, by means of a further change of variables which consists of replacing ω with $-\omega$, we get

$$
f(\boldsymbol{x}) = \frac{1}{(2\pi)^2} \int_0^{2\pi} \left(\int_{-\infty}^{0} |\omega| \hat{f}(\omega\boldsymbol{\theta}) \exp(i\omega\boldsymbol{\theta} \cdot \boldsymbol{x}) d\omega \right) d\phi. \qquad (11.15)
$$

By adding equation (11.13) and equation (11.15), we obtain the desired representation of $f(\boldsymbol{x})$

$$f(\boldsymbol{x}) = \frac{1}{2(2\pi)^2} \int_0^{2\pi} \left(\int_{-\infty}^{+\infty} |\omega| \hat{f}(\omega\boldsymbol{\theta}) \exp(i\omega\boldsymbol{\theta} \cdot \boldsymbol{x}) d\omega \right) d\phi. \tag{11.16}$$

Let us assume now that f is a solution of equation (11.2). From the Fourier slice theorem, equation (11.3), it follows that, if we denote by $\hat{g}(\omega, \boldsymbol{\theta})$ the Fourier transforms of the measured projections, then $\hat{f}(\omega\boldsymbol{\theta}) = \hat{g}(\omega, \boldsymbol{\theta})$. By inserting this relation in equation (11.16) we get

$$f(\boldsymbol{x}) = \frac{1}{2(2\pi)^2} \int_0^{2\pi} \left(\int_{-\infty}^{+\infty} |\omega| \hat{g}(\omega, \boldsymbol{\theta}) \exp(i\omega\boldsymbol{\theta} \cdot \boldsymbol{x}) d\omega \right) d\phi \tag{11.17}$$

and this formula is the key of the *filtered backprojection* (FBP) algorithm.

Indeed, if we define filtered projections $G(s, \boldsymbol{\theta})$ as follows

$$G(s, \boldsymbol{\theta}) = \frac{1}{2\pi} \int_{-\infty}^{+\infty} |\omega| \hat{g}(\omega, \boldsymbol{\theta}) \exp(i\omega s) d\omega, \tag{11.18}$$

then equation (11.17) can be written in the following way

$$f = \frac{1}{4\pi} R^{\#} G. \tag{11.19}$$

We find that f is obtained by applying the backprojection operator to the filtered projections, defined in equation (11.18).

In conclusion, the inversion procedure can be sectioned in the following steps:

- for each value of $\boldsymbol{\theta}$ compute the Fourier transform $\hat{g}(\omega, \boldsymbol{\theta})$ of $g(s, \boldsymbol{\theta})$
- multiply $\hat{g}(\omega, \boldsymbol{\theta})$ by the ramp filter $|\omega|$
- compute the inverse Fourier transform of $|\omega|\hat{g}(\omega, \boldsymbol{\theta})$ to obtain the filtered projections $G(s, \boldsymbol{\theta})$
- apply the backprojection operator to $G(s, \boldsymbol{\theta})$.

Figure 11.3 illustrates this scheme. In this figure we also show the difference between the original sinogram and the filtered one as well as the difference between the reconstruction provided by FBP and that obtained by applying the backprojection operator to the original sinogram.

The filtered backprojection algorithm provides a further insight into the numerical instability of tomography. The ill-posedness of the problem was already established by means of the computation of the singular values in section 9.5, because it was found that the singular values of the Radon transform tend to zero. As a consequence, the Picard criterion, equation (10.26), is not satisfied by arbitrary data functions. This criterion is now replaced by the condition that

$$\int_{-\infty}^{+\infty} |\omega|^2 |\hat{g}(\omega, \boldsymbol{\theta})|^2 d\omega < \infty \tag{11.20}$$

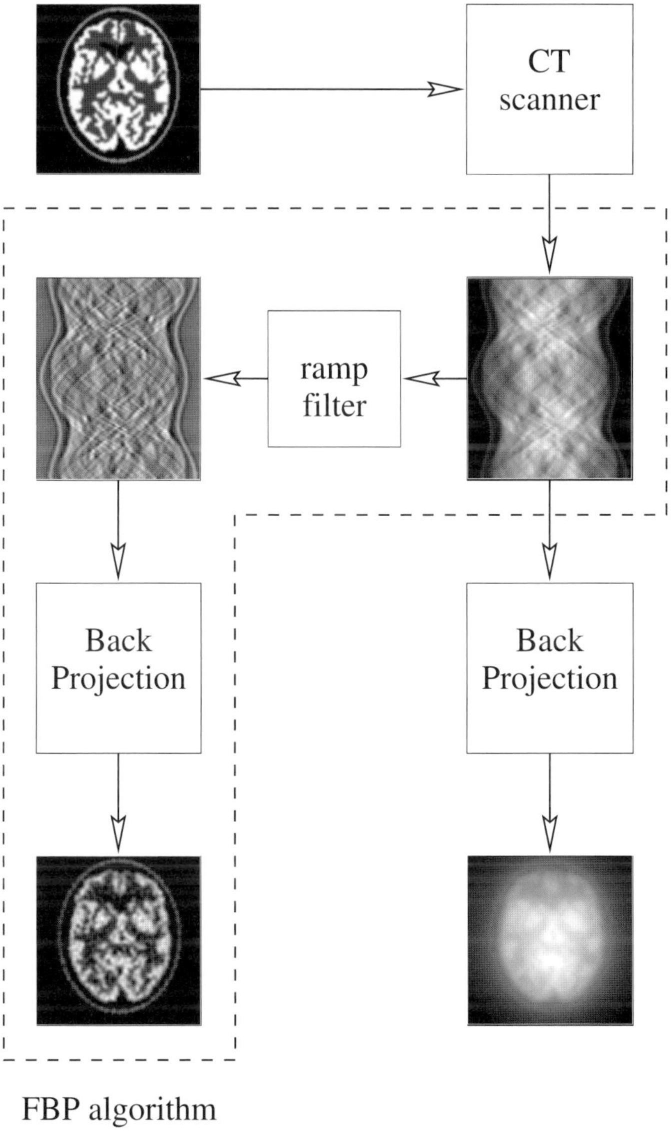

Figure 11.3. Scheme of the FBP algorithm and comparison of the results obtained by applying the backprojection operator to the original sinogram (to the right) and to the filtered one (to the left)

which also is not satisfied by arbitrary square integrable functions. In other words the ramp filter can amplify the high frequency noise in such a way that the restored function is not acceptable from the physical point of view.

In order to avoid this effect it is necessary to regularize the second step of the FBP algorithm or, in other words, to introduce a low-pass filter $\hat{W}_{\Omega}(|\omega|)$ characterized by a cut-off frequency Ω. Then the second step must be modified as follows:

- multiply $\hat{g}(\omega, \boldsymbol{\theta})$ by the ramp filter $|\omega|$ and a suitable low-pass filter $W_{\Omega}(|\omega|)$.

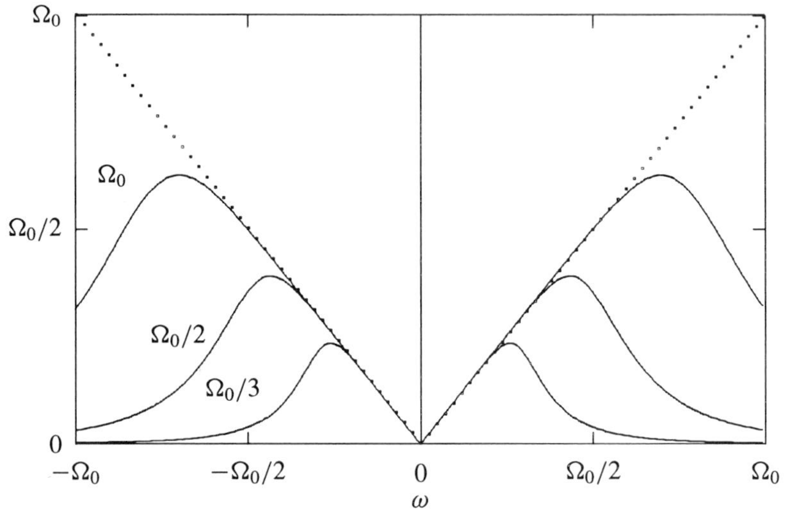

Figure 11.4. Picture of the filter $F_{\Omega}(|\omega|) = |\omega| W_{\Omega}(|\omega|)$ on an interval $[-\Omega_0, \Omega_0]$, when $W_{\Omega}(|\omega|)$ is a Butterworth filter, equation (5.59), of order $n = 10$, for various values of Ω (given in terms of Ω_0). The dotted line represents the pure ramp filter.

It follows that the global filter $F_{\Omega}(|\omega|) = |\omega| W_{\Omega}(|\omega|)$ has in general, a behaviour similar to that plotted in figure 11.4.

The filtered projections, including the low pass filter, are now given by

$$G_{\Omega}(s, \boldsymbol{\theta}) = \frac{1}{2\pi} \int_{-\infty}^{+\infty} |\omega| W_{\Omega}(|\omega|) \hat{g}(\omega, \boldsymbol{\theta}) e^{-i\omega s} d\omega \tag{11.21}$$

and the backprojection operator must be applied to these filtered projections.

If we denote by f_{Ω} the approximation of the unknown object $f^{(0)}$ which can be obtained in this way

$$f_{\Omega} = \frac{1}{4\pi} R^{\#} G_{\Omega} \tag{11.22}$$

and if we assume the usual model for the noisy data, i.e. $g = Rf^{(0)} + w$, then it is easy to show that

$$f_\Omega(\boldsymbol{x}) = \frac{1}{(2\pi)^2}\int W_\Omega(|\boldsymbol{\omega}|)\hat{f}^{(0)}(\boldsymbol{\omega})e^{i\boldsymbol{x}\cdot\boldsymbol{\omega}}d\boldsymbol{\omega} + \frac{1}{4\pi}\left(R^{\#}w_\Omega\right)(\boldsymbol{x}) \qquad (11.23)$$

where $w_\Omega(s, \boldsymbol{\theta})$ is the noise contribution to $G_\Omega(s, \boldsymbol{\theta})$. Therefore the result is a noisy and bandlimited approximation of $f^{(0)}(\boldsymbol{x})$.

11.3 Implementation of the discrete FBP

The implementation of the FBP algorithm seems to be obvious as concerns the first three steps: replace the Fourier transform with the discrete Fourier transform of the discretized projections and use the FFT algorithm. However, a few points must be considered carefully. The first point is that, if we multiply the DFT of a discrete projection by the ramp filter and then we compute the inverse DFT of the result, we are in fact computing a cyclic convolution and this is affected by the periodicity effects discussed in section 2.6. The second point is that, if we put the ramp filter equal to zero at zero frequency, we are in fact putting the filter equal to zero in a neighbourhood of $\omega = 0$, whose size coincides with the sampling distance in frequency domain. As a result, the zero frequency component of the object is not correctly estimated so that the restored object is smaller than the original one, because its integral has been understimated.

In order to clarify these points, let us assume again that the support of the unknown object $f^{(0)}$ is interior to the disc of radius a, so that the support of all projections is interior to the interval $[-a, a]$. Let $\boldsymbol{\theta}_j (j = 0, 1, \cdots, p-1)$ be the directions of the projections and let $s_m (m = 0, 1, \cdots, M-1)$ be the sampling points of each projection, given by

$$s_m = -a + m\delta_s; \quad \delta_s = \frac{2a}{M}. \qquad (11.24)$$

Then, as discussed in section 2.6, this discretization implies a frequency cut-off Ω and a sampling distance δ_ω in frequency domain given by

$$\Omega = \frac{\pi}{\delta_s} = \frac{\pi M}{2a}, \quad \delta_\omega = \frac{\pi}{a}. \qquad (11.25)$$

If we put $g_{k,m} = g(s_m, \boldsymbol{\theta}_k)$, then the DFT $\hat{g}_{k,l}$ of $g_{k,m}$, with fixed k, must be multiplied by $\omega_l = (\pi l/a)$ for $l = 0, 1, \cdots, M/2$ and by $\omega_l = (M-l)\pi/a$ for $l = M/2+1, \cdots, M-1$ (notice that $\omega_{M/2} = \Omega$). This is the apparently obvious implementation of the second step of the FBP algorithm (multiplication by the ramp filter).

In order to understand the result of this procedure let us come back to functions of continuous variables, and let us denote by $H_\Omega(s)$ the PSF associated with the ramp filter truncated to the band $[-\Omega, \Omega]$ (which is also called the Ramachandran–Lakshminarayanan filter [3])

$$H_\Omega(s) = \frac{1}{2\pi}\int_{-\Omega}^{\Omega} |\omega| e^{-is\omega} d\omega \tag{11.26}$$
$$= \frac{\Omega^2}{2\pi}\left\{2\mathrm{sinc}\left(\Omega\frac{s}{\pi}\right) - \mathrm{sinc}^2\left(\Omega\frac{s}{2\pi}\right)\right\}.$$

If we multiply the Fourier transform of a projection by the truncated ramp filter (this is equivalent to take in equation (11.21) $W_\Omega(|\omega|)$ as a perfect low-pass filter) we obtain the convolution of the projection with the PSF $H_\Omega(s)$, i.e.

$$\tilde{G}_\Omega(s, \theta_k) = \int_{-\infty}^{+\infty} H_\Omega(s - s')g(s', \theta_k)ds'. \tag{11.27}$$

This is the filtered projection we should compute. However, the procedure outlined above in the discrete case does not compute approximate values of this convolution product but the cyclic convolution of M samples of $g(s, \theta_k)$ with M approximate samples of $H_\Omega(s)$ (obtained from the inverse DFT of the discrete ramp filter). The result of this computation can be completely unsatisfactory.

In order to get approximate values of (11.27) we can use, as shown in section 2.6, the zero padding, i.e. we can add $M/2$ zeros to the right and to the left of the samples $g_{k,m}$ in order to get a vector of length $2M$. This procedure is equivalent to doubling the interval, i.e. to considering the interval $[-2a, 2a]$, without changing the sampling distance δ_s. As a consequence the cut-off frequency Ω does not change while the sampling distance in frequency domain is divided by 2.

This approximation, however, may not yet be satisfactory because it has been recognized that, if one applies the backprojection operator to these filtered projections, one does not obtain the correct zero frequency component (DC component) of the object. To this purpose the following procedure has been proposed: sample both the projection and the PSF of the filter (11.26) on the interval $[-2a, 2a]$ and then compute their cyclic convolution. The main difference with respect to the previous method is that the DFT of $H_{\Omega,m} = H_\Omega(s_m)$ is not zero at zero frequency. It is also interesting to remark that the values of $H_\Omega(s)$ at the sampling points of the projections have a very simple expression:

$$H_\Omega(m\delta_s) = \begin{cases} 0 & m \text{ even}; m \neq 0 \\ \Omega^2/2\pi & m = 0 \\ -2\Omega^2/\pi^3 m^2 & m \text{ odd}. \end{cases} \tag{11.28}$$

This property can be used for a direct implementation of the convolution product, possibly based on specially designed hardware.

The filtered projections obtained by means of one of the previous methods are characterized by a frequency cut-off which is determined by the sampling distance of the projections. According to the previous discussion of tomography two other kinds of cut-off must be considered and precisely:

- cut-off required for suppressing the effects due to insufficient sampling of the Fourier transform of the solution at high frequency, when the set of projections is finite
- cut-off required for suppressing noise propagation.

As shown in section 11.1, the first cut-off is given by equation (11.7). Therefore one can try to get this cut-off by an appropriate sampling of the projections, i.e. by choosing M in such a way that the cut-off (11.7) and the cut-off (11.25) coincide. In this way we obtain the relation

$$\frac{p}{M} = \frac{\pi}{2}. \tag{11.29}$$

In conclusion: *the number of sampling points for each projection must be approximately equal to the number of projections.*

A valuable discussion of the aliasing artifacts due to the undersampling of the projections and/or to the finite set of projections is given in the book of Kak and Slaney [1].

As concerns noise propagation we already observed that one can control its effect by a suitable low-pass filtering (using for instance the Butterworth filter). To this purpose a widely used filter is that proposed by Shepp and Logan [3], which consists of replacing the ramp-filter by the following one

$$\hat{S}_{\Omega}(\omega) = \frac{2\Omega}{\pi}\left|\sin\left(\frac{\pi|\omega|}{2\Omega}\right)\right|, \quad |\omega| \leq \Omega. \tag{11.30}$$

In practice, the value of Ω to be used in this filter is determined by condition (11.29). The values of the PSF of this filter at the sampling points of the projections have also a very simple expression

$$S_{\Omega}(m\delta_s) = -\frac{4}{\pi^3}\frac{\Omega^2}{4m^2-1} \tag{11.31}$$

and, therefore, also in this case a direct implementation of the convolution product, possibly based on specially designed hardware, can be used. The difference between the Shepp–Logan and the Ramachandran–Lakshminarayanan filter is shown in figure 11.5. As we can see, the Shepp–Logan filter is smaller at high frequencies, thus reducing the effect of noise propagation.

Finally, as concerns the backprojection, i.e. the last step of the FBP algorithm, its discretization can be performed in several ways [4]. In principle, it can always be expressed in terms of a matrix. If we indicate by $\{k, m\}$ the indices

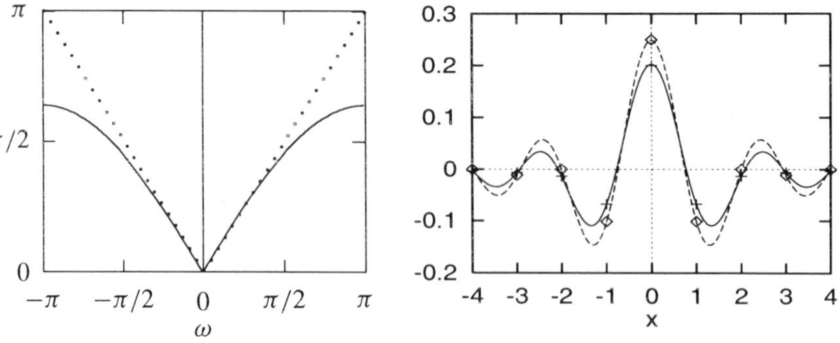

Figure 11.5. Plot of the Ramachandran–Lakshminarayanan (dotted line) and Shepp–Logan (full line) filters (left panel). In the right panel the corresponding PSF are shown with indication of the sampling points.

characterizing the bins of the projections and by $\{i, j\}$ the indices characterizing the pixels in the object domain, we can write

$$f_{i,j} = \sum_{k,m} R^{\#}_{i,j;k,m} g_{k,m}. \tag{11.32}$$

The most simple approximation consists of taking the matrix element equal to 1 if the straight line corresponding to the centre of the bin $\{k, m\}$ crosses the pixel $\{i, j\}$ and zero elsewhere.

11.4 Resolution and super-resolution in image restoration

Filtered backprojection provides a bandlimited approximation of the unknown object. If we look at the structure of the FBP solution, as given, for instance, in equation (11.23), we see that its Fourier transform is a reliable estimate of the Fourier transform of the object inside a disc with a certain radius Ω while it is negligible outside the same disc. In the case of a very large number of projections, the value of Ω depends on the noise level affecting the measured data while, when the number p of projections is not too large, the value of Ω depends essentially on p and is given approximately by equation (11.7).

Similar remarks apply to the linear regularization methods for image deconvolution, at least when the transfer function $\hat{K}(\omega)$ has circular symmetry and does not have zeros. Indeed, in such a case, the window functions $\hat{W}(\omega)$ of these methods, as defined in section 5.5, may be approximately one inside a disc with a certain radius Ω and negligible outside the same disc. The estimate of Ω depends mainly on the transfer function and on the noise affecting the data. In order to be more concrete we consider a specific example.

Nearfield acoustic holography, in the case of planar surfaces (see section 3.6), is described by the PSF given in equation (3.54), with the corresponding TF given by equations (3.55) and (3.56). In this example the TF has circular symmetry and is never zero.

The unknown field amplitude on the boundary plane $x_3 = 0$, can be estimated by means of one of the basic regularization methods such as Tikhonov or Landweber. Since all methods provide, in general, quite similar results (see the discussion in the next chapter), we can take the Tikhonov method as their representative and discuss its application to the problem under consideration.

Assume that the regularization parameter has been determined using, for instance, the Miller method, equation (5.77) (this choice also corresponds to the Wiener filter method discussed in section 7.5). If we denote by $\tilde{f}_0$ the corresponding regularized solution, then the relationship between $\tilde{f}_0$ and the unknown object $f^{(0)}(\boldsymbol{x})$ is given by (see section 5.5)

$$\widehat{\tilde{f}}_0(\boldsymbol{\omega}) = \hat{W}_0(|\boldsymbol{\omega}|)\hat{f}^{(0)}(\boldsymbol{\omega}) + \hat{W}_0(|\boldsymbol{\omega}|)\frac{\hat{w}(\boldsymbol{\omega})}{\hat{K}(\boldsymbol{\omega})} \tag{11.33}$$

where the second term is the noise contribution to $\tilde{f}_0$ and $\hat{W}_0(|\boldsymbol{\omega}|)$ is the Tikhonov window function given by

$$\hat{W}_0(|\boldsymbol{\omega}|) = \frac{|\hat{K}(|\boldsymbol{\omega}|)|^2}{|\hat{K}(|\boldsymbol{\omega}|)|^2 + (\varepsilon/E)^2}. \tag{11.34}$$

Let Ω be defined by

$$\hat{W}_0(\Omega) = \frac{1}{2}. \tag{11.35}$$

It follows that $\hat{W}_0(|\boldsymbol{\omega}|)$ is close to 1 for $|\boldsymbol{\omega}| < \Omega$ (remember that $|\hat{K}(\boldsymbol{\omega})|$ is exactly 1 for $|\boldsymbol{\omega}| < k = 2\pi/\lambda$) and negligible for $|\boldsymbol{\omega}| > \Omega$. Therefore, if we neglect the noise term in equation (11.33) we see that $\widehat{\tilde{f}}_0(\boldsymbol{\omega})$ is approximately equal to $\hat{f}^{(0)}(\boldsymbol{\omega})$ for $|\boldsymbol{\omega}| < \Omega$ and approximately zero for $|\boldsymbol{\omega}| > \Omega$. We conclude that $\tilde{f}_0(\boldsymbol{x})$ provides, in practice, an Ω-bandlimited approximation of $f_0(\boldsymbol{x})$.

For the problem under consideration it is easy to compute Ω. If we observe that condition (11.35) is equivalent to the following one

$$|\hat{K}(\Omega)| = \frac{\varepsilon}{E}, \tag{11.36}$$

from equations (3.55) and (3.56) we get

$$\Omega = k\left[1 + \frac{1}{(ka)^2}\ln^2\left(\frac{E}{\varepsilon}\right)\right]^{1/2}. \tag{11.37}$$

We point out that:

- Ω depends on the PSF through the distance a between the two planes (in particular $\Omega \cong k$ when a is very large);

- Ω depends on the noise through the quantity E/ε.

A similar analysis applies to most problems in image restoration. When the PSF does not have circular symmetry, the region, where the window function is close to one, in general is not a disc in the frequency domain. But it may contain a disc with a certain radius Ω so that, by a further filtering, we can obtain a window function which is negligible when $|\omega| > \Omega$. In conclusion, we always obtain a (noisy) bandlimited approximation of $f^{(0)}(\boldsymbol{x})$.

If the FT of $\tilde{f}_0$ is exactly zero for $|\omega| > \Omega$ and exactly coincides with the FT of $f^{(0)}$ for $|\omega| < \Omega$, then the relationship between $\tilde{f}_0$ and $f^{(0)}$ is that provided by an ideal imaging system with a circular pupil, as discussed in section 3.4. It follows that

$$\tilde{f}_0(\boldsymbol{x}) = \frac{\Omega}{2\pi} \int \frac{J_1(\Omega|\boldsymbol{x} - \boldsymbol{x}'|)}{|\boldsymbol{x} - \boldsymbol{x}'|} f^{(0)}(\boldsymbol{x}')d\boldsymbol{x}'. \tag{11.38}$$

Obviously the exact relationship between $\tilde{f}_0$ and $f^{(0)}$ is not this one since it is given by equation (11.33). This approximate relationship, however, allows one to understand what features of $f^{(0)}$ can be recovered in $\tilde{f}_0$. In particular it allows one to apply the Rayleigh criterion, briefly discussed in section 3.4, to this particular problem: two narrow peaks in the objects $f^{(0)}$ cannot be detected in the restored image $\tilde{f}_0$ if their distance is smaller than the quantity $\delta = 1.22\pi/\Omega$. A similar conclusion is reached by observing that $\tilde{f}_0$ can be represented by means of sampling expansions (section 2.2). If we use a square lattice, then the size of the cells is $\delta = \pi/\Omega$ while, if we use a 120° rhombic lattice, then the size of the cells is $\delta = 2\pi/\sqrt{3}\Omega = 1.15\pi/\Omega$. Any one of these values of δ can be used as a measure of the resolution achievable in the restoration of $f^{(0)}$ by means of $\tilde{f}_0$.

The conclusion of the previous analysis is that the Tikhonov regularization method provides an approximately bandlimited estimate of the unknown object, characterized by a bandwidth Ω, and that the quality of this estimate can be described by a finite resolution distance δ, inversely proportional to the bandwidth Ω.

As discussed in part I, however, for specific problems one can use additional constraints in the restoration of the object, such as positivity, etc. Then we say that a particular inversion method is *super-resolving* (or also, that it provides *super-resolution*) with respect to the Tikhonov method, if the restored images provided by this method are characterized by a resolution which is better than that estimated by means of the Tikhonov method.

Remark 11.2. In the theory of optical instruments and wave propagation the term super-resolution may have a different meaning.

As discussed in section 3.4 the resolution achievable with an optical instrument is of the order of $\lambda/2$, where λ is the wavelength of the light. The same resolution is achieved in inverse diffraction from far-field data. Indeed, in such a case equation (11.37) gives $\Omega = k = 2\pi/\lambda$ and therefore $\delta = 1.22\lambda/2$.

Since the resolution limit $\lambda/2$ is considered as the natural one, one says that in a particular imaging method, related to wave-propagation phenomena, super-resolution is achieved if a resolution better or much better than $\lambda/2$ is obtained. In this sense inverse diffraction from near field data is super-resolving because, as follows from equation (11.37), one can obtain a resolution much better than $\lambda/2$ if $a << \lambda$. This fact has applications in near-field acoustic holography (NAH) and scanning near-field optical microscopy (see section 3.6), which are considered super-resolving techniques.

An inversion method which, applied to far-field data, provides a resolution better than $\lambda/2$, is super-resolving both in this sense and in our sense.

Since the resolution distance δ is inversely proportional to the bandwidth Ω, *a super-resolving inversion method is a method which recovers the Fourier transform of the object outside the band Ω provided by the Tikhonov method.* In other words, super-resolution is a synonym of out-of-band extrapolation [5].

To recover the Fourier transform of the object outside the effective band provided by the usual regularization methods one needs additional information about the object. Indeed, out-of-band extrapolation is feasible, at least in principle, if the Fourier transform of the object is analytic and this condition is satisfied when the object has a bounded support.

At first sight this condition seems to be always satisfied because all objects are spacelimited. However we will show that, since the problem of analytic continuation is ill-posed, a much stronger condition is required for obtaining a significant out-of band extrapolation: the diameter of the support of the object must be of the order of the resolution distance. In a wave propagation problem with far-field data, this means that the linear dimensions of the object must be of the order of the wavelength λ of the radiation.

Remark 11.3. In tomography the situation is a little different because the projections of a spacelimited object are also spacelimited and therefore are not bandlimited. As shown in section 11.1, a bandlimiting of the solution is needed in the case of a finite set of projections and this can be obtained using a number of sampling points in the interval $[-a, a]$ approximately equal to the number p of projections (see equation (11.29)). The band-width is given by $\Omega = p/a$ and therefore the resolution δ is proportional to a/p. If the support of the object is contained in a smaller disc of radius $a' < a$, then we can restrict the projections to the interval $[-a', a']$ and obtain a bandwidth $\Omega' = p/a' > \Omega$ using about p sampling points in $[-a', a']$. This procedure, however, is not a super-resolving method with respect the previous one, because different data are used in the two cases.

11.5 Out-of-band extrapolation

In this section we formulate the problem of out-of-band extrapolation in terms of integral equations and we discuss the main features of this problem, indicating when out-of-band extrapolation, and therefore super-resolution, is feasible. We consider mainly the 2D case, which is the most important one in imaging, even if more quantitative results can be obtained mainly in the 1D case.

Let $f(\boldsymbol{x})$ be a function whose support is a bounded domain $\mathcal{D}$. We assume that this domain is sufficiently regular and has a finite measure (area) $m(\mathcal{D})$. As already remarked in section 2.2, the Fourier transform $\hat{f}(\boldsymbol{\omega})$ is an entire analytic function. Moreover, let us suppose that $\hat{f}(\boldsymbol{\omega})$ is known over a bounded domain $\mathcal{B}$ in frequency space and that $\mathcal{B}$ has also a finite measure (area) $m(\mathcal{B})$. Then, the problem of out-of-band extrapolation is the problem of determining $f(\boldsymbol{x})$ from the limited knowledge of its Fourier transform in $\mathcal{B}$.

The solution of this problem, when it exists, is unique. Indeed, if we have two solutions f_1, f_2, the Fourier transform of their difference, $f_1 - f_2$, is an entire analytic function which is zero over $\mathcal{B}$. Since an analytic function which is zero over a set with non-zero measure is zero everywhere, then $f_1 = f_2$.

An equivalent formulation of the problem of out-of-band extrapolation is the following: determine the function $f(\boldsymbol{x})$, being given a function $g(\boldsymbol{x})$ whose Fourier transform coincides with the Fourier transform of $f(\boldsymbol{x})$ on $\mathcal{B}$ and is zero outside $\mathcal{B}$, i.e.

$$\hat{g}(\boldsymbol{\omega}) = \chi_{\mathcal{B}}(\boldsymbol{\omega})\hat{f}(\boldsymbol{\omega}) \tag{11.39}$$

where $\chi_{\mathcal{B}}(\boldsymbol{\omega})$ is the characteristic function of $\mathcal{B}$.

If we denote by $H_{\mathcal{B}}(\boldsymbol{x})$ the inverse Fourier transform of $\chi_{\mathcal{B}}(\boldsymbol{\omega})$

$$H_{\mathcal{B}}(\boldsymbol{x}) = \frac{1}{(2\pi)^2}\int \chi_{\mathcal{B}}(\boldsymbol{\omega})e^{i\boldsymbol{x}\cdot\boldsymbol{\omega}}d\boldsymbol{\omega} \tag{11.40}$$

then, by means of the convolution theorem, from equation (11.39) we obtain

$$g(\boldsymbol{x}) = \int_{\mathcal{D}} H_{\mathcal{B}}(\boldsymbol{x}-\boldsymbol{x}')f(\boldsymbol{x}')d\boldsymbol{x}', \tag{11.41}$$

where we have used *a priori* knowledge about the support of $f(\boldsymbol{x})$ by restricting the integral to the domain $\mathcal{D}$.

This is a Fredholm integral equation of the first kind, corresponding to the integral operator

$$(Af)(\boldsymbol{x}) = \int_{\mathcal{D}} H_{\mathcal{B}}(\boldsymbol{x}-\boldsymbol{x}')f(\boldsymbol{x}')d\boldsymbol{x}' \tag{11.42}$$

which transforms functions defined over $\mathcal{D}$ into functions defined everywhere. If we consider, as usual, square integrable functions, then A is an operator from $L^2(\mathcal{D})$ into $L^2(\mathrm{R}^2)$.

The operator (11.42) is a particular case of the operators considered in section 11.4 and characterized in equation (9.62). If we define the L^2-norm of the kernel as in equation (9.63) (with $K(\boldsymbol{x}, \boldsymbol{x}') = H_{\mathcal{B}}(\boldsymbol{x} - \boldsymbol{x}')$) then, by means of the Parseval equality, we get

$$\|K\|^2 = m(\mathcal{D}) \int |H_{\mathcal{B}}(\boldsymbol{x})|^2 d\boldsymbol{x} = \frac{1}{(2\pi)^2} m(\mathcal{D}) m(\mathcal{B}). \tag{11.43}$$

It follow that A is an integral operator with a square-integrable kernel, so that it has a singular value decomposition with the properties described in section 9.4. We denote by $\{\sigma_j; u_j, v_j\}$ its singular system.

The uniqueness of the solution of the problem of out-of-band extrapolation implies that the null space of the operator A contains only the zero element. On the other hand, the functions in the range of the operator A are $\mathcal{B}$-bandlimited, as follows from equation (11.39). As a consequence, the null space of the operator A^*, which is given by

$$\left(A^* g\right)(\boldsymbol{x}) = \int H_{\mathcal{B}}^*(\boldsymbol{x} - \boldsymbol{x}') g(\boldsymbol{x}) d\boldsymbol{x}, \quad \boldsymbol{x} \in \mathcal{D}, \tag{11.44}$$

(see equation (9.67)) is the orthogonal complement of the subspace of the $\mathcal{B}$-bandlimited functions. In other words, the null space of A^* contains all square-integrable functions whose Fourier transform is zero inside $\mathcal{B}$.

From general properties of the singular system of an integral operator with a square-integrable kernel, we can conclude that:

- the singular values σ_j tend to zero when $j \to \infty$
- the singular functions u_j form an orthonormal basis in the subspace of the $\mathcal{B}$-bandlimited functions of $L^2(\mathrm{R}^2)$
- the singular functions v_j form an orthonormal basis in $L^2(\mathcal{D})$.

Remark 11.4. The operator A can also be written in terms of projection operators. Indeed, the function $H_{\mathcal{B}}(\boldsymbol{x} - \boldsymbol{x}')$ is the integral kernel of the bandlimiting projection operator $P^{(\mathcal{B})}$ defined in section 4.5. If we also introduce a spacelimiting projection operator $P^{(\mathcal{D})}$ defined by

$$\left(P^{(\mathcal{D})} f\right)(\boldsymbol{x}) = \begin{cases} f(\boldsymbol{x}) & ; \quad \boldsymbol{x} \in \mathcal{D} \\ 0 & ; \quad \boldsymbol{x} \notin \mathcal{D} \end{cases} \tag{11.45}$$

and we observe that $P^{(\mathcal{D})} f = f$ when the support of f is interior to $\mathcal{D}$, then equation (11.42) can be written as follows

$$Af = P^{(\mathcal{B})} P^{(\mathcal{D})} f. \tag{11.46}$$

To be pedantic, this form is not completely equivalent to equation (11.42). Indeed, in equation (11.46) A is considered as an operator in $L^2(\mathrm{R}^2)$ and therefore has *a null space which contains all square-integrable functions which are zero on $\mathcal{D}$. Then the singular functions v_j are zero outside $\mathcal{D}$ and constitute an orthonormal basis in the subspace of the square integrable functions which are also zero outside $\mathcal{D}$. This is equivalent to say that they constitute a basis in $L^2(\mathcal{D})$.*

The singular functions of the operator A are related to the *generalized prolate spheroidal functions* introduced by Slepian [6]. The relationship can be obtained by computing the operator A^*A. From the symmetry property $H_{\mathcal{B}}^*(\boldsymbol{x}) = H_{\mathcal{B}}(-\boldsymbol{x})$ and the identity

$$\int H_{\mathcal{B}}(\boldsymbol{x}-\boldsymbol{x}')H_{\mathcal{B}}(\boldsymbol{x}')d\boldsymbol{x}' = H_{\mathcal{B}}(\boldsymbol{x}), \tag{11.47}$$

which derives from the Fourier convolution theorem and the obvious identity $\chi_{\mathcal{B}}^2(\boldsymbol{\omega}) = \chi_{\mathcal{B}}(\boldsymbol{\omega})$, we obtain

$$\left(A^*Af\right)(\boldsymbol{x}) = \int_{\mathcal{D}} H_{\mathcal{B}}(\boldsymbol{x}-\boldsymbol{x}')f(\boldsymbol{x}')d\boldsymbol{x}'; \quad \boldsymbol{x} \in \mathcal{D}. \tag{11.48}$$

This self-adjoint and positive-definite operator is precisely the integral operator investigated by Slepian. Its eigenvalues $\lambda_1 \geq \lambda_2 \geq \lambda_3 \geq \cdots$ are positive and its eigenfunctions $\psi_1, \psi_2, \psi_3, \cdots$ are a basis in $L^2(\mathcal{D})$. These functions, which are defined on $\mathcal{D}$, can be extended to functions defined everywhere by means of the eigenvalue equation

$$\psi_j(\boldsymbol{x}) = \frac{1}{\lambda_j}\int_{\mathcal{D}} H_{\mathcal{B}}(\boldsymbol{x}-\boldsymbol{x}')\psi_j(\boldsymbol{x}')d\boldsymbol{x}', \tag{11.49}$$

using the fact that the kernel is defined everywhere. The extended eigenfunctions, normalized with respect to the norm of $L^2(\mathrm{R}^2)$, can be called *generalized prolate spheroidal functions*, even if Slepian uses this term only in the case where both $\mathcal{B}$ and $\mathcal{D}$ are discs. The eigenfunctions corresponding to this case are also called *circular prolate functions* [7].

In conclusion we have the following general results:

- the singular values σ_j of the operator A are the square roots of the eigenvalues of the Slepian operator (11.48)
- the singular functions u_j coincide with the generalized prolate spheroidal functions

$$u_j(\boldsymbol{x}) = \psi_j(\boldsymbol{x}) \tag{11.50}$$

- the singular functions v_j are proportional to the generalized prolate spheroidal functions

$$v_j(\boldsymbol{x}) = \frac{1}{\sigma_j}\psi_j(\boldsymbol{x}). \tag{11.51}$$

The last property derives from the fact that the ψ_j are $\mathcal{B}$-bandlimited, so that $A^*\psi_j = \psi_j$. We also observe that, since the u_j are orthogonal in $L^2(\mathrm{R}^2)$ while the v_j are orthogonal in $L^2(\mathcal{D})$, the ψ_j have a double orthogonality property, in the sense that they are orthogonal with respect to two different scalar products.

It is possible to give now the solution of the problem of out-of-band extrapolation by solving the integral equation (11.41) in terms of the singular system of the operator A. The result is the following

$$f(\boldsymbol{x}) = \sum_{j=1}^{\infty} \frac{1}{\sqrt{\lambda_j}}(g, u_j)v_j(\boldsymbol{x}) \tag{11.52}$$

where

$$(g, u_j) = \int g(\boldsymbol{x})\psi_j^*(\boldsymbol{x})d\boldsymbol{x} = \frac{1}{(2\pi)^2}\int_{\mathcal{B}} \hat{f}(\boldsymbol{\omega})\hat{\psi}_j^*(\boldsymbol{\omega})d\boldsymbol{\omega}. \tag{11.53}$$

This expression shows that $f(\boldsymbol{x})$ can be uniquely determined from the values of its Fourier transform on the set $\mathcal{B}$.

The series (11.52) is convergent if and only if we know the exact values of $\hat{f}(\boldsymbol{\omega})$ on $\mathcal{B}$. This is equivalent to require that the the coefficients (11.53) satisfy the Picard criterion (10.26). If we only have approximate values of $\hat{f}(\boldsymbol{\omega})$, then the series (11.52) in general is not convergent and, even if it converges, it can give completely unreliable values of $f(\boldsymbol{x})$. In other words, the problem of out-of-band extrapolation is ill-posed and therefore one can only have estimates of $f(\boldsymbol{x})$ by the use of regularization methods. A consequence of this fact is that it will not be possible to extrapolate $\hat{f}(\boldsymbol{\omega})$ everywhere but only over a bounded set $\mathcal{B}' \supset \mathcal{B}$. If the set $\mathcal{B}'$ is not significantly broader than $\mathcal{B}$, then no significant improvement in resolution can be obtained by means of out-of-band extrapolation.

From the previous remark it is obvious that, in order to estimate the feasibility of out-of-band extrapolation, it should be necessary to compute the eigenvalues and the eigenfunctions of the Slepian operator (11.48). This is not easy in the general case of arbitrary sets $\mathcal{B}$ and $\mathcal{D}$. Results are available in the rather frequent case where both $\mathcal{B}$ and $\mathcal{D}$ are discs, with radii respectively Ω and R [6] [7]. In such a case the operator A is given by (see equation (11.38))

$$(Af)(\boldsymbol{x}) = \frac{\Omega}{2\pi}\int_{|\boldsymbol{x}'|\le R} \frac{J_1(\Omega|\boldsymbol{x}-\boldsymbol{x}'|)}{|\boldsymbol{x}-\boldsymbol{x}'|} f(\boldsymbol{x}')d\boldsymbol{x}'. \tag{11.54}$$

A simple change of variables, transforming the disc of radius R into the disc of radius 1, shows that the circular prolate functions depend only on the parameter

$$c = R\Omega \tag{11.55}$$

which is the so-called *space-bandwidth product*. We see in a moment that this parameter is important for evaluating the amount of out-of-band extrapolation.

The case were a clear picture of the behaviour of the singular values is available is that where both $\mathcal{B}$ and $\mathcal{D}$ are squares of sides 2Ω and $2R$ respectively. Using again a change of variables which transforms the square of side $2R$ into the square of side 2 and introducing the parameter c, as defined in (11.55), we find

$$(Af)(\boldsymbol{x}) = \int_{-1}^{1} dx_1 \int_{-1}^{1} dx_2 \frac{\sin c(x_1 - x_1')}{\pi(x_1 - x_1')} \frac{\sin c(x_2 - x_2')}{\pi(x_2 - x_2')} f(x_1', x_2'). \quad (11.56)$$

The singular system of this operator can be given in terms of the eigenvalues and eigenfunctions of the Slepian–Pollack operator [8]

$$(A_{SP}f)(x) = \int_{-1}^{1} \frac{\sin c(x - x')}{\pi(x - x')} f(x')dx', \quad |x| \leq 1. \quad (11.57)$$

The eigenfunctions are the well-known *prolate spheroidal wave functions* (PSWF) denoted by $\psi_k(c, x)$, $k = 0, 1, 2, \cdots$. If we denote by λ_k the eigenvalue of A_{SP} associated with $\psi_k(c, x)$, then the singular system of the operator (11.56) is given by

$$\sigma_{j,k} = \sqrt{\lambda_j \lambda_k}; \quad j, k = 0, 1, 2, \cdots \quad (11.58)$$

$$u_{j,k}(\boldsymbol{x}) = \psi_j(c, x_1)\psi_k(c, x_2) \quad (11.59)$$

$$v_{j,k}(\boldsymbol{x}) = \frac{1}{\sqrt{\lambda_j \lambda_k}} \psi_j(c, x_1)\psi_k(c, x_2). \quad (11.60)$$

Using the properties of the eigenvalues of the Slepian–Pollack operator it is possible to understand when out-of-band extrapolation is feasible and, in particular, how it depends on the parameter c. If the space-bandwidth product is sufficiently large, the prolate eigenvalues $\lambda_k = \lambda_k(c)$ have a rather typical behaviour as a function of k: they are approximately equal to 1 for $k < 2c/\pi$ and tend to zero very rapidly for $k > 2c/\pi$. The quantity

$$S = \frac{2c}{\pi} = \frac{2\Omega R}{\pi} \quad (11.61)$$

is called the *Shannon number* [9] and has a very simple meaning: it is the number of sampling points, spaced by the Nyquist distance π/Ω, interior to the interval $[-R, R]$.

Thanks to this behaviour, if we consider a regularized solution provided by a truncated SVD, for any reasonable threshold value the number of terms will be of the order of S^2, i.e. of the order of the number of sampling points interior to the square of side $2R$. Since in such a case the sampling distance coincides with the resolution distance, it is clear that no improvement of resolution has been obtained. The result is interpreted by saying that no significant amount of out-of-band extrapolation can be achieved when c is large, i.e. when the size of the support of f is large with respect to the resolution distance π/Ω.

The situation is different when c is not much larger than 1, i.e. when the linear dimensions of the support of the object f are comparable with the resolution distance. Clearly this condition can be satisfied in microscopy or astronomy.

In such a case, using tables of the prolate eigenvalues, it is possible to show [5] that the number of terms in a truncated SVD solution can be greater than the Shannon number S, even by factors of 2 or 3. Therefore super-resolution can be achieved as well as a considerable amount of out-of-band extrapolation.

11.6 The Gerchberg method and its generalization

In the previous sections the analysis of super-resolution in image restoration has been divided in two parts

- identification of a band $\mathcal{B}$ where the Tikhonov regularized solution provides a reliable approximation of the Fourier transform of the unknown object
- extrapolation of the Fourier transform of the object out of the band $\mathcal{B}$, using information about the support of the object.

This procedure is useful from the theoretical point of view because it allows an analysis of the effect of the support of the object on the amount of out-of-band extrapolation. In such a way it is possible to give arguments showing that super-resolution is feasible when the diameter of the support of the object is of the order of the resolution distance, which is proportional to the inverse of the diameter of the band $\mathcal{B}$. It is not difficult to realize that the analysis implies that super-resolution should also be possible when the support $\mathcal{D}$ is the union of disjoint sets $\mathcal{D}_1, \mathcal{D}_2, \cdots, \mathcal{D}_p$, each with a diameter of the order of the resolution distance δ, the distance between any pair of these sets being much greater than δ.

An object satisfying these conditions is essentially a *nearly black object* in the sense of Donoho *et al* [10], a concept introduced for characterizing the class of the objects which are restored in an excellent way by the maximum entropy method [11]. Many applications of this technique, which is not treated in this book, have been developed: in spectroscopy [12], in interferometric astronomy [11] and in infra-red absorption spectroscopy [13]. The regularizing properties of maximum entropy are discussed in the book of Engl *et al* [14].

The analysis of section 11.5 can be hardly put in practice because the computation of the singular system of the operator (11.42) may be difficult, especially when $\mathcal{B}$ or $\mathcal{D}$ does not have a simple geometrical structure. However the very simple iterative scheme proposed by Gerchberg [15] in the 1D case, i.e. in the case where the PSWF can be used, can be easily extended to the general case of an arbitrary number of variables and arbitrary sets $\mathcal{B}$ and $\mathcal{D}$. Moreover it turns out that the method is a very clever implementation of the Landweber method.

In order to derive the Gerchberg method from the Landweber method for out-of-band extrapolation, it is convenient to write the integral operator (11.42) in the form (11.46). We have

$$A = P^{(\mathcal{B})} P^{(\mathcal{D})}, \quad A^* = P^{(\mathcal{D})} P^{(\mathcal{B})}, \quad A^* A = P^{(\mathcal{D})} P^{(\mathcal{B})} P^{(\mathcal{D})}. \tag{11.62}$$

Since A is the product of two non-commuting projection operators, its norm is certainly less than 1: $\|A\| < 1$. It follows that the relaxation parameter τ can take values between 0 and 2 and that we can choose $\tau = 1$, this choice being that corresponding to the original Gerchberg method.

The iterative scheme of the Landweber method, combined with equations (11.62), provides

$$f_{k+1} = f_k + P^{(\mathcal{D})} \left(P^{(\mathcal{B})} g - P^{(\mathcal{B})} P^{(\mathcal{D})} f_k \right) \tag{11.63}$$

where g is the $\mathcal{B}$-bandlimited data function providing the approximate Fourier transform of the object on the band $\mathcal{B}$ (see equation (11.39)).

If we take as an initial guess $f_0 = 0$, we find

$$f_1 = P^{(\mathcal{D})} P^{(\mathcal{B})} g = P^{(\mathcal{D})} g \tag{11.64}$$

where we have used the property

$$P^{(\mathcal{B})} g = g. \tag{11.65}$$

Since equation (11.64) implies that

$$P^{(\mathcal{D})} f_1 = f_1, \tag{11.66}$$

by induction, we can derive that

$$P^{(\mathcal{D})} f_k = f_k \tag{11.67}$$

for any k. Equations (11.65) and (11.67) as well as the linearity of the projection operators show that the iterative scheme (11.63) is equivalent to the following one

$$f_{k+1} = P^{(\mathcal{D})} \left[g + \left(I - P^{(\mathcal{B})} \right) f_k \right]. \tag{11.68}$$

The implementation of this method is precisely the Gerchberg algorithm:

- compute $f_1(\boldsymbol{x}) = \left(P^{(\mathcal{D})} g \right)(\boldsymbol{x})$
- from $f_k(\boldsymbol{x})$ compute $\hat{f}_k(\boldsymbol{\omega})$
- compute $\hat{g}_{k+1}(\boldsymbol{\omega})$ given by

$$\hat{g}_{k+1}(\boldsymbol{\omega}) = \hat{g}(\boldsymbol{\omega}) + [1 - \chi_{\mathcal{B}}(\boldsymbol{\omega})] \hat{f}_k(\boldsymbol{\omega}) \tag{11.69}$$

- from $\hat{g}_{k+1}(\boldsymbol{\omega})$ compute $g_{k+1}(\boldsymbol{x})$
- compute $f_{k+1}(\boldsymbol{x}) = \left(P^{(\mathcal{D})} g_{k+1} \right)(\boldsymbol{x})$

A few remarks follow. The computation of the projection operator $P^{(\mathcal{D})}$ is easy because it amounts to zeroing a function outside $\mathcal{D}$. Analogously the third step, equation (11.69), amounts to replacing $\hat{f}_k(\omega)$ on $\mathcal{B}$ by the data values $\hat{g}(\omega)$, without modifying $\hat{f}_k(\omega)$ outside $\mathcal{B}$ (since $\hat{g}(\omega)$ is zero outside $\mathcal{B}$). It follows that the implementation of the method requires essentially the computation of one direct and one inverse Fourier transform.

Since the Gerchberg method is simply a version of the Landweber method, its convergence and regularization properties are a direct consequence of the same properties of the Landweber method. Moreover it is obvious that, at the step k, it provides, in a way which is very cheap from the computational point of view, a filtered version of the singular function expansion (11.52), (11.53), since we must have

$$f_k(\boldsymbol{x}) = \sum_{j=1}^{\infty} \frac{1-(1-\lambda_j)^{k-1}}{\sqrt{\lambda_j}} (g, u_j) v_j(\boldsymbol{x}). \tag{11.70}$$

The use of the Gerchberg method for super-resolution, however, is related to the separation, mentioned at the beginning of this section, between the step of image deconvolution and the step of out-of-band extrapolation. This separation, useful from the theoretical point of view, may not be convenient in practice where it should be more convenient an algorithm which provides at the same time image restoration over the band $\mathcal{B}$ and out-of-band extrapolation. This algorithm is provided by a generalization of the Gerchberg method, which can be obtained again from the Landweber method.

To this purpose, let us consider the convolution operator

$$(Af)(\boldsymbol{x}) = \int K(\boldsymbol{x}-\boldsymbol{x}') f(\boldsymbol{x}') d\boldsymbol{x}' \tag{11.71}$$

and let us assume that we look for a solution of the usual equation $g = Af$, with support interior to a bounded set $\mathcal{D}$. Then we can write the basic equation as follows

$$g = AP^{(\mathcal{D})} f. \tag{11.72}$$

If we write the Landweber method for the operator $A_{\mathcal{D}} = AP^{(\mathcal{D})}$, we obtain

$$f_{k+1} = f_k + \tau P^{(\mathcal{D})} \left[A^* g - A^* A P^{(\mathcal{D})} f_k \right]. \tag{11.73}$$

In the case $f_0 = 0$, we can prove again by induction that $P^{(\mathcal{D})} f_k = f_k$, for any k, so that equation (11.73) is equivalent to the following one

$$f_{k+1} = P^{(\mathcal{D})} \left[f_k + \tau \left(A^* g - A^* A f_k \right) \right]. \tag{11.74}$$

When written in this way the method looks like a projected Landweber method. This form is important for its implementation which is just the same of the projected Landweber method and which is repeated here for convenience of the reader:

- compute $f_1(\boldsymbol{x}) = \left(P^{(\mathcal{D})} A^* g\right)(\boldsymbol{x})$
- from $f_k(\boldsymbol{x})$ compute $\hat{f}_k(\boldsymbol{\omega})$
- compute $\hat{h}_{k+1}(\boldsymbol{\omega})$ given by

$$\hat{h}_{k+1}(\boldsymbol{\omega}) = \tau \hat{K}^*(\boldsymbol{\omega})\hat{g}(\boldsymbol{\omega}) + \left(1 - \tau|\hat{K}(\boldsymbol{\omega})|^2\right) \hat{f}_k(\boldsymbol{\omega}) \tag{11.75}$$

- from $\hat{h}_{k+1}(\boldsymbol{\omega})$ compute $h_{k+1}(\boldsymbol{x})$
- compute $f_{k+1}(\boldsymbol{x}) = \left(P^{(\mathcal{D})} h_{k+1}\right)(\boldsymbol{x})$.

We remark that, even if the method looks like a projected Landweber method for the operator A, it is the Landweber method for the operator $A_{\mathcal{D}} = AP^{(\mathcal{D})}$ and therefore has the same convergence and regularization properties of this method. Moreover it provides a filtered version of the solution of the problem as given in terms of the singular system of the operator $A_{\mathcal{D}}$. Indeed, while the operator A is a convolution operator so that it does not have a singular value decomposition, the operator $A_{\mathcal{D}}$ has a singular value decomposition, as discussed in section 9.4, remark 9.1.

It is also important to observe that the projected Landweber method can be used for introducing the constraint of positivity in addition to the constraint on the support of the object. Since the projection operator P_+ onto the convex set of all non-negative functions—see equation (4.58)—commutes with the orthogonal projection $P^{(\mathcal{D})}$, as one can easily verify

$$P_+ P^{(\mathcal{D})} = P^{(\mathcal{D})} P_+, \tag{11.76}$$

it follows that the projection operator onto the convex set of the non-negative functions with support interior to $\mathcal{D}$ is given by

$$P_{\mathcal{C}} = P_+ P^{(\mathcal{D})}. \tag{11.77}$$

This operator can be easily computed, so that the corresponding projected Landweber method can also be easily implemented.

We illustrate the method by means of an application to the problem of far-field acoustic holography, already considered in section 4.5. In such a case we know that we can use the inverse filtering method because the problem of determining the generalized solution is well-posed. The resolution achievable is of the order of $\lambda/2$, λ being the wavelength of the monochromatic acoustic radiation.

We consider again a binary object which consists of a grid $2.5\lambda \times 2.5\lambda$ with vertical and horizontal bars 0.5λ wide. Therefore the size of this object, which is shown in figure 11.6(a), is not much larger than the resolution distance. If we apply the inverse filtering method to the noisy image of the object at a distance 5λ from the boundary plane (see figure 11.6(c)), the structure of the grid is not clearly resolved, as shown in figure 11.6(d). The basic reason is illustrated in figure 11.6(b): the bandlimiting due to wave propagation annihilates an important part of the Fourier spectrum of the object.

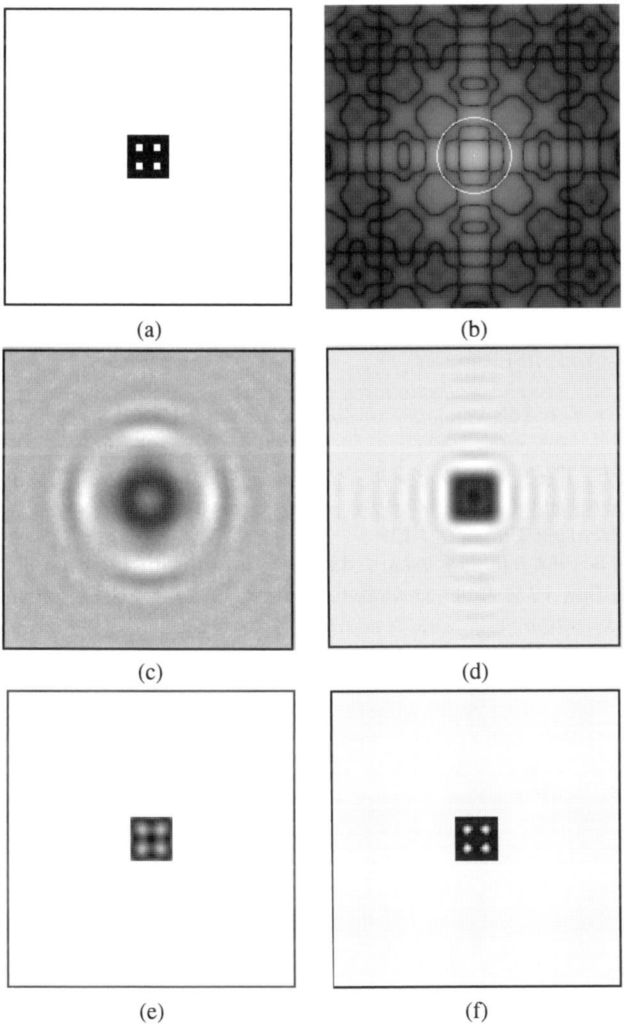

Figure 11.6. Example of super-resolution in far-field acoustic holography. (a) The object: a grid $2.5\lambda \times 2.5\lambda$ with bars 0.5λ wide. (b) The modulus of the FT of the object in (a); the white circle indicates the band of the far-field data. (c) The modulus of the noisy image at the distance 5λ from the object plane. (d) The reconstruction obtained by means of the inverse filtering method. (e) The reconstruction obtained by means of the iterative method with the constraint of bounded support. (f) The reconstruction obtained by means of the iterative method with the constraints of bounded support and positivity.

The result obtained by means of the iterative method (11.74) when the iterates are constrained to be zero outside the square $2.5\lambda \times 2.5\lambda$, is shown in figure 11.6(e). The improvement with respect to the inverse filtering solution is visible, even if the size of the object is larger than the resolution distance $\lambda/2$ by a factor 5. Finally in figure 11.6(f) we give the result obtained when also the positivity constraint on the support of the object is used in the iterative method. The restoration obtained is now quite satisfactory.

References

[1] Kak A C and Slaney M 1988 *Principles of Computerized Tomographic Imaging.* (New York: IEEE Press)

[2] Natterer F 1986 *The Mathematics of Computerized Tomography* (New York: Wiley)

[3] Krestel E (ed) 1990 *Imaging Systems for Medical Diagnostics* (Berlin: Siemens Aktiengesellschaft)

[4] Herman G T (ed) 1979 *Image Reconstruction from Projections* (Berlin: Springer)

[5] Bertero M and De Mol C 1996 *Progress in Optics* vol XXXVI ed E Wolf (Amsterdam: Elsevier) p 129

[6] Slepian D 1964 *Bell Syst. Tech. J.* **43** 3009

[7] Frieden B R 1971 *Progress in Optics* vol IX, ed E Wolf (Amsterdam: Elsevier) p 311

[8] Slepian D and Pollack H O 1961 *Bell Syst. Tech. J.* **40** 43

[9] Toraldo di Francia G 1969 *J. Opt. Soc. Am.* **59** 799

[10] Donoho D L, Johnstone I M Hoch J C and Stern A S 1991 *J. R. Stat. Soc.* **54** 41

[11] Gull S F and Daniell G J 1978 *Nature* **272** 686

[12] Sibisi S, Skilling J, Brereton R G, Laue E D and Staunton J 1984 *Nature* **311** 446

[13] Frieden B R 1972 *J. Opt. Soc. Am.* **62** 1202

[14] Engl H W, Hanke M and Neubauer A 1996 *Regularization of Inverse Problems* (Dordrecht: Kluwer)

[15] Gerchberg R W 1974 *Opt. Acta* **21** 709

Chapter 12

Comments and concluding remarks

In this book we have presented a selection of the methods which are most frequently used for the solution of linear inverse problems. Our presentation should make clear that, for the same problem, we can have at our disposal several methods. This richness generates a quite natural question: how can we decide which method is the most appropriate one for the specific problem we have to solve? In this concluding chapter we discuss possible answers to this question and possible strategies for obtaining an answer.

12.1 Does there exist a general-purpose method?

The methods for the solution of linear inverse problems, discussed in the previous chapters, can be set in the following groups:

- *The generalized-inverse method*
- *Methods for stable approximations of the generalized solution:*
  - Tikhonov regularization
  - Landweber method
  - Steepest descent
  - Conjugate gradient
- *Methods with additional constraints*
  - Projected Landweber method
  - Expectation Maximization

The order of the methods in this list derives from the following considerations. The first five methods do not require specific properties of the solution (such as positivity, for instance) and provide a solution or approximate solutions of the least-squares equation

$$A^*Af = A^*g. \tag{12.1}$$

The last two methods provide, respectively, approximate solutions of constrained least-squares problems, and approximate solutions of a maximum likelihood problem with the constraint of positivity.

In the case of image deconvolution we have also discussed the van Cittert method, which can be used for the approximate solution of the basic equation

$$Af = g \tag{12.2}$$

when the operator A is positive semidefinite. In such a case one can also use the *Lavrentev method* which consists of considering the following family of approximate solutions, similar to the Tikhonov regularized solutions

$$f_\mu = (A + \mu I)^{-1}\, g. \tag{12.3}$$

Moreover, one can also apply directly to equation (12.2) both the steepest descent and the conjugate gradient method. The disadvantages of the van Cittert method have already been discussed in chapter 6. A general comment on all these methods for the approximate solution of equation (12.2) is that they are much more affected by noise propagation than those for the approximate solution of equation (12.1). The reason resides in the fact that the data of equation (12.1) are obtained by applying the operator A^* to the original image g and this operation produces a filtering of the image, hence a reduction of the high-frequency noise.

Let us now return to the methods listed at the beginning of this section. The method of the generalized inverse, which coincides with the inverse filtering method in the case of deconvolution problems, produces the minimal norm solution $f^\dagger(\boldsymbol{x})$ of equation (12.1). Conditions for the existence of this solution were discussed in section 4.5 and section 10.1. For image deconvolution $f^\dagger(\boldsymbol{x})$ is given by

$$f^\dagger(\boldsymbol{x}) = \frac{1}{(2\pi)^2} \int_{\mathcal{B}} \frac{\hat{g}(\boldsymbol{\omega})}{\hat{K}(\boldsymbol{\omega})} e^{i\boldsymbol{x}\cdot\boldsymbol{\omega}} d\boldsymbol{\omega} \tag{12.4}$$

where $\mathcal{B}$ is the band of the imaging system, while for a linear inverse problem described by the singular system of the imaging operator it is given by

$$f^\dagger(\boldsymbol{x}) = \sum_{j=1}^{p} \frac{1}{\sigma_j} \left(g, u_j\right)_{\mathcal{Y}} v_j(\boldsymbol{x}). \tag{12.5}$$

Then the first two methods of the first group provide approximations of $f^\dagger(\boldsymbol{x})$ which can obtained by means of a suitable filtering of the representations (12.4) and (12.5). The general structure of these approximate solutions, here denoted by $\tilde{f}$, is the following:

$$\tilde{f}(\boldsymbol{x}) = \frac{1}{(2\pi)^2} \int_{\mathcal{B}} \frac{\hat{W}(\boldsymbol{\omega})}{\hat{K}(\boldsymbol{\omega})} \hat{g}(\boldsymbol{\omega}) e^{i\boldsymbol{x}\cdot\boldsymbol{\omega}} d\boldsymbol{\omega} \tag{12.6}$$

in the case of equation (12.4) and

$$\tilde{f}(\boldsymbol{x}) = \sum_{j=1}^{p} \frac{W_j}{\sigma_j} \left(g, u_j\right)_{\mathcal{Y}} v_j(\boldsymbol{x}) \tag{12.7}$$

in the case of equation (12.5). The two methods correspond to different choices of the window functions $\hat{W}(\omega)$ or W_j. In the case of Tikhonov method they are given by

$$\hat{W}_\mu(\omega) = \frac{|\hat{K}(\omega)|^2}{|\hat{K}(\omega)|^2 + \mu} \tag{12.8}$$

$$W_{\mu,j} = \frac{\sigma_j^2}{\sigma_j^2 + \mu} \tag{12.9}$$

where $\mu > 0$ is the regularization parameter, while in the case of the Landweber method they are given by:

$$\hat{W}_k(\omega) = 1 - \left(1 - \tau|\hat{K}(\omega)|^2\right)^k \tag{12.10}$$

$$W_{k,j} = 1 - \left(1 - \tau\sigma_j^2\right)^k \tag{12.11}$$

where k is the number of iterations and τ is the relaxation parameter.

Other linear methods, which can be reduced to the general form (12.6) or (12.7), such as low-pass filtering for image deconvolution or truncated singular function expansions for the linear inverse problems of part II, were also discussed (see section 5.4 and section 10.3, respectively).

The remaining methods in the list are nonlinear, the steepest descent and the conjugate gradient still providing approximate solutions of the least-squares equation (12.1).

The common features of all methods mentioned (except the generalized inverse) are the following:

- they are based on some *a priori* information about the solution of the inverse problem, the minimal one being that the energy of the unknown object cannot be too large;
- they have the *semiconvergence property*, which implies the possibility of adapting the value of the regularization parameter or of the number of iterations to obtain an optimal approximation of the object.

All methods have been applied to the example of section 5.6, the restoration of an image blurred by uniform motion. In the case of the projected Landweber method, the constraint of positivity was used. The results obtained by means of the various methods have been already given but, since they are scattered throughout the book, for convenience of the reader they are collected in table 12.1. In this table we give both the minimum restoration error and the

Table 12.1. The values of the minimum restoration error and of the corresponding optimal parameter (regularization parameter, μ_{opt}, or number of iterations, k_{opt}) for the methods listed above, when applied to the example of figure 5.15.

	Generalized inverse	Tikhonov
Rest. error	235%	13.3%
μ_{opt}	-	7.0×10^{-3}

	Landweber	Steepest descent	Conjugate gradient	Landweber with positivity	EM
Rest. error	13.2%	13.3%	13.6%	11.9%	12.1%
k_{opt}	45	41	11	42	62

value of the regularization parameter (or of the number of iterations) which is needed for obtaining this minimum error.

A glance at this table is sufficient to realize that all methods (except the generalized inverse) provide essentially the same result. Indeed the various restorations look quite similar because a variation of about $1 \div 2\%$ in the restoration error is not visible. In any case, a few more specific comments seem to be appropriate.

First, all methods of the first group of the above list provide the same restoration error, about 13.3%. These methods do not require additional constraints on the solution and can be applied to any linear inverse problem. Therefore they look like equivalent general-purpose methods. The decision of choosing one method over another cannot be based on the quality of the solution (they provide the same solution) but must be based on an analysis of the questions raised by the implementation of the various methods.

For instance, in image deconvolution where the FFT algorithm can be used, the Tikhonov method may be much faster than the iterative methods. On the other hand, in cases where the imaging matrix is large and sparse, it may not be convenient to use the Tikhonov method which requires the computation of an inverse matrix. Iterative methods, which only require a matrix-vector multiplication, should be considered and, among them, the conjugate gradient method appears to be the most efficient one because, in general, it requires a rather small number of iterations. On the other hand the choice of the number of iterations is critical for the CG method, so that accurate criteria for the estimation of this number (stopping rules) are needed. For this reason, in some cases, a less efficient method may be convenient if for this method the choice of the correct number of iterations is less dramatic.

The second comment concerns the constraint of positivity. In the example of table 12.1 the object to be restored is certainly non-negative. However it represents a rather complex situation and its values are, in general, considerably larger than zero. In such a situation, the unconstrained methods already produce restorations which are essentially non-negative and therefore the use of the positivity constraint does not improve the restoration in a significant way (even if a small improvement is obtained).

In this book several examples have been shown where positivity or similar constraints may be useful. We would like to conclude by discussing a further example not yet considered and which is intended to show the relevance of these constraints in some specific cases.

We consider a binary object 128×128, shown in figure 12.1(a), contaminated by out-of-focus blur, with COC diameter $D = 13$ (see section 3.3), and by white Gaussian noise with $s = 2$. The two levels of the object are 0 and 255. As shown in figure 12.1(b) the inscription cannot be identified from its blurred and noisy image. Indeed, the difference between g and $f^{(0)}$ is 79%. The relative RMS-error produced by the Gaussian noise is about 4%.

The condition number of the problem is rather large, $\alpha = 2.8 \times 10^6$, so that the restoration error produced by the inverse filtering method is also very large. In table 12.2 we report the results obtained by means of the methods used in the previous example. In the present example we have also used the projected Landweber method with positivity and upper bound on the solution (i.e. 255). For all different types of the Landweber method the value of the relaxation parameter is $\tau = 1.8$. Some of the restorations we have obtained are shown in figure 12.1 while the behaviour of the restoration error, as a function of the number of iterations, for the various iterative methods we have considered, is given in figure 12.2.

These results confirm some of the conclusions deduced from the previous example. The four methods in the second group of the list are essentially equivalent from the point of view of the restoration error. Even if this error is still rather large, about 47%, the inscription is readable and therefore, from the point of view of the interpretation of the image, these methods are quite satisfactory. The restored images, however, are not very nice from an aesthetic point of view. If we wish to improve the quality of the image, then the best results are provided by the projected Landweber methods, while the EM method provides a restoration which is intermediate between that of the general-purpose methods and that of the projected Landweber methods. The reason for this behaviour of the EM method may reside in the fact that the noise is Gaussian and rather small, as we discussed in section 7.3. A further comparison of these methods is given in figure 12.3, where the restorations of a piece of a line of the object in figure 12.1 are shown.

Figure 12.1. Comparison of: (a) the object; (b) the image (out-of focus blur); (c) the restoration provided by the Tikhonov method; (d) by the Landweber method with positivity; (e) by the EM method; (f) by the Landweber method with positivity and upper bound.

Table 12.2. Restoration errors in the example of figure 12.1.

	Generalized inverse	Tikhonov
Rest. error	705%	46.8%
μ_{opt}	-	1.2×10^{-3}

	Landweber	Steepest descent	Conjugate gradient
Rest. error	46.2%	46.8%	46.5%
k_{opt}	346	307	28
	Landweber with pos.	Landweber with pos. and upper bound	EM
Rest. error	31.6%	23.9%	43.0%
k_{opt}	1129	3867	445

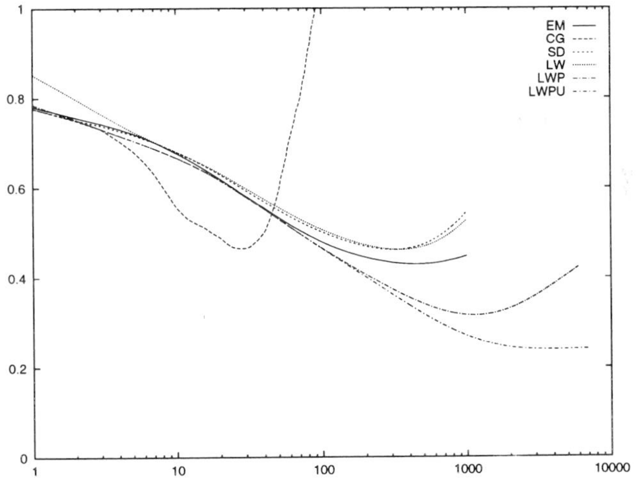

Figure 12.2. Behaviour of the restoration error for the iterative methods used in the restoration of the object of figure 12.1. The symbols have the following meaning: LW = Landweber, SD = Steepest descent, CG = conjugate gradient, LWP = Landweber with positivity, LWPU = Landweber with positivity and upper bound, EM = expectation maximization.

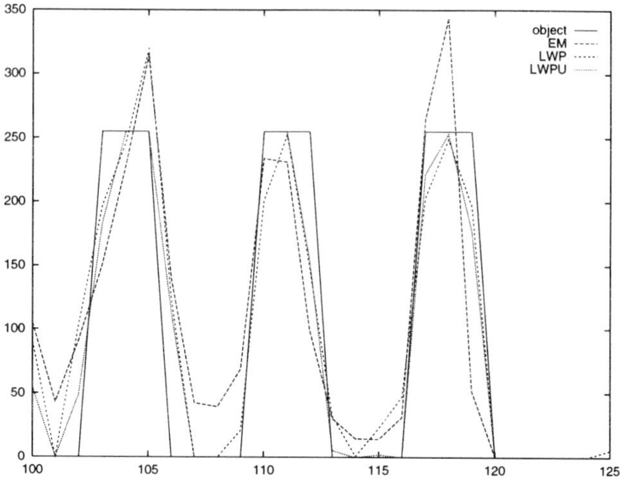

Figure 12.3. Plot of a piece of a line of the restorations of the object shown in figure 12.1. The full line corresponds to the profile of the object. The symbols have the following meaning: LWP = Landweber with positivity, LWPU = Landweber with positivity and upper bound, EM = expectation maximization.

12.2 In praise of simulation

In the usual problems of mathematical physics and applied mathematics such as those of diffusion, wave propagation, scattering, etc, one has, in general, to solve well-posed boundary value or initial value problems for partial differential equations. Since the solution exists, is unique and depends continuously on the data, the main problem is to find efficient algorithms for computing accurate approximations of the solution.

In the case of inverse problems the situation is different. When formulated in terms of functions depending on continuous variables an inverse problem, is, in general, ill-posed so that its solution does not exist in the case of noisy data. Of course one can consider a regularized version of the problem such as that provided by the Tikhonov method and, in such a case, one can look again for accurate and efficient numerical methods for computing the solution. Such an effort, however, may not be very useful in those cases where the regularized solutions are intrinsically poor approximations of the true one, because of large noise and/or small information content of the data. The last situation occurs when the effective band of the imaging system is small so that the corresponding resolution distance is large (see section 11.4) or when the number of singular values of the imaging operator, significantly different from zero, is also small. In these cases one does not need very accurate numerical methods and a rough discretization of the problem can be good enough.

Another typical feature of inverse problems, already discussed in the previous section, is that one has at one's disposal many different approximations of the solution provided by the various methods. As a consequence, it may happen that, to restore the images of different objects, provided by the same imaging system, one may be obliged to use different methods. An example is provided by astronomic images: different methods could be used for restoring diffuse objects such as a galaxy or point-like objects such as a star field over a black background. Moreover, since the regularization algorithms contain a free parameter, namely the regularization parameter, once the algorithm has been chosen one still has to check the various methods which can be used for the choice of the value of the parameter.

For all the above reasons, the use of numerical simulations is vital when one has to solve a practical inverse problem and therefore has to decide which method is the best for this purpose.

A program of numerical simulations consists of the following steps:

- step 1: definition of the mathematical model of the imaging system and computation of the matrix which approximates the imaging operator;
- step 2: choice of typical representatives of the class or of the classes of objects which should be restored;
- step 3: generation of noisy images of the representatives, chosen in step 2, using realistic models of the noise;
- step 4: selection of appropriate restoration methods and, for each one, computation of the optimal solution, using some measure of the quality of the restoration;
- step 5: for any given method compare the optimal solution obtained in step 4 with those provided by the various methods for the choice of the regularization parameter.

The final result of such a program of numerical simulations should be a decision about the method to be used for each class of objects. These simulations should also indicate what kind of information can be obtained and what kind of information cannot be obtained from the restoration of the blurred and noisy images.

We briefly discuss now the various steps mentioned above.

Step 1 is obvious; it is the basic point for the formulation of the problem of image formation. In some cases, as in X-ray tomography, this step may be rather easy while it may be very difficult in others (for instance, in the case of emission tomography). In image deconvolution this step is equivalent to the identification of the PSF by means of measurements or of computations based on a sufficiently accurate model of the imaging system. If one has only a poor knowledge of the PSF one can attempt to improve it by means of methods of *blind deconvolution*, a term introduced by Stockham, Cannon and Ingebretsen [1] to denote problems where both the PSF and the object are unknown. One has now at one's disposal many methods of blind-deconvolution: spectral methods [1], iterative methods [2] or maximum likelihood methods [3].

When the PSF is known, some important quantities obtainable from the PSF must be estimated. The first is the effective bandwidth Ω which provides the resolution achievable by means of the general-purpose methods such as Tikhonov, CG, etc. Moreover the values of $\hat{K}_{max}$ and $\hat{K}_{min}$, i.e. the maximum and the minimum value of the modulus of the TF, must be computed. If the maximum value is reached at $\omega = 0$, then it may be convenient to normalize the TF (and therefore the PSF) in such a way that $\hat{K}_{max} = 1$. This is the normalization we have used in all simulations presented in this book.

From $\hat{K}_{max}$ and $\hat{K}_{min}$ we obtain the condition number which provides an estimate of the noise propagation from the noisy image to the inverse filter solution. Finally the value of $\hat{K}_{max}$ can be used for the choice of the relaxation parameter if the Landweber or the projected Landweber method is used.

Step 2 is the true starting point of a program of numerical simulations. In fact the objects imaged by the imaging system under consideration are not completely unknown. We have at least some kind of *a priori* information about their qualitative features. Even if these features cannot be expressed in a quantitative mathematical form (such as a constraint, for instance), they can be used for generating typical examples of the possible objects to be imaged. These representatives or test objects can be digital or physical. By digital we mean a computer-generated test object, i.e. an array of numbers providing a representation of a possible object; by physical we mean a manufactured and perfectly known test object. While in the first case the image will be computer-generated, in the second case one can obtain a true image of the test object. It is obvious that it is not always possible to proceed in this way. It is certainly not possible in the case of astronomical images but it is possible, for instance, in the case of medical imaging, where the use of the so-called *phantoms* is very frequent.

The previous remarks can be clarified by a few examples. The first example is provided by the problem of restoring the HST images before installation of the corrective optics (see section 3.4). In order to test and compare various restoration methods several different brightness distributions were generated corresponding to typical celestial objects [4] and made available to the scientific community via anonymous-ftp. In figure 12.4 we show two of these distributions: one corresponds to a cluster of 470 stars (already used in section 6.2) while the other corresponds to a galaxy with a simple elliptical shape and no structures.

Tomography is another problem where test objects are frequently used. A well-known example is the Shepp–Logan phantom [5] which provides a model of a head section consisting of ten ellipses with different densities. This model has been extensively used for testing the accuracy of the reconstruction algorithms in X-ray tomography because it is believed that the reconstruction of the human head in particular requires great numerical accuracy and freedom from artifacts.

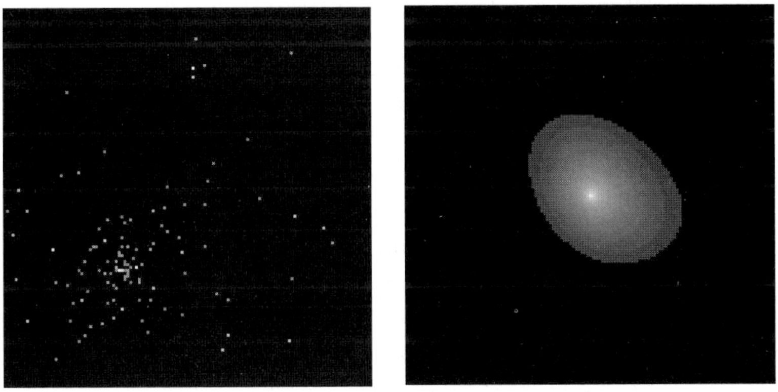

(a) (b)

Figure 12.4. Two test objects used for comparing restoration methods in the case of HST imaging: (a) a star cluster; (b) an elliptic galaxy (images produced by AURA/STScI).

A different test object is used for emission tomography. This is the so-called Hoffman phantom [6] which simulates cerebral blood-flow and metabolic pictures of the human brain. Two sections of this phantom are shown in figure 12.5. This is the digital version of the phantom. A physical version has also been manufactured. It can be used for data acquisition from PET or SPECT machines.

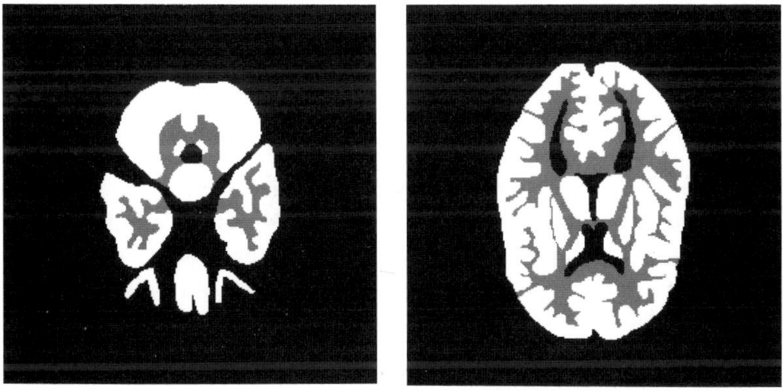

Figure 12.5. Two slices of the 3D Hoffman phantom which is used for testing the algorithms of emission tomography.

When the test objects have been chosen, it may be useful to compute for each object its Euclidean norm, which is proportional to the square root of its energy. This quantity can be used in simulation experiments, for instance for estimating the effect on energy of various choices of the regularization parameter.

The third step is naturally divided into two parts. The first one is the computation of the noise-free image $g^{(0)}$, which is simply the computation of a matrix-vector multiplication or of a convolution product, using the results of step 1 and step 2. The second part is the generation of a realistic noise to corrupt the noise-free image $g^{(0)}$.

To this purpose an analysis of the noise affecting the real data is required. For instance, both in X-ray tomography and in emission tomography it is assumed that the noise is mainly due to fluctuations in the counting process so that it can be accurately described by a Poisson distribution. Poisson noise due to the counting process is also the main noise in the case of HST images where two other kinds of noise are also present: the so-called background noise, which is also Poisson distributed, and the read-out noise which can be described by a Gaussian random process [7].

In general a routine for the generation of Gaussian distributed and one for the generation of Poisson distributed random numbers are sufficient to cover the most important cases of noise. The Gauss distribution can be obtained from the uniform distribution (which is the usual one for the generators of sequences of random numbers) by means of the transformation method, while the Poisson distribution can be obtained by means of the rejection method [8].

In the case of Gaussian additive noise, one has to estimate the variance of the distribution (the expectation value, in general, is zero), generate a sequence of random numbers corresponding to this distribution and add the numbers of the sequence to the various samples of the noise-free image. In the case of Poisson noise, one has first to normalize the noise-free image in such a way that the sum of the values of the image corresponds to the expected total number of counts. Then the value of the image at any given pixel must be considered as the expectation value of a Poisson process. The value obtained from the generator of Poisson random numbers, with that expectation value, must be taken as the value of the noisy image at that sampling point.

Once the noisy image g has been generated, the difference $g - g^{(0)}$ provides the realization of the noise term w. This image of the noise is completely random if the noise is object independent; otherwise, as in the case of Poisson noise, it still keeps a reminiscence of the noise-free image. In any case the Euclidean norm of the noise term w provides the exact value of the discrepancy which is basic for estimating the regularization parameter or the number of iterations.

We point out that in many circumstances it may be useful to generate several noisy images corresponding to the same noise-free image, i.e. images which correspond to different realizations of the noise process. Indeed, by applying the same restoration algorithm to these images one can obtain a clear picture of the noise propagation from the image to the restored object and possibly identify parts of the object domain where the error due to the noise is more important than in others.

At this point we have all the elements we need for performing the numerical experiments of steps 4 and 5. Step 4 consists of computing the best restorations which can be obtained by means of a given method. This expression can be a little ambiguous because it depends on the measure we choose for the quality of the restoration. The usual one is the Euclidean norm of the difference between the restored object and the true one (possibly divided by the norm of the true object, so that we obtain a relative restoration error). In the case of image deconvolution this relative restoration error is equivalent to the MSEIF, defined in equation (4.29). Finally step 5 provides information about the efficiency of the various methods which have been proposed for the estimation of the regularization parameter, number of iterations, etc, i.e. of the methods which must be used in the case of the real images produced by the real imaging system.

This is essentially the story of all the numerical experiments we performed for illustrating this book.

References

[1] Stockham T G, Cannon T M and Ingebretsen R B 1975 *Proc. IEEE* **63** 678

[2] Ayres G R and Dainty J C 1988 *Opt. Lett.* **13** 547

[3] Lagendijk R L and Biemond J 1991 *Iterative Identification and Restoration of Images* (Boston: Kluwer)

[4] Hanish R J 1993 *Restoration-Newsletter of STScI's Image Restoration Project* **1** 76

[5] Shepp L A and Logan B F 1974 *IEEE Trans. Nucl. Sci.* **21** 21

[6] Hoffman E, Cutler P, Digby W and Mazziotta J 1990 *IEEE Trans. Nucl. Sci.* **37** 616

[7] Snyder D L, Hammond A M and White R L 1993 *J Opt Soc Am* **A-10** 1014

[8] Press W H, Teukolsky S A, Vetterling W T and Flannery B P 1992 *Numerical Recipes* (Cambridge: Cambridge University Press)

PART 3

MATHEMATICAL APPENDICES

Appendix A

Euclidean and Hilbert spaces of functions

The appropriate mathematical framework for the analysis of inverse problems is the theory of function spaces and of operators in these spaces. The most frequent approach consists of the formulation of an inverse problem in terms of a first kind operator equation in a Hilbert space setting. This is the abstract framework adopted, for instance, in the book of Groetsch [1] and in that of Engl, Hanke and Neubauer [2]. The first is essentially an introductory book while the second provides a more complete presentation of the most important mathematical results which have been proved in regularization theory.

In this appendix we sketch a few properties of Hilbert spaces, which are essentially infinite-dimensional linear spaces equipped with a scalar (inner) product. A complete theory can be found in any book on functional analysis (see, for instance, [3]).

We assume that the reader is familiar with the definition and basic properties of a linear space. Examples of linear spaces of functions are: the class of all continuous functions defined on a given domain; the class of all integrable functions defined on a given domain, etc. Since in the problems considered in this book we must use very often complex valued functions, we mainly consider linear spaces where multiplication by complex numbers is defined. If α is a complex number, its complex conjugate will be denoted by α^*.

We will denote by $\mathcal{X}$ a linear space of functions and we will use symbols such as $f, g, h, \cdots$ for indicating elements of $\mathcal{X}$. Then we say that $\mathcal{X}$ is *Euclidean* if a *scalar product* has been introduced in $\mathcal{X}$, i.e. a complex valued functional defined on the pairs of elements f, h of $\mathcal{X}$, denoted by (f, h)—sometimes by $(f, h)_{\mathcal{X}}$ if it is necessary to specify the space $\mathcal{X}$—and having the following properties:

- $(f, f) \geq 0$; $(f, f) = 0$ if and only if $f = 0$ (zero element of the linear space $\mathcal{X}$);
- $(f, h) = (h, f)^*$;
- $(\alpha f, h) = \alpha(f, h)$, for any complex number α;
- $(f^{(1)} + f^{(2)}, h) = (f^{(1)}, h) + (f^{(2)}, h)$.

The last two conditions imply that (f, h) is a linear functional of the first argument. From the second and third conditions one easily derives that

$$(f, \alpha h) = \alpha^*(f, h). \tag{A.1}$$

Moreover it is possible to prove (see, for instance, [3]) that, as a consequence of these conditions, the scalar product of two functions satisfies the so-called *Schwarz inequality*

$$|(f, h)|^2 \leq (f, f)(h, h). \tag{A.2}$$

The definition of a scalar product allows the definition of orthogonality of two functions f, h of $\mathcal{X}$: they are orthogonal if $(f, h) = 0$. Then, if $\mathcal{S}$ is a subset of elements of $\mathcal{X}$, the *orthogonal complement* of $\mathcal{S}$, denoted by $\mathcal{S}^\perp$, is the set of all functions of $\mathcal{X}$ which are orthogonal to all functions of $\mathcal{S}$.

In order to describe the structure of the orthogonal complement, we need the definition of linear subspace. A subset $\mathcal{L}$ of $\mathcal{X}$ is called a *linear subspace* if it has the structure of a linear space, i.e. if any linear combination of elements of $\mathcal{L}$ is also an element of $\mathcal{L}$. Then it is easy to show that *the orthogonal complement $\mathcal{S}^\perp$ of a subset $\mathcal{S}$ of $\mathcal{X}$ is a linear subspace of $\mathcal{X}$* .

When a scalar product has been introduced in $\mathcal{X}$, one can define the length of a function of $\mathcal{X}$. This length is called the *norm* of f, is denoted by $\|f\|$ and is defined by

$$\|f\| = (f, f)^{1/2}. \tag{A.3}$$

From the properties of the scalar product it follows that the norm has the following basic properties:

- $\|f\| \geq 0$; $\|f\| = 0$ if and only if $f = 0$ (null element of the linear space $\mathcal{X}$);
- $\|\alpha f\| = |\alpha| \|f\|$, for any complex number α;
- $\|f + h\| \leq \|f\| + \|h\|$ (triangle inequality).

The third property follows from Schwarz inequality, equation (A.2) [3]. It is called triangle inequality because, if we consider the triangle with sides f, h and $f + h$, the inequality states that the length of the side $f + h$ does not exceed the sum of the lengths of the two other sides. To this purpose we remark that, if f and h are orthogonal, then

$$\|f + h\|^2 = \|f\|^2 + \|h\|^2 \tag{A.4}$$

and this relation is the extension of *Pythagoras' theorem* to Euclidean spaces.

Equation (A.4) can be generalized to the case of n mutually orthogonal elements $f^{(1)}, f^{(2)}, \cdots, f^{(n)}$

$$\|f^{(1)} + f^{(2)} + \cdots + f^{(n)}\|^2 = \|f^{(1)}\|^2 + \cdots + \|f^{(n)}\|^2, \tag{A.5}$$

and an application of this relation is the following. Consider a set of n orthonormal functions $v_1, v_2, \cdots, v_n$, i.e. functions satisfying the conditions:

$$(v_k, v_j) = \delta_{k,j}, \tag{A.6}$$

and a function f which is a linear combination of the functions v_k

$$f = \sum_{k=1}^{n} c_k v_k. \tag{A.7}$$

The coefficients c_k are the components of f with respect to the orthonormal functions v_k and, as follows from equation (A.6), are given by

$$c_k = (f, v_k). \tag{A.8}$$

Then, from equation (A.5) we get

$$\|f\|^2 = \sum_{k=1}^{n} |c_k|^2. \tag{A.9}$$

This is a particular case of the so-called *Parseval equality*, which will be given in a moment.

The introduction of a norm allows also the introduction of a *metric*: the *distance* of the element f from the element h is the norm of $f - h$:

$$d(f, h) = \|f - h\|. \tag{A.10}$$

From the properties of the norm one can easily derive the following basic properties of the distance:

- $d(f, h) \geq 0$; $d(f, h) = 0$, if and only if $f = h$;
- $d(f, h) = d(h, f)$;
- $d(f, h) \leq d(f, g) + d(g, h)$ for any g (triangle inequality)

The concept of convergence is deeply related to the concept of distance: *a sequence f_n $(n = 1, 2, 3, \cdots)$ is said to be convergent to an element f of $\mathcal{X}$ if $d(f_n, f) = \|f_n - f\| \to 0$ when $n \to \infty$.*

Given a subset $\mathcal{S}$ of $\mathcal{X}$, a function f is a *limit point* of $\mathcal{S}$ if it is the limit of a sequence contained in $\mathcal{S}$. A set $\mathcal{S}$ is said to be *closed* if it contains all its limit points. If $\mathcal{S}$ is not closed, it is possible to get a closed set by adding to $\mathcal{S}$ all limit points which do not belong to $\mathcal{S}$. The new set is called the *closure* of $\mathcal{S}$ and denoted by $\bar{\mathcal{S}}$. It is the smallest closed set containing $\mathcal{S}$.

The notion of *completeness* of a Euclidean space is strictly related to the notion of convergence. A *Cauchy sequence* is a sequence f_n, $(n = 1, 2, 3, \cdots)$, such that $\|f_n - f_m\| \to 0$ when $n, m \to \infty$. From the triangle inequality it follows that any convergent sequence is also a Cauchy sequence. The converse, in general, is not true. A Euclidean space such that any Cauchy sequence is convergent to an element of the space is said to be *complete* and is called a *Hilbert space* [3].

A Hilbert space is *separable* if there exists a countable basis, i.e. a countable orthonormal set of functions v_j $(j = 1, 2, 3, \cdots)$ such that any function f of the space can be represented by

$$f = \sum_{j=1}^{\infty}(f, v_j)v_j, \tag{A.11}$$

the convergence of the series being in the sense that the sequence of the partial sums converges to f

$$\|f - \sum_{j=1}^{n}(f, v_j)v_j\| \to 0, \quad n \to \infty. \tag{A.12}$$

Then it is possible to prove that the following equality, called the *Parseval equality*, holds true

$$\|f\|^2 = \sum_{j=1}^{\infty}|(f, v_j)|^2 \tag{A.13}$$

as well as the following one, called the *generalized Parseval equality*

$$(f, h) = \sum_{j=1}^{\infty}(f, v_j)(h, v_j)^*. \tag{A.14}$$

The set of the functions v_j is called an orthonormal basis of the Hilbert space $\mathcal{X}$ and the numbers (f, v_j) are the components of f with respect to this basis. As follows from equation (A.13), the series of their squares is convergent. Then the following question arises: given a sequence of complex numbers $c_1, c_2, \cdots$, does there exists a function f of the space $\mathcal{X}$ such that these numbers are its components with respect to the basis $v_1, v_2, v_3, \cdots$? The answer is positive if and only if the numbers satisfy the condition

$$\sum_{j=1}^{\infty}|c_j|^2 < \infty. \tag{A.15}$$

The proof derives from the completeness of the Hilbert space. This result implies the Picard criterion which is discussed in section 10.1.

The classical example of a separable Hilbert space is the space of all square-integrable functions defined on a domain $\mathcal{D}$. This space is denoted by $L^2(\mathcal{D})$. The rigorous definition of this space requires Lebesgue theory of measure and integration. Even a short account of this theory is beyond the scope of this book. The interested reader can consult any textbook on the theory of functions of real variables (see, for instance, [4]). We only give an important concept which can be introduced without using the complete theory of measure.

A *set of measure zero* is a set which can be included in open sets of arbitrarily small measure. For instance, a point is a set of measure zero because it is the centre of balls of arbitrarily small radius ε. Finite sets of points or countable sets of points provide other examples. In two dimensions a regular curve is also a set of measure zero and so on.

The following definition is based on the concept of set of measure zero: *a property is said to hold almost everywhere (abbreviated a.e.) if the set of points where it fails to hold is a set of measure zero.* One says, for instance, that $f = g$ a.e. if the set of points $\boldsymbol{x}$ where $f(\boldsymbol{x}) \neq g(\boldsymbol{x})$ has measure zero. Similarly one says that $f(\boldsymbol{x}) \neq 0$ a.e. if the set points $\boldsymbol{x}$ where $f(\boldsymbol{x}) = 0$ has zero measure (the functions (2.5) and (2.7) introduced in section 2.1 are examples of functions different from zero a.e.). One also says that the sequence f_n converges to f a.e. if the set of points $\boldsymbol{x}$ where $f_n(\boldsymbol{x})$ does not converge to $f(\boldsymbol{x})$ has zero measure and so on.

If f, h are functions of $L^2(\mathcal{D})$, it is possible to define their scalar product as follows

$$(f, h) = \int_{\mathcal{D}} f(\boldsymbol{x}) h^*(\boldsymbol{x}) d\boldsymbol{x}. \tag{A.16}$$

This is the canonical scalar product of a space of square integrable functions and is defined for any pair of these functions because the product of two square integrable functions is always integrable. Then it is easy to verify that the properties of the scalar product, except the first one, are satisfied. Indeed the second part of this property does not hold true if we consider $L^2(\mathcal{D})$ as a space of functions: from the condition $(f, f) = 0$ one only derives that $f = 0$ a.e. and therefore the element f such that $(f, f) = 0$ is not unique. Uniqueness can be obtained if we consider two square-integrable functions to be equivalent if they are equal a.e. and if we do not distinguish between equivalent functions. Then, the zero element of the space $L^2(\mathcal{D})$ is the class of all functions which are zero a.e. in $\mathcal{D}$. Accordingly any element of $L^2(\mathcal{D})$ is a class of functions and precisely the class of all functions which are equal a.e. to a given function $f(\boldsymbol{x})$. This can be taken as a representative of the class.

A quite natural extension of the classical L^2-space is provided by a *weighted* L^2*-space*. In the most simple case a weighted L^2-space can be defined as follows: let $C(\boldsymbol{x})$ be a positive function (a.e.) defined on the domain $\mathcal{D}$, then $L^2_C(\mathcal{D})$ is the space of all functions $f(\boldsymbol{x})$ such that

$$\int_{\mathcal{D}} C(\boldsymbol{x}) |f(\boldsymbol{x}|^2 d\boldsymbol{x} < \infty. \tag{A.17}$$

It is easy to verify that this is a Euclidean space with a scalar product defined by

$$(f, h) = \int_{\mathcal{D}} C(\boldsymbol{x}) f(\boldsymbol{x}) h^*(\boldsymbol{x}) d\boldsymbol{x}. \tag{A.18}$$

In other words, $L^2_C(\mathcal{D})$ is the space of all functions f such that $C^{1/2} f$ is square integrable. Therefore it may contain functions which are not square integrable. If the function $C(\boldsymbol{x})$ satisfies a.e. in $\mathcal{D}$ the conditions: $0 < a \leq C(\boldsymbol{x}) \leq b < \infty$, then $C^{1/2} f$ is square integrable if and only if f is square integrable. In such a case $L^2_C(\mathcal{D})$ contains the same elements of $L^2(\mathcal{D})$.

More general weighted spaces can be defined in terms of positive definite kernels, i.e. functions $C(\boldsymbol{x}, \boldsymbol{x}')$ such that

$$\int\int_{\mathcal{D}} C(\boldsymbol{x}, \boldsymbol{x}') f(\boldsymbol{x}) f^*(\boldsymbol{x}') d\boldsymbol{x} d\boldsymbol{x}' > 0 \qquad \text{(A.19)}$$

for any $f \neq 0$ a.e. such that the integral is convergent. In the linear space of all functions satisfying this condition one can introduce the following scalar product

$$(f, h) = \int\int_{\mathcal{D}} C(\boldsymbol{x}, \boldsymbol{x}') f(\boldsymbol{x}) h^*(\boldsymbol{x}') d\boldsymbol{x} d\boldsymbol{x}'. \qquad \text{(A.20)}$$

In this book we do not need this kind of Euclidean space but we need its discrete version which will be discussed in appendix C.

Appendix B

Linear operators in function spaces

This appendix contains a summary of the main properties of linear operators and, in particular, of projection operators.

Consider a pair of Euclidean spaces of functions, say $\mathcal{X}$ and $\mathcal{Y}$. An operator A from $\mathcal{X}$ into $\mathcal{Y}$, denoted by $A : \mathcal{X} \rightarrow \mathcal{Y}$, is a mapping which assigns to each element f of a certain subset $\mathcal{D}(A)$ of $\mathcal{X}$ an element g of another subset $\mathcal{R}(A)$ of $\mathcal{Y}$. We write

$$A(f) = g \tag{B.1}$$

and we say that g is the image of f. The set $\mathcal{D}(A)$ is called the *domain* of the operator A. The set $\mathcal{R}(A)$, where every element g is the image of at least one element in $\mathcal{D}(A)$, is called the *range* of the operator A. If $\mathcal{D}(A) = \mathcal{X}$ one says that the operator is defined everywhere on $\mathcal{X}$. When $\mathcal{Y} = \mathcal{X}$ one says that A is an operator in $\mathcal{X}$.

An operator A is *linear* if for any pair of complex numbers α_1, α_2 and for any pair of functions in its domain $\mathcal{D}(A)$, $f^{(1)}$, $f^{(2)}$, one has

$$A(\alpha_1 f^{(1)} + \alpha_2 f^{(2)}) = \alpha_1 A(f^{(1)}) + \alpha_2 A(f^{(2)}). \tag{B.2}$$

The notation used in (B.1) recalls that the concept of operator is a particular case of the more general concept of function. In the case of linear operators, however, a simplified notation is used: one usually writes Af instead of $A(f)$ so that equation (B.1) becomes

$$Af = g. \tag{B.3}$$

In the applications to inverse problems in imaging $\mathcal{X}$ is the space of the objects while $\mathcal{Y}$ is the space of the images. Therefore one is interested both to the case where the two spaces coincide and to the case where the two spaces are distinct. For instance, in the first part of this book both $\mathcal{X}$ and $\mathcal{Y}$ are spaces of square-integrable functions and the convolution operators are examples of linear operators in $\mathcal{X}$. On the other hand the Radon transform, introduced in section 8.2, defines a linear operator from a space $\mathcal{X}$ into a space $\mathcal{Y}$. The space

$\mathcal{X}$ is, for instance, the space of all square integrable functions defined on the disc of radius a (space of the density functions) while the space $\mathcal{Y}$ is the space of all square integrable functions defined on the rectangle $|s| \leq a$, $0 \leq \phi \leq \pi$ (space of the sinograms).

A linear operator, defined everywhere on $\mathcal{X}$, is *bounded* if there exists a constant M such that, for every f in $\mathcal{X}$

$$\|Af\|_{\mathcal{Y}} \leq M\|f\|_{\mathcal{X}}. \tag{B.4}$$

Here we consider the case where the two spaces are distinct so that it is convenient to specify the space as a suffix of the norm.

The convolution operators considered in section 2.3 and corresponding to bounded transfer functions are examples of bounded operators. The integral operators, with square integrable kernels, investigated in section 9.4, are also examples of bounded operators.

The smallest constant M such that (B.4) holds true is called the *norm of the operator A* and denoted by $\|A\|$. It follows that

$$\|A\| = \sup_{f \in \mathcal{X}} \frac{\|Af\|_{\mathcal{Y}}}{\|f\|_{\mathcal{X}}}. \tag{B.5}$$

A linear operator A is *continuous* if it has the following property: let f_n be a sequence convergent to f in $\mathcal{X}$, i.e. $\|f_n - f\|_{\mathcal{X}} \to 0$, then Af_n is a sequence convergent to Af in $\mathcal{Y}$, i.e. $\|Af_n - Af\|_{\mathcal{Y}} \to 0$. In particular, if $f_n \to 0$ (null element of $\mathcal{X}$), then $Af_n \to 0$ (null element of $\mathcal{Y}$). Equation (B.4) shows that the last property holds true for a bounded operator. One can prove that *a linear operator A is continuous if and only if it is bounded* [3].

Given a linear and bounded operator A, its adjoint operator A^* is the unique operator such that

$$(Af, g)_{\mathcal{Y}} = (f, A^*g)_{\mathcal{X}}. \tag{B.6}$$

If A is an operator from $\mathcal{X}$ into $\mathcal{Y}$, then A^* is an operator from $\mathcal{Y}$ into $\mathcal{X}$. In the case of a convolution operator A^* is given by equation (2.37). It can be proved that the adjoint A^* of a bounded operator A has the same norm of A, i.e. $\|A^*\| = \|A\|$ [3].

A bounded operator A in $\mathcal{X}$ is *self-adjoint* if $A^* = A$. A convolution operator is self-adjoint if and only if its PSF satisfies the condition $K^*(\boldsymbol{x}) = K(-\boldsymbol{x})$. This condition implies that the transfer function $\hat{K}(\boldsymbol{\omega})$ is real valued.

The *null space* of a linear operator A, denoted by $\mathcal{N}(A)$, is the set of all elements f such that $Af = 0$, i.e.

$$\mathcal{N}(A) = \{f \in \mathcal{X};\quad Af = 0\}. \tag{B.7}$$

It is easy to prove that $\mathcal{N}(A)$ is a linear subspace and also that it is closed if the operator A is continuous. In the case of convolution operators, the null space was also called the subspace of the *invisible objects* (see section 2.3).

There exist important relationships between null space and range of an operator, on one side, and null space and range of its adjoint on the other side. More precisely, by recalling the definition of orthogonal complement of a set (introduced in appendix A) the following relationships hold true

$$\mathcal{N}(A) = \mathcal{R}(A^*)^{\perp}, \quad \mathcal{N}(A^*) = \mathcal{R}(A)^{\perp}. \tag{B.8}$$

Proof B.1. We prove the first relation (B.8). From the definition (B.6) of the adjoint operator it follows that, if $f \in \mathcal{N}(A)$, i.e. $Af = 0$, then the l.h.s. of the equation (B.6) is zero for any g, so that f is orthogonal to all elements A^*g, i.e. to $\mathcal{R}(A^*)$. This result implies that $\mathcal{N}(A)$ is contained in $\mathcal{R}(A^*)^{\perp}$. On the other hand, if $f \in \mathcal{R}(A^*)^{\perp}$, then the r.h.s. of equation (B.6) is zero for any g and therefore Af is orthogonal to all elements of $\mathcal{Y}$. But this implies $Af = 0$, i.e. $f \in \mathcal{N}(A)$. It follows that $\mathcal{R}(A^*)^{\perp}$ is contained in $\mathcal{N}(A)$. By combining the two inclusions, we get $\mathcal{N}(A) = \mathcal{R}(A^*)^{\perp}$. The second relation in (B.8) is obtained from the first one by simply exchanging A with A^*. □.

In section 2.3 we observed that the invisible objects are orthogonal to the noise-free images, i.e. the elements of $\mathcal{R}(A)$. This implies that $\mathcal{N}(A) = \mathcal{R}(A)^{\perp}$. This property, however, is not true in general for operators in a Euclidean space $\mathcal{X}$. The general relationships are those described by equation (B.8). It holds true for a convolution operator because, in such a case, $\mathcal{N}(A) = \mathcal{N}(A^*)$. This is true also for a self-adjoint operator, so that for such an operator the equations (B.8) are replaced by the following one

$$\mathcal{N}(A) = \mathcal{R}(A)^{\perp}. \tag{B.9}$$

Given a Euclidean space $\mathcal{X}$, a *projection operator* P is a linear operator in $\mathcal{X}$, satisfying the following conditions

- $P^* = P$
- $P^2 = P$.

These conditions imply that P is bounded and that $\|P\| = 1$ [3].

If P is a projection operator, then also $Q = I - P$ (where I is the identity operator in $\mathcal{X}$) is a projection operator; indeed it is easy to verify that $Q^* = Q$ and $Q^2 = Q$.

A projection operator P is associated with two linear subspaces defined as follows

$$\mathcal{M} = \{f \in \mathcal{X};\, Pf = f\}, \qquad \mathcal{N} = \{f \in \mathcal{X};\, Pf = 0\}. \tag{B.10}$$

The subspace $\mathcal{N}$ is precisely the null space $\mathcal{N}(P)$ of the projection operator P while the subspace $\mathcal{M}$ is the range $\mathcal{R}(P)$, as one can prove by the use of the second property. Therefore, from equation (B.9) it follows that $\mathcal{N}$ is the orthogonal complement of $\mathcal{M}$ and conversely. This result can also be obtained by observing that, for any f in $\mathcal{X}$ one has the following decomposition

$$f = Pf + Qf \tag{B.11}$$

with $Pf \in \mathcal{M}$ and $Qf \in \mathcal{N}$, and that the two components are orthogonal because $PQ = 0$.

The previous remarks imply that P is the *orthogonal projection* onto the subspace $\mathcal{M}$. Conversely, given a subspace $\mathcal{M}$, it is possible to define the orthogonal projection onto $\mathcal{M}$. This result follows from a well-known theorem (*decomposition theorem*) which holds true in a complete Euclidean space (Hilbert space) [3]: *let $\mathcal{M}$ be a closed linear subspace of $\mathcal{X}$ and $\mathcal{N}$ its orthogonal complement; then for any element f of $\mathcal{X}$ there exists the unique decomposition*

$$f = f_{\mathcal{M}} + f_{\mathcal{N}} \tag{B.12}$$

with $f_{\mathcal{M}} \in \mathcal{M}$ and $f_{\mathcal{N}} \in \mathcal{N}$.

The function $f_{\mathcal{M}}$ is the component of f along $\mathcal{M}$ while $f_{\mathcal{N}}$ is the component of f orthogonal to $\mathcal{M}$. Therefore $f_{\mathcal{M}}$ is the orthogonal projection of f onto $\mathcal{M}$. In figure B.1 a two-dimensional picture of this decomposition is given.

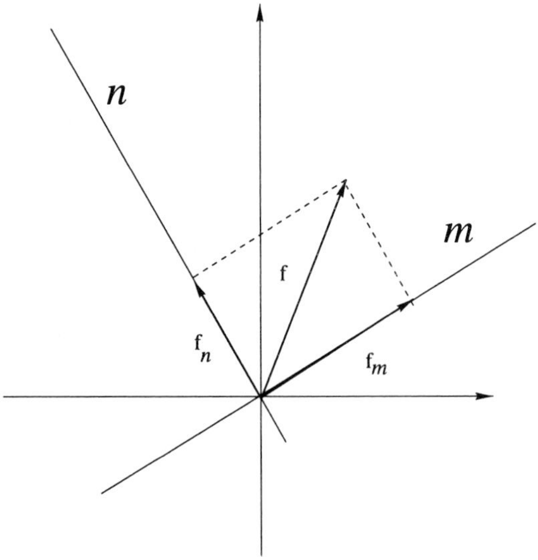

Figure B.1. Two-dimensional representation of the decomposition theorem

Now, if we assign to any f its projection $f_{\mathcal{M}}$, we define an operator $P_{\mathcal{M}}$

as follows

$$P_{\mathcal{M}} f = f_{\mathcal{M}}. \tag{B.13}$$

It is easy to verify that this operator satisfies the two basic properties and therefore is a projection operator. Moreover the subspace $\mathcal{M}$ is precisely the subspace of functions f such that $P_{\mathcal{M}} f = f$.

We provide a few examples of projection operators.

If $\mathcal{M}$ is the one-dimensional linear subspace generated by the function v such that $\|v\| = 1$ then the projection operator onto $\mathcal{M}$ is given by

$$Pf = (f, v)v. \tag{B.14}$$

More generally, if $\mathcal{M}$ is the N-dimensional subspace generated by the orthonormal functions $v_1, v_2, \cdots, v_N$, then the projection operator onto $\mathcal{M}$ is given by

$$Pf = (f, v_1)v_1 + \cdots + (f, v_N)v_N. \tag{B.15}$$

Projection operators frequently used in image processing are the spacelimiting operator and the bandlimiting operator. The *spacelimiting operator* associated with a bounded domain $\mathcal{D}$ is defined by

$$(P^{(\mathcal{D})} f)(\boldsymbol{x}) = \chi_{\mathcal{D}}(\boldsymbol{x}) f(\boldsymbol{x}) \tag{B.16}$$

where $\chi_{\mathcal{D}}(\boldsymbol{x})$ is the characteristic function of the domain $\mathcal{D}$. This operator projects onto the subspace of all functions which are zero outside the domain $\mathcal{D}$.

Analogously the *bandlimiting operator* associated with a bounded domain $\mathcal{B}$ in frequency space is defined by

$$(P^{(\mathcal{B})} f)(\boldsymbol{x}) = \frac{1}{(2\pi)^q} \int \chi_{\mathcal{B}}(\boldsymbol{\omega}) \hat{f}(\boldsymbol{\omega}) e^{i(\boldsymbol{x}\cdot\boldsymbol{\omega})} d\boldsymbol{\omega} \tag{B.17}$$

where $\chi_{\mathcal{B}}(\boldsymbol{\omega})$ is the characteristic function of the domain $\mathcal{B}$. This operator projects onto the subspace of all functions whose Fourier transform is zero outside the domain $\mathcal{B}$.

Appendix C

Euclidean vector spaces and matrices

The discretization of an inverse problem very often implies the use of finite-dimensional linear spaces equipped with weighted scalar products. The need to introduce a weighted scalar product in the discrete image space is discussed, for instance, in section 7.2, while the need for a weighted scalar product in the discrete object space is discussed in section 9.1.

The use of weighted scalar products can cause difficulties because the usual algorithms of matrix diagonalization or singular value decomposition apply to problems where the canonical scalar product is used. The scope of this appendix is twofold. First we prove, in the canonical case, some results used in section 9.2 for deriving the SVD of a matrix. Secondly, we discuss the case of weighted scalar products and we show how this case can be reduced to the canonical one.

An N-dimensional linear space $\mathcal{X}_N$ will be called *vector space*. An element of this space can always be represented by a vector $\boldsymbol{f}$ with components $f_1, f_2, \cdots, f_N$. For instance, if $\mathcal{X}_N$ is the space of the polynomials of degree $\leq N-1$, then the component f_n of the vector $\boldsymbol{f}$, corresponding to a given polynomial, is the coefficient of x^{n-1}.

It is always possible to give a Euclidean structure to a vector space $\mathcal{X}_N$ by introducing a suitable scalar product. Indeed, it is possible to introduce an infinity of different scalar products, whose general structure will be discussed in the following. We will call the *canonical scalar product* that defined as follows

$$(\boldsymbol{f} \cdot \boldsymbol{h}) = \sum_{n=1}^{N} f_n h_n^*. \tag{C.1}$$

This definition applies to the case of complex valued vectors. In the case of real-valued vectors the complex conjugation is obviously omitted. The vector space equipped with this scalar product will be called the *canonical Euclidean space* and will be denoted by $\mathcal{E}_N$. If it is necessary to specify the dimension of the space, we will denote by $(\boldsymbol{f} \cdot \boldsymbol{h})_N$ the scalar product of $\mathcal{E}_N$.

The scalar product (C.1) is the scalar product defined in equation (2.43) (except for a shift in the values of the index) and used in the theory of the DFT.

It is also the scalar product used in the theory of the diagonalization or singular value decomposition of a matrix, as discussed in section 9.2, as well as in the corresponding numerical algorithms.

Consider now a linear operator A from $\mathcal{E}_N$ into $\mathcal{E}_M$. It is well known that to such a linear operator is associated a matrix $M \times N$, i.e. a matrix with M rows and N columns, which will be denoted by $\mathbf{A}$. The notation $A_{m,n}$ or $(\mathbf{A})_{m,n}$ will be used for its matrix elements. Moreover we will denote by p the *rank* of the matrix.

Remark C.1. We recall the definition of the rank of a matrix and the properties of the matrix related to this concept.

A minor of order r of $\mathbf{A}$ is the determinant of a $r \times r$ submatrix of $\mathbf{A}$. Then the matrix $\mathbf{A}$ is said to have rank p if at least one of its minors of order p is different from zero while every minor of order $p+1$, if any, is zero. Since this definition is invariant with respect to an exchange of rows and columns, a matrix $\mathbf{A}$ and its transposed $\mathbf{A}^T$, as well as its adjoint $\mathbf{A}^*$, have the same rank. Moreover, from the definition it is obvious that

$$p \leq \min\{M, N\}. \tag{C.2}$$

The rank is related to other important properties of the matrix. Consider first the equation

$$\mathbf{A}f = 0. \tag{C.3}$$

The definition of rank implies that there exist p, and not more than p, linearly independent rows of the matrix, so that equation (C.3) is equivalent to a homogeneous linear system with p equations and N unknowns. If $p < N$, it is possible to give arbitrary values to $N-p$ components of f and, as a consequence, equation (C.3) has $N-p$ linearly independent solutions. In other words, the null space $\mathcal{N}(\mathbf{A})$ of the matrix (see appendix B) has dimension $N-p$. A similar analysis applies to $\mathbf{A}^T$. In conclusion

$$dim\mathcal{N}(\mathbf{A}) = N - p, \quad dim\mathcal{N}(\mathbf{A}^T) = M - p. \tag{C.4}$$

Consider now the range of the matrix $\mathbf{A}$, $\mathcal{R}(\mathbf{A})$, i.e. the set of all vectors g which are the image of some vector f, $g = \mathbf{A}f$. Since the matrix $\mathbf{A}$ has p, and not more than p, linearly independent columns, it follows that $\mathcal{R}(\mathbf{A})$ contains at most p linearly independent vectors (the columns of $\mathbf{A}$ are the images of the vectors having one component equal to 1 and the others equal to 0). Therefore the range of $\mathbf{A}$ has dimension p. The same analysis applies to $\mathbf{A}^T$, so that

$$dim\mathcal{R}(\mathbf{A}) = dim\mathcal{R}(\mathbf{A}^T) = p. \tag{C.5}$$

Sometimes this property is taken as the definition of the rank.

If we observe that, in a finite dimensional space, the sum of the dimensions of a linear subspace and of its orthogonal complement coincides with the dimension of the space, then for a real matrix, equation (C.5) follows from equation (C.4) and equation (B.8). Indeed if the matrix elements of $\mathbf{A}$ *are real valued then the transposed matrix coincides with the adjoint matrix*

$$\mathbf{A}^* = \mathbf{A}^T \tag{C.6}$$

because, for any $\boldsymbol{f} \in \mathcal{E}_N$ *and* $\boldsymbol{g} \in \mathcal{E}_M$ *we have*

$$(\mathbf{A}\boldsymbol{f} \cdot \boldsymbol{g})_M = \left(\boldsymbol{f} \cdot \mathbf{A}^T \boldsymbol{g}\right)_N . \tag{C.7}$$

We can prove now the properties, stated in section 9.2, of the following matrices

$$\bar{\mathbf{A}} = \mathbf{A}^*\mathbf{A}, \quad \tilde{\mathbf{A}} = \mathbf{A}\mathbf{A}^*. \tag{C.8}$$

It is obvious that these are square matrices and that $\bar{\mathbf{A}}$ is $N \times N$ while $\tilde{\mathbf{A}}$ is $M \times M$.

- *The two matrices are self-adjoint*: In the case of $\bar{\mathbf{A}}$ we have

$$\bar{\mathbf{A}}^* = (\mathbf{A}^*\mathbf{A})^* = \mathbf{A}^*(\mathbf{A}^*)^* = \mathbf{A}^*\mathbf{A} = \bar{\mathbf{A}}. \tag{C.9}$$

  In a similar way we can prove that also $\tilde{\mathbf{A}}$ is self-adjoint.
- *The two matrices are positive semi-definite*: In the case of $\bar{\mathbf{A}}$ we have

$$\left(\bar{\mathbf{A}}\boldsymbol{f} \cdot \boldsymbol{f}\right)_N = (\mathbf{A}^*\mathbf{A}\boldsymbol{f} \cdot \boldsymbol{f})_N = (\mathbf{A}\boldsymbol{f} \cdot \mathbf{A}\boldsymbol{f})_M = \|\mathbf{A}\boldsymbol{f}\|_M^2 \geq 0. \tag{C.10}$$

  A similar proof holds in the case of $\tilde{\mathbf{A}}$. We observe that a positive semi-definite matrix has non-negative eigenvalues. Indeed, if $\bar{\mathbf{A}}\boldsymbol{v} = \lambda \boldsymbol{v}$, then

$$0 \leq \left(\bar{\mathbf{A}}\boldsymbol{v}, \boldsymbol{v}\right)_N = \lambda \|\boldsymbol{v}\|_N^2 \tag{C.11}$$

  and therefore $\lambda \geq 0$.
- *The two matrices have rank p*: We will prove that

$$\mathcal{N}(\bar{\mathbf{A}}) = \mathcal{N}(\mathbf{A}), \quad \mathcal{N}(\tilde{\mathbf{A}}) = \mathcal{N}(\mathbf{A}^*); \tag{C.12}$$

  then the property follows from equations (C.4).
  In order to prove the first of the relations (C.12) let us observe that, if $\mathbf{A}\boldsymbol{f} = 0$, then $\bar{\mathbf{A}} = 0$ so that $\mathcal{N}(\mathbf{A})$ is contained in $\mathcal{N}(\bar{\mathbf{A}})$. On the other hand, if $\bar{\mathbf{A}}\boldsymbol{f} = 0$, then

$$0 = \left(\bar{\mathbf{A}}\boldsymbol{f} \cdot \boldsymbol{f}\right)_N = \|\mathbf{A}\boldsymbol{f}\|_M^2 \tag{C.13}$$

  so that $\mathbf{A}\boldsymbol{f} = 0$. This implies that $\mathcal{N}(\bar{\mathbf{A}})$ is contained in $\mathcal{N}(\mathbf{A})$. By combining the two inclusions, we get the first of the relations (C.12). The second is obtained in a similar way.

In the final part of this appendix we consider the case of weighted scalar products. Indeed the most general scalar product in a space $\mathcal{X}_N$ of real-valued vectors has the following structure. Let $\mathbf{C}$ be a $N \times N$ *symmetric and positive-definite matrix*, i.e. a matrix whose matrix elements $C_{m,n}$ satisfy the following conditions:

$$C_{m,n} = C_{n,m} \tag{C.14}$$

and

$$\sum_{m,n=1}^{N} C_{m,n} f_n f_m^* > 0 \tag{C.15}$$

for any vector $\boldsymbol{f} \neq 0$. We also assume that this matrix is real-valued. Such a matrix $\mathbf{C}$ is non-singular, so that the inverse matrix $\mathbf{C}^{-1}$ exists. Then it is easy to verify that the following equation

$$(\boldsymbol{f} \cdot \boldsymbol{h})_{N,\mathbf{C}} = \sum_{m,n=1}^{N} C_{m,n} f_n h_m^* \tag{C.16}$$

defines a scalar product because $(\boldsymbol{f} \cdot \boldsymbol{h})_{N,\mathbf{C}}$ satisfies the basic properties of a scalar product given in appendix A. As a consequence the norm and the distance can be defined respectively by equation (A.3) and equation (A.10). We also remark that the scalar product (C.16) can be written in terms of the canonical scalar product as follows

$$(\boldsymbol{f} \cdot \boldsymbol{h})_{N,\mathbf{C}} = (\mathbf{C}\boldsymbol{f} \cdot \boldsymbol{h})_N = (\boldsymbol{f} \cdot \mathbf{C}\boldsymbol{h})_N. \tag{C.17}$$

Equation (C.16) is essentially the discrete version of equation (A.20). Such a scalar product will be called a *weighted scalar product* and the vector space $\mathcal{X}_N$, equipped with this scalar product will be denoted by $\mathcal{E}_{N,\mathbf{C}}$.

It is obvious that, even if $\mathcal{E}_N$ and $\mathcal{E}_{N,\mathbf{C}}$ coincide as linear spaces, their geometrical structure is different: two vectors, which are orthogonal in $\mathcal{E}_N$, in general are not orthogonal in $\mathcal{E}_{N,\mathbf{C}}$ and, conversely, an orthonormal basis of $\mathcal{E}_N$ is still a basis of $\mathcal{E}_{N,\mathbf{C}}$, because the vectors are linearly independent, but it is not orthonormal; the orthogonal complements of a subset $\mathcal{S}$ in $\mathcal{E}_N$ and $\mathcal{E}_{N,\mathbf{C}}$ are different linear subspaces and so on. Since the orthogonality relations are very important in the theory of matrices, one understands that the need of introducing a weighted scalar product in the vector space of the images and/or of the objects can generate difficulties.

There always exists, however, a suitable change of variables transforming a weighted Euclidean space into a canonical one. This change of variables is elementary in the case of a weighted scalar product defined by a diagonal matrix $C_{m,n} = c_n \delta_{m,n}$, with $c_n > 0$:

$$(\boldsymbol{f} \cdot \boldsymbol{h})_{N,\mathbf{C}} = \sum_{n=1}^{N} c_n f_n h_n^*. \tag{C.18}$$

Indeed, by introducing the vectors $\bar{f}$ and $\bar{h}$ with components $\bar{f}_n = \sqrt{c_n} f_n$ and $\bar{h}_n = \sqrt{c_n} h_n$, respectively, we find that the scalar product of f, h coincides with the canonical scalar product of $\bar{f}, \bar{h}$: $(f \cdot h)_{N,\mathbf{C}} = (\bar{f} \cdot \bar{h})_N$.

This procedure can be extended to the general case by the use of the so-called *Choleski factorization* of a matrix (see [5], pp 89–91): *if* $\mathbf{C}$ *is positive definite, there exists an upper triangular matrix* $\mathbf{H}$, *with positive diagonal elements, such that*

$$\mathbf{C} = \mathbf{H}^T \mathbf{H}. \tag{C.19}$$

Since efficient and simple algorithms are available for this factorization, given the matrix $\mathbf{C}$, which defines the scalar product (C.16), we can easily compute its Choleski factor $\mathbf{H}$. Then, given two vectors f and h of $\mathcal{E}_{N,\mathbf{C}}$, we can perform the following change of variables

$$\bar{f} = \mathbf{H} f, \quad \bar{h} = \mathbf{H} h. \tag{C.20}$$

From equations (C.17) and (C.19) it follows that

$$\begin{aligned} (f \cdot h)_{N,\mathbf{C}} &= (\mathbf{C} f \cdot h)_N = (\mathbf{H}^T \mathbf{H} f \cdot h)_N \\ &= (\mathbf{H} f \cdot \mathbf{H} h)_N = (\bar{f} \cdot \bar{h})_N \end{aligned} \tag{C.21}$$

and therefore: *the weighted scalar product of the vectors* f, h *coincides with the canonical scalar product of the vectors* $\bar{f}, \bar{h}$.

We show how this change of variables can be applied to the SVD of a matrix defining an imaging operator in vector spaces equipped with weighted scalar products. Assume that the object space is $\mathcal{E}_{N,\mathbf{B}}$ while the image space is $\mathcal{E}_{M,\mathbf{C}}$, with weighting matrices having respectively the following Choleski factorizations

$$\mathbf{B} = \mathbf{F}^T \mathbf{F}, \quad \mathbf{C} = \mathbf{G}^T \mathbf{G}. \tag{C.22}$$

Consider a matrix $\mathbf{A}$ defining a linear operator from $\mathcal{E}_{N,\mathbf{B}}$ into $\mathcal{E}_{M,\mathbf{C}}$. Such a matrix defines also an operator from $\mathcal{E}_N$ into $\mathcal{E}_M$. Its action, as an operator, is independent of the choice of the scalar product. In particular the null space and the range are independent of this choice. What is changing is the adjoint matrix. Indeed, while in the case of the pair $\{\mathcal{E}_N, \mathcal{E}_M\}$ the adjoint is just the Hermitian conjugate matrix, in the case of the pair $\{\mathcal{E}_{N,\mathbf{B}}, \mathcal{E}_{M,\mathbf{C}}\}$ the adjoint is associated with a different matrix which will be denoted by $\mathbf{A}_w^*$ (weighted adjoint). This matrix can be computed using the general definition (B.6). From the relation, which holds true for any pair of vectors, f, g

$$\begin{aligned} (\mathbf{A} f \cdot g)_{M,\mathbf{C}} &= (\mathbf{C}\mathbf{A} f \cdot g)_M = (f \cdot \mathbf{A}^* \mathbf{C} g)_N \\ &= (\mathbf{B}^{-1}\mathbf{B} f \cdot \mathbf{A}^* g)_N = (f \cdot \mathbf{B}^{-1}\mathbf{A}^*\mathbf{C} g)_{N,\mathbf{B}} \\ &= (f \cdot \mathbf{A}_w^* g)_{N,\mathbf{B}} \end{aligned} \tag{C.23}$$

it follows that

$$\mathbf{A}_w^* = \mathbf{B}^{-1}\mathbf{A}^*\mathbf{C}. \tag{C.24}$$

By the use of this adjoint matrix we can repeat the analysis of section 9.2: introduce the matrices $\bar{\mathbf{A}}_w = \mathbf{A}_w^*\mathbf{A}$ and $\tilde{\mathbf{A}}_w = \mathbf{A}\mathbf{A}_w^*$; prove that these matrices have the same positive eigenvalues etc. In such a way we obtain a weighted singular system of the matrix $\mathbf{A}$, $\{\sigma_{w,k}; \boldsymbol{u}_{w,k}, \boldsymbol{v}_{w,k}\}$, which consists of the solutions of the shifted eigenvalue problem

$$\mathbf{A}\boldsymbol{v}_{w,k} = \sigma_{w,k}\boldsymbol{u}_{w,k}, \quad \mathbf{A}_w^*\boldsymbol{u}_{w,k} = \sigma_{w,k}\boldsymbol{v}_{w,k}. \tag{C.25}$$

In conclusion, the theory can be developed along the lines indicated in section 9.2 but the singular system cannot be computed using the standard algorithms because these apply only to the case of canonical Euclidean spaces.

However if we consider the change of variables associated with the Choleski factorizations (C.22)

$$\boldsymbol{g}_0 = \mathbf{G}\boldsymbol{g}, \quad \boldsymbol{f}_0 = \mathbf{F}\boldsymbol{f}, \tag{C.26}$$

then the basic equation $\boldsymbol{g} = \mathbf{A}\boldsymbol{f}$ becomes

$$\boldsymbol{g}_0 = \mathbf{A}_0\boldsymbol{f}_0 \tag{C.27}$$

where

$$\mathbf{A}_0 = \mathbf{G}\mathbf{A}\mathbf{F}^{-1}. \tag{C.28}$$

The relationship between $\mathbf{A}_0^*$ and $\mathbf{A}_w^*$ is given by

$$\mathbf{A}_0^* = \mathbf{F}\mathbf{A}_w^*\mathbf{G}^{-1}, \tag{C.29}$$

as follows from equations (C.28), (C.24) and (C.22).

Since the modified vectors $\boldsymbol{f}_0, \boldsymbol{g}_0$ are elements of canonical spaces and the matrix $\mathbf{A}_0$ represents the effect of the imaging system on these modified vectors, we can apply to $\mathbf{A}_0$ the standard theory of section 9.2. Moreover we can also use the standard algorithms for computing the singular system of $\mathbf{A}_0$

$$\mathbf{A}_0\bar{\boldsymbol{v}}_k = \bar{\sigma}_k\bar{\boldsymbol{u}}_k, \quad \mathbf{A}_0^*\bar{\boldsymbol{u}}_k = \bar{\sigma}_k\bar{\boldsymbol{v}}_k. \tag{C.30}$$

By comparing equations (C.30) and (C.25) and by taking into account equations (C.28) and (C.29) we find that

$$\sigma_{w,k} = \bar{\sigma}_k, \quad \boldsymbol{u}_{w,k} = \mathbf{G}^{-1}\bar{\boldsymbol{u}}_k, \quad \boldsymbol{v}_k = \mathbf{F}^{-1}\bar{\boldsymbol{v}}_k. \tag{C.31}$$

In conclusion, in order to compute the singular system of a matrix in weighted Euclidean spaces the following procedure can be used:

- compute the Choleski factorization of the weighting matrices, equation (C.22);
- compute the matrix $\mathbf{A}_0$, equation (C.28);
- compute the singular system of $\mathbf{A}_0$ by means of standard algorithms;
- apply the inverse change of variables to the singular vectors of $\mathbf{A}_0$, equation (C.31).

In steps 2 and 4 we need the inverse of a triangular matrix; it is well known that this inverse matrix can be computed easily and very efficiently [5].

Appendix D

Properties of the DFT and the FFT algorithm

In this appendix we prove the properties of the DFT stated in section 2.4 and we provide the foundations of the FFT algorithm. Some of the proofs are simplified by the use of methods from matrix theory.

Indeed, the DFT of a vector, can be written as the result of a matrix-vector multiplication. If we introduce the $N \times N$ matrix F defined by

$$F_{m,n} = e^{-i\frac{2\pi}{N}mn}; \quad m, n = 0, 1, \cdots, N-1 \tag{D.1}$$

(here we use indices running from 0 to $N-1$, and not from 1 to N, because this is the choice in the standard definition of the DFT) then the vector $\hat{\boldsymbol{f}}$, defined in equation (2.39) is given by

$$\hat{\boldsymbol{f}} = \mathsf{F}\boldsymbol{f}. \tag{D.2}$$

The inversion formula (2.40) implies the existence of the inverse matrix F^{-1}.

Proof D.1. The inversion formula (2.40). *From the definition of the DFT, equation (2.39), we get*

$$\frac{1}{N}\sum_{m=0}^{N-1}\hat{f}_m \exp\left(i\frac{2\pi}{N}nm\right) = \frac{1}{N}\sum_{m=0}^{N-1}\left(\sum_{l=0}^{N-1} f_l \exp\left(-i\frac{2\pi}{N}ml\right)\right)\exp\left(i\frac{2\pi}{N}nm\right)$$
$$= \sum_{l=0}^{N-1} f_l\left(\frac{1}{N}\sum_{m=0}^{N-1}\exp\left[i\frac{2\pi}{N}(n-l)m\right]\right). \tag{D.3}$$

Then equation (2.40) is a consequence of the following orthogonality property

$$\frac{1}{N}\sum_{m=0}^{N-1}\exp\left[i\frac{2\pi}{N}(n-l)m\right] = \delta_{nl} \tag{D.4}$$

where δ_{nl} is the usual Kronecker symbol. Equation (D.4) is obvious in the case $n = l$. In the case $n \neq l$ it is a consequence of the fact that the sum of the Nth roots of 1 is zero. □

The inversion formula (2.40) implies that the inverse matrix $\mathbf{F}^{-1}$ exists and is related to the adjoint (hermitian conjugate) matrix $\mathbf{F}^*$ by

$$\mathbf{F}^{-1} = \frac{1}{N}\mathbf{F}^*. \tag{D.5}$$

This relation can also be obtained by observing that equation (D.4) can be written in matrix notation as follows

$$\mathbf{F}^*\mathbf{F} = \mathbf{F}\mathbf{F}^* = N\mathbf{I} \tag{D.6}$$

where $\mathbf{I}$ is the $N \times N$ unity matrix. This equation also provides a very simple proof of the generalized Parseval equality.

Proof D.2. The generalized Parseval equality (2.42). From equations (D.2), (D.6), using the scalar product (2.43) and the property (B.6) of the adjoint matrix we have

$$(\hat{\boldsymbol{f}} \cdot \hat{\boldsymbol{h}}) = (\mathbf{F}\boldsymbol{f} \cdot \mathbf{F}\boldsymbol{h}) = (\boldsymbol{f} \cdot \mathbf{F}^*\mathbf{F}\boldsymbol{h}) = N(\boldsymbol{f} \cdot \boldsymbol{h}) \tag{D.7}$$

and this is precisely equation (2.42). By taking $\boldsymbol{h} = \boldsymbol{f}$, we get equation (2.41). □

In section 2.4 the extension of the vector $\boldsymbol{f}$ to a periodic sequence of period N has been considered as well as a translation of this sequence. It has been shown that a translation of the sequence is equivalent to a cyclic permutation of the components of the vector $\boldsymbol{f}$. It follows that a p-translation is equivalent to the multiplication of $\boldsymbol{f}$ by a *permutation matrix* $\mathbf{R}_p$ defined by

$$(\mathbf{R}_p f)_n = \begin{cases} f_{n+p} & ; \quad n = 0, 1, \cdots, N-p-1 \\ f_{n+p-N} & ; \quad n = N-p, N-p+1, \cdots, N-1. \end{cases} \tag{D.8}$$

The structure of the matrix $\mathbf{R}_p$ is the following

$$\mathbf{R}_p = \begin{pmatrix} \mathbf{0} & \mathbf{I}_{N-p} \\ \mathbf{I}_p & \mathbf{0} \end{pmatrix} \tag{D.9}$$

where $\mathbf{I}_p$ and $\mathbf{I}_{N-p}$ are the unity matrices $p \times p$ and $(N-p) \times (N-p)$ respectively.

Equation (2.51) can now be proved by computing the DFT of the vector $\mathbf{R}_p \boldsymbol{f}$.

Proof D.3. DFT and translations. The DFT of $\mathbf{R}_p f$ *is given by*

$$(\mathbf{FR}_p f)_m = \sum_{n=0}^{N-1} \exp\left(-i\frac{2\pi}{N}mn\right)(\mathbf{R}_p f)_n \tag{D.10}$$

$$= \sum_{n=0}^{N-1-p} \exp\left(-i\frac{2\pi}{N}mn\right) f_{n+p} + \sum_{n=N-p}^{N-1} \exp\left(-i\frac{2\pi}{N}mn\right) f_{n+p-N}.$$

By introducing the index $n' = n + p$ *in the first summation and the index* $n' = n + p - N$ *in the second summation we get*

$$(\mathbf{FR}_p f)_m = \sum_{n'=p}^{N-1} \exp\left[-i\frac{2\pi}{N}m(n'-p)\right] f_{n'}$$

$$+ \sum_{n'=0}^{p-1} \exp\left[-i\frac{2\pi}{N}m(n'+N-p)\right] f_{n'}$$

$$= \exp\left(i\frac{2\pi}{N}pm\right)(\mathbf{F}f)_m \tag{D.11}$$

and this is just equation (2.51). □

One of the basic properties of the DFT is that the matrix **F** diagonalizes any cyclic matrix **A**. This is the content of the cyclic convolution theorem, as given in equation (2.61).

Proof D.4. The cyclic-convolution theorem. By computing the DFT of both sides of equation (2.59) we have

$$\hat{g}_m = \sum_{n=0}^{N-1}\left(\sum_{l=0}^{N-1} K_{n-l} f_l\right) \exp\left(-i\frac{2\pi}{N}mn\right)$$

$$= \sum_{l=0}^{N-1}\left(\sum_{n=0}^{N-1} K_{n-l} \exp\left(-i\frac{2\pi}{N}mn\right)\right) f_l. \tag{D.12}$$

Then, using equation (2.51) we have

$$\hat{g}_m = \sum_{l=0}^{N-1} \hat{K}_m \exp\left(-i\frac{2\pi}{N}ml\right) f_l = \hat{K}_m \hat{f}_m \tag{D.13}$$

which is just equation (2.61). □

From this result and the definition (2.62) of a cyclic matrix it follows that

$$(\mathbf{A}f)_m = \hat{K}_m (\mathbf{F}f)_m \tag{D.14}$$

or also

$$\mathbf{FA} = \mathbf{\Lambda F} \tag{D.15}$$

where $\mathbf{\Lambda}$ denotes the diagonal matrix whose diagonal elements are just the $\hat{K}_m$. From equation (D.15) and equation (D.6), if $\mathbf{V}$ is the matrix defined by

$$\mathbf{V} = \frac{1}{\sqrt{N}} \mathbf{F}^* \tag{D.16}$$

one obtains

$$\mathbf{A} = \mathbf{V\Lambda V}^* \tag{D.17}$$

and this is precisely the diagonal form of the matrix $\mathbf{A}$.

As it is known, the columns of the matrix $\mathbf{V}$ are the eigenvectors of $\mathbf{A}$ and therefore they coincide with the orthonormal vectors (2.46). It follows that equation (2.64) holds true.

The Fast Fourier Transform (FFT) algorithm is a very efficient way to compute the DFT with a substantial reduction of computer time due to the reduction of the number of operations. It applies in general to the case $N = r^p$ [6] but it is used mainly in the case $r = 2$ because in such a case it can be easily implemented in computers with binary arithmetic. While the trivial evaluation of DFT by means of equation (2.39) requires N^2 operation, the FFT algorithm only requires $2N \log_2 N$ operations in the case $N = 2^p$, without requiring more data storage. In the following we will assume that N is a power of 2.

If we introduce the complex number $W = \exp(-i2\pi/N)$, the DFT formula and the inversion formula become respectively

$$\hat{f}_m = \sum_{n=0}^{N-1} f_n W^{mn}; \quad m = 0, 1, \cdots, N-1 \tag{D.18}$$

and

$$f_n = \frac{1}{N} \sum_{m=0}^{N-1} \hat{f}_m (W^*)^{nm}; \quad n = 0, 1, \cdots, N-1. \tag{D.19}$$

We derive the FFT algorithm in the case of the DFT; the same algorithm is also applicable, with some minor modifications, to the inversion formula (D.19).

If we introduce the binary decompositions ($N = 2^p$; $m, n < N$)

$$\begin{aligned} m &= 2^{p-1} m_{p-1} + 2^{p-2} m_{p-2} + \cdots + 2m_1 + m_0 \\ n &= 2^{p-1} n_{p-1} + 2^{p-2} n_{p-2} + \cdots + 2n_1 + n_0 \end{aligned} \tag{D.20}$$

where $n_i, m_i = 0, 1$, then m, n are represented by binary numbers of length p

$$m = (m_{p-1}, m_{p-2}, \cdots, m_1, m_0)$$
$$n = (n_{p-1}, n_{p-2}, \cdots, n_1, n_0). \tag{D.21}$$

The representations (D.20) can be used for computing the product mn, i.e. the exponent of W in (D.18). This product will contain multiples of $2^p = N$, which can be neglected because their contribution to W^{mn} is 1. Therefore, for the computation of W^{mn} we can replace mn by the following sum

$$mn \sim c_0 + c_1 + \cdots + c_{p-1} \tag{D.22}$$

where

$$\begin{aligned} c_0 &= 2^{p-1} m_0 n_{p-1} \\ c_1 &= (2^{p-1} m_1 + 2^{p-2} m_0) n_{p-2} \\ &\cdots\cdots\cdots \\ c_{p-1} &= (2^{p-1} m_{p-1} + 2^{p-2} m_{p-2} + \cdots + 2m_1 + m_0) n_0. \end{aligned} \tag{D.23}$$

If we insert these formulae in equation (D.18) and if we write $\hat{f}(m_{p-1}, \cdots, m_0)$ in place of $\hat{f}_m$ and $f(n_{p-1}, \cdots, n_0)$ in place of f_n, we obtain

$$\begin{aligned} &\hat{f}(m_{p-1}, \cdots, m_0) \\ &\quad = \sum_{n_0=0}^{1} \sum_{n_1=0}^{1} \cdots \sum_{n_{p-1}=0}^{1} f(n_{p_1}, \cdots, n_0) W^{c_0} W^{c_1} \ldots W^{c_{p-1}}. \end{aligned} \tag{D.24}$$

One finds that the computation of one component of $\hat{f}$ is reduced to the computation of $p = \log_2 N$ summations, each one containing two terms. Since N is the number of components, the total number of operations required is $2N \log_2 N$.

The computation can be performed in p steps by introducing intermediate vectors as follows

$$\begin{aligned} f^{(1)}(m_0, n_{p-2}, \cdots, n_1, n_0) &= \sum_{n_{p-1}=0}^{1} f(n_{p-1}, n_{p-2}, \cdots, n_1, n_0) W^{c_0} \\ f^{(2)}(m_0, m_1, \cdots, n_1, n_0) &= \sum_{n_{p-2}=0}^{1} f^{(1)}(m_0, n_{p-2}, \cdots, n_1, n_0) W^{c_1} \\ &\cdots\cdots\cdots \\ f^{(p)}(m_0, m_1, \cdots, m_{p-2}, m_{p-1}) &= \sum_{n_0=0}^{1} f^{(p-1)}(m_0, m_1, \cdots, m_{p-2}, n_0) W^{c_{p-1}}. \end{aligned} \tag{D.25}$$

By comparing with equation (D.24), we see that $\hat{f}(m_{p-1}, m_{p-2}, \cdots, m_0) = f^{(p)}(m_0, m_1, \cdots, m_{p-1})$. Now, if we observe that the indexing convention used

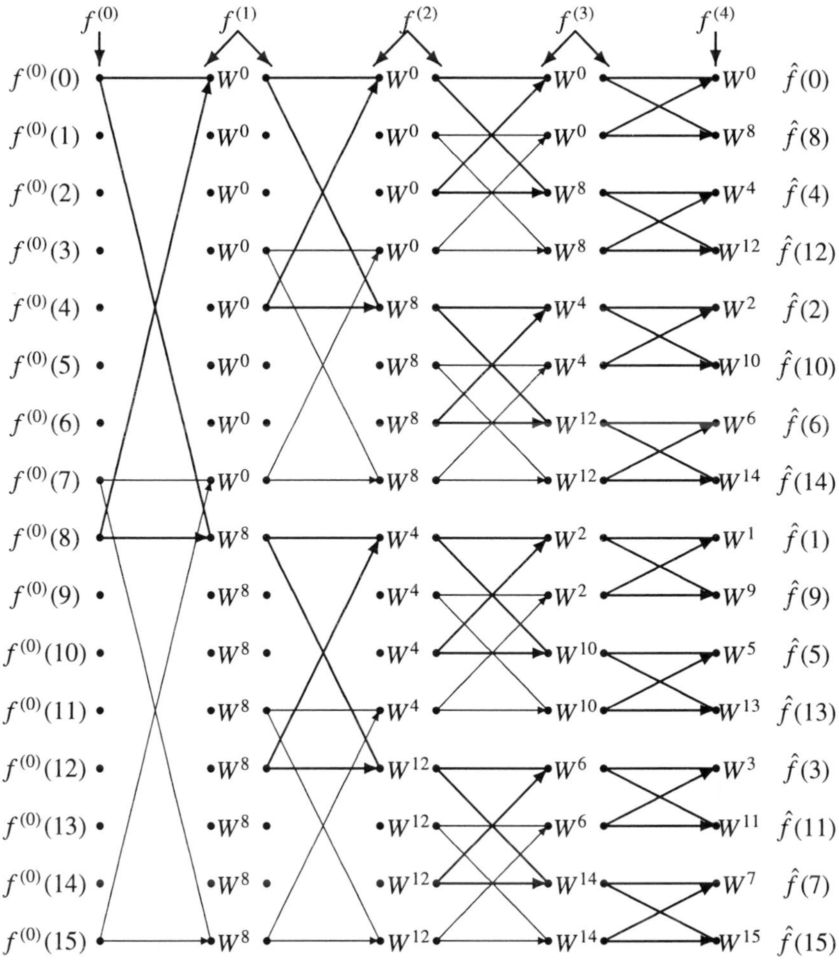

Figure D.1. Butterfly scheme of the FFT algorithm in the case $N = 2^4 = 16$

indicates the locations where the components of the vector are stored, we find that the previous procedure locates the components of $\hat{f}$ according to an inverted binary representation of the index because m_{p-1} is replaced by m_0, m_{p-2} by m_1, and so on. This order is called *bit-reversal*. By reordering the vector $f^{(p)}$ we obtain $\hat{\boldsymbol{f}}$.

In order to clarify the FFT algorithm we give its graphical representation by means of the well-known *butterfly scheme*. In figure D.1 we plot this scheme in the case $N = 16$, i.e. $p = 4$. From this figure it is easy to deduce that the FFT algorithm consists of p steps and that, at each step, the number of operations is $2N$, so that the total number of operations is $2N \log_2 N$, as already observed.

Many excellent implementations of this algorithm are available in the most important computer languages (see, for instance, [5]). These algorithms contain the computation of the quantities W^{c_i}. In methods where the FFT routine is called many times (typically, iterative methods) this implies a useless repetition of operations. In these cases it is convenient to write an initializing routine where the quantities W^{c_i}, which depend only on the number of points N, are computed with high accuracy, using, for instance, double-precision values of the trigonometric functions.

Appendix E

Minimization of quadratic functionals

In this appendix we consider the problem of minimizing the quadratic functional which is basic in regularization theory ($\mu \geq 0$)

$$\Phi_\mu(f; g) = \|Af - g\|_{\mathcal{Y}}^2 + \mu \, \|f\|_{\mathcal{X}}^2 ; \tag{E.1}$$

for $\mu = 0$ we obtain the discrepancy functional, i.e. the functional of the least-squares problem.

We assume that, in general, the spaces $\mathcal{X}$ and $\mathcal{Y}$ are different in order to cover the problems considered in the second part of the book. In the case of discrete problems, these spaces can be weighted Euclidean spaces (see appendix C). We consider this case in detail at the end of this appendix.

The main scope of the appendix is to derive the Euler equation which must be satisfied by the minimum points of the functional. To this purpose, let us denote by f_μ any one of these points and let us observe that, since the functional is convex and bounded from below, all minima are global. This means that, for any f in $\mathcal{X}$

$$\Phi(f_\mu; g) \leq \Phi(f; g). \tag{E.2}$$

As a consequence, if we consider the variation of the functional in the direction characterized by a function (vector) h, we have

$$\Phi(f_\mu; g) \leq \Phi(f_\mu + \alpha h; g) \tag{E.3}$$

for any complex number α (and, of course, also for any h).

From the properties of the scalar product we have

$$\begin{aligned}
\Phi(f_\mu + \alpha h; g) = \; & \left\|Af_\mu - g\right\|_{\mathcal{Y}}^2 + \mu \|f_\mu\|_{\mathcal{Y}}^2 \\
& + \alpha \left\{ \left(Ah, Af_\mu - g\right)_{\mathcal{Y}} + \mu \left(h, f_\mu\right)_{\mathcal{X}} \right\} \\
& + \alpha^* \left\{ \left(Af_\mu - g, Ah\right)_{\mathcal{Y}} + \mu \left(f_\mu, h\right)_{\mathcal{X}} \right\} \\
& + |\alpha|^2 \left(\|Ah\|_{\mathcal{Y}}^2 + \mu \|h\|_{\mathcal{X}}^2 \right) .
\end{aligned} \tag{E.4}$$

Since the term quadratic in $|\alpha|$ is non-negative, it is obvious that condition (E.3) can be satisfied for any α and any h if and only if the term linear in α is zero. Using the definition of the adjoint operator this condition can be written as follows

$$\alpha\left(h, A^*Af_\mu + \mu f_\mu - A^*g\right)_{\mathcal{X}} + \alpha^*\left(A^*Af_\mu + \mu f_\mu - A^*g, h\right)_{\mathcal{X}} = 0 \qquad \text{(E.5)}$$

and it implies that

$$\left(A^*Af_\mu + \mu f_\mu - A^*g, h\right)_{\mathcal{X}} = 0. \qquad \text{(E.6)}$$

Finally since this orthogonality condition must be true for any element h of $\mathcal{X}$, it follows that

$$\left(A^*A + \mu I\right) f_\mu = A^*g. \qquad \text{(E.7)}$$

This is precisely the so-called Euler equation of the functional (E.1).

The previous analysis implies that, if f_μ is a minimum point of $\Phi_\mu(f; g)$, then it must be a solution of equation (E.7). Conversely, if f_μ is a solution of equation (E.7), then from equation (E.4) one derives that f_μ satisfies condition (E.3) and therefore is a minimum point of $\Phi_\mu(f; g)$.

If $\mu > 0$, equation (E.7) has a unique solution and therefore the functional (E.1) has a unique minimum point. Indeed, the operator at the l.h.s. of equation (E.7) is positive definite, as follows from the following inequality

$$\left(A^*Af + \mu f, f\right)_{\mathcal{X}} = \|Af\|_{\mathcal{Y}}^2 + \mu \|f\|_{\mathcal{X}}^2 \geq \mu \|f\|_{\mathcal{X}}^2 . \qquad \text{(E.8)}$$

Then, by applying Schwarz inequality to the scalar product in the first member we obtain

$$\left\|\left(A^*A + \mu I\right) f\right\|_{\mathcal{X}} \geq \mu \|f\|_{\mathcal{X}} . \qquad \text{(E.9)}$$

This inequality has the following implications

- the equation $(A^*A + \mu I)f = 0$ has the unique solution $f = 0$, i.e. the solution of equation (E.7) is unique;
- the inverse operator $(A^*A + \mu I)^{-1}$ is bounded (and its norm is bounded by μ^{-1}) so that equation (E.7) has a solution for any g.

The solution can be written in the form

$$f_\mu = \left(A^*A + \mu I\right)^{-1} A^*g \qquad \text{(E.10)}$$

which is the form used both in chapter 5 and in chapter 10.

The previous computations provide also the expression of the so-called *gradient* $\nabla_f \Phi_\mu$ of the functional (E.1) which is defined by

$$\Phi_\mu(f + h; g) = \Phi_\mu(f; g) + Re\left(\nabla_f \Phi_\mu, h\right)_{\mathcal{X}} + O\left(\|h\|_{\mathcal{X}}^2\right). \qquad \text{(E.11)}$$

By comparing this equation with equation (E.4) we obtain

$$\nabla_f \Phi_\mu(f; g) = 2\left(A^*Af + \mu f - A^*g\right). \qquad \text{(E.12)}$$

We consider now the limiting case $\mu = 0$. The limit of the functional (E.1) is the discrepancy functional

$$\lim_{\mu \to 0} \Phi_\mu(f; g) = \varepsilon^2(f; g), \tag{E.13}$$

so that equation (E.7) with $\mu = 0$ becomes the equation of the least-squares solutions

$$A^* A f = A^* g. \tag{E.14}$$

If the null space of the operator A contains only the null element, then the inverse of the operator $A^* A$ exists and the least-squares solution is unique.

This situation never occurs in the case of bandlimited imaging systems but it may occur for some of the problems considered in Part II. In such a case, if the data g satisfy Picard criterion, the unique least-squares solution, i.e. the generalized solution, is given by

$$f^\dagger = (A^* A)^{-1} A^* g. \tag{E.15}$$

In the case of a non-trivial null space $\mathcal{N}(A)$, given any solution $\tilde{f}$ of equation (E.14), all others are given by $f = \tilde{f} + v$ where v is an arbitrary element of $\mathcal{N}(A)$. Now, if we denote by $f^\dagger$ the generalized solution, i.e. the least-squares solution of minimal norm, then, for any v

$$\| f^\dagger \|_{\mathcal{X}} \leq \| f^\dagger + v \|_{\mathcal{X}}. \tag{E.16}$$

Since we have, for any complex number α

$$\| f^\dagger + \alpha v \|_{\mathcal{X}}^2 = \| f^\dagger \|_{\mathcal{X}}^2 + \alpha \left(v, f^\dagger \right)_{\mathcal{X}} + \alpha^* \left(f^\dagger, v \right)_{\mathcal{X}} + |\alpha|^2 \|v\|_{\mathcal{X}}^2 \tag{E.17}$$

condition (E.16) is satisfied if and only if

$$\left(f^\dagger, v \right)_{\mathcal{X}} = 0 \tag{E.18}$$

for any v in $\mathcal{N}(A)$. We conclude that *the generalized solution is the unique least-squares solution which is orthogonal to the null-space of the operator A.*

From equation (E.12) with $\mu = 0$ we also obtain the expression of the gradient of the discrepancy functional

$$\nabla_f \varepsilon^2(f, g) = 2 \left(A^* A f - A^* g \right). \tag{E.19}$$

We conclude this appendix with a few remarks about the case of discrete problems in weighted Euclidean spaces. Assume that $\mathcal{X} = \mathcal{E}_{N,\mathbf{B}}$ and $\mathcal{Y} = \mathcal{E}_{M,\mathbf{C}}$ (see appendix C). In such a case the matrix $\mathbf{A}^*$ in equation (E.7) is the matrix denoted by $\mathbf{A}_w^*$ in equation (C.24). From this expression we get, multiplying both sides of equation (E.7) by the matrix $\mathbf{B}$, that this equation takes the following form

$$\left(\mathbf{A}^* \mathbf{C} \mathbf{A} + \mu \mathbf{B} \right) \boldsymbol{f}_\mu = \mathbf{A}^* \mathbf{C} \boldsymbol{g} \tag{E.20}$$

where $\mathbf{A}^*$ is the Hermitian conjugate of the matrix $\mathbf{A}$. Then, the inequality (E.8) implies that the matrix on the l.h.s. has an inverse so that

$$f_\mu = \left(\mathbf{A}^*\mathbf{C}\mathbf{A} + \mu\mathbf{B}\right)^{-1}\mathbf{A}^*\mathbf{C}g. \tag{E.21}$$

If we take $\mathbf{C} = \mathbf{S}_\nu^{-1}$ and $\mathbf{B} = \mathbf{S}_\phi^{-1}$ we obtain equation (7.72) and equation (7.73).

Appendix F

Contraction and non-expansive mappings

In this appendix we give a few results about contraction mappings. These operators are usually defined in a metric space setting. However, since in this book we only consider Hilbert spaces, we give the definition in this case. More results can be found in [7].

An operator T (in general, nonlinear) in a Hilbert space $\mathcal{X}$ is said to be a *contraction mapping* if there exists a constant $\rho < 1$ such that, for any pair of elements f_1, f_2 of $\mathcal{X}$

$$\|T(f_2) - T(f_1)\| \leq \rho \|f_1 - f_2\|. \tag{F.1}$$

An example of contraction mapping is provided in section 6.3. If the operator T is linear, then condition (F.1) is equivalent to say that the norm of the operator is strictly smaller than 1.

Condition (F.1) implies that the operator T is continuous. Indeed, if f_k is a sequence converging to f, from (F.1) we get $\|T(f_k) - T(f)\| \leq \rho \|f_k - f\|$ and therefore the sequence $T(f_k)$ converges to $T(f)$.

A fixed point of the operator T is any element f of $\mathcal{X}$ such that

$$T(f) = f. \tag{F.2}$$

If T is a contraction mapping, then there exists at most one fixed point. Indeed, if f_1, f_2 are two fixed points of T, using again condition (F.1) we get $\|f_2 - f_1\| = \|T(f_2) - T(f_1) \leq \rho \|f_2 - f_1\|$ and therefore $\|f_2 - f_1\| = 0$ since $\rho < 1$. This implies $f_2 = f_1$. If T is also linear, then the unique fixed point is $f = 0$.

As concerns the existence of the fixed point of T, this is assured by the *contraction principle: for any starting element f_0, the sequence of the successive approximations*

$$f_{k+1} = T(f_k) \tag{F.3}$$

is convergent and its limit is the unique fixed point of T.

The proof is based on the completeness of the Hilbert space $\mathcal{X}$ and essentially consists of proving that the sequence f_k is a Cauchy sequence (see appendix A). First observe that from condition (F.1) we have

$$\|f_{k+1} - f_k\| \leq \rho \|f_k - f_{k-1}\| \leq \cdots \leq \rho^k \|f_1 - f_0\|. \tag{F.4}$$

Then, for any $p > 1$, using the triangle inequality, we obtain

$$\begin{aligned} \|f_{k+p} - f_k\| &\leq \|f_{k+p} - f_{k+p-1}\| + \cdots + \|f_{k+1} - f_k\| \\ &\leq (\rho^{k+p-1} + \rho^{k+p-2} + \cdots + \rho^k)\|f_1 - f_0\| < \frac{\rho^k}{1-\rho}\|f_1 - f_0\| \end{aligned} \tag{F.5}$$

and therefore $\|f_{k+p} - f_k\| \to 0$, $k \to \infty$ for any p. This result implies that the sequence f_k is a Cauchy sequence. Since the space is complete the sequence has a limit f. If remains only to prove that f is a fixed point of T. Using again the triangle inequality, we have

$$\begin{aligned} \|T(f) - f\| &\leq \|T(f) - f_{k+1}\| + \|f_{k+1} - f\| \\ &= \|T(f) - T(f_k)\| + \|f_{k+1} - f\| \leq \rho\|f_k - f\| + \|f_{k+1} - f\|. \end{aligned} \tag{F.6}$$

Since the r.h.s. tends to zero when $k \to \infty$, we obtain $T(f) = f$. From the uniqueness of the fixed point we conclude that all sequences of successive approximations converge to this fixed point.

An operator T in a Hilbert space $\mathcal{X}$ is said to be a *non-expansive mapping* if it satisfies the condition

$$\|T(f_2) - T(f_1)\| \leq \|f_2 - f_1\|. \tag{F.7}$$

As is obvious, this condition is much weaker than condition (F.1). In any case, it is still possible to prove that any *non-expansive mapping* has at least one fixed point. Then, it is not difficult to prove that the set of the fixed points of T is a closed and convex set of $\mathcal{X}$. However, it is not possible to prove, in general, that the sequence of the successive approximations is convergent, i.e. the contraction principle does not apply to non-expansive mappings. This is the origin of the difficulty in proving the convergence of the projected Landweber method.

Indeed, the operator T defined by (see section 6.1)

$$T(f) = f + \tau(A^*g - A^*Af) \tag{F.8}$$

is, in general a non-expansive mapping for the problems considered in this book. It is a contraction mapping only for discrete problems such that the operator A^*A is a full-rank matrix.

These results can be derived by observing that

$$T(f_2) - T(f_1) = (I - \tau A^*A)(f_2 - f_1). \tag{F.9}$$

Then we obtain that

$$\|T(f_2) - T(f_1)\| \leq \rho \|f_2 - f_1\| \tag{F.10}$$

where

$$\rho = \sup_{\omega} \left|1 - \tau |\hat{K}(\omega)|^2\right| \tag{F.11}$$

for convolution operators, while

$$\rho = \sup_{k} \left|1 - \tau \sigma_k^2\right| \tag{F.12}$$

for operators with a singular value decomposition.

Since $\hat{K}(\omega)$ is zero at infinity we find that $\rho = 1$ for a convolution operator; analogously $\rho = 1$ for operators such that $\sigma_k \to 0$ or such that A^*A has the zero eigenvalue. Equation (F.12) implies that $\rho < 1$ if and only if the eigenvalues of the operator A^*A have a positive lower bound and this is true only in the case of a full-rank matrix.

The projected Landweber method, discussed in section 6.2, is the method of successive approximations for the operator $P_C T$ where P_C is the projection operator onto the closed and convex set C. If both T and P_C are non-expansive, then also $P_C T$ is non-expansive. Indeed the projection operator P_C is always non-expansive, as follows from its definition, which we recall for convenience of the reader.

For closed and convex sets in a Hilbert space $\mathcal{X}$, the basic result is the following [7]: *let C denote a closed and convex set of $\mathcal{X}$ and let f be any element of $\mathcal{X}$; then, there exists a unique element f_C of C which has minimal distance from f, i.e.*

$$\|f_C - f\| = \min_{h \in C} \|h - f\|. \tag{F.13}$$

According to this theorem, any closed and convex set C defines a mapping of $\mathcal{X}$ into itself by associating to each element f the unique element f_C which has minimal distance from it. This element is called the *convex projection* of f onto C and the mapping P_C defined by

$$P_C f = f_C \tag{F.14}$$

is called the projection operator onto the set C. This operator is, in general, nonlinear. It is linear if and only if the set C is a closed subspace of $\mathcal{X}$. In such a case P_C coincides with the projection operator onto this subspace (see appendix B).

A basic property of the convex projections is the following one [7]: *if f_1, f_2 is any pair of elements of $\mathcal{X}$ and $f_{1,C}$, $f_{2,C}$ the corresponding convex projections onto the closed and convex set C, then*

$$\|f_{1,C} - f_{2,C}\| \leq \|f_1 - f_2\|. \tag{F.15}$$

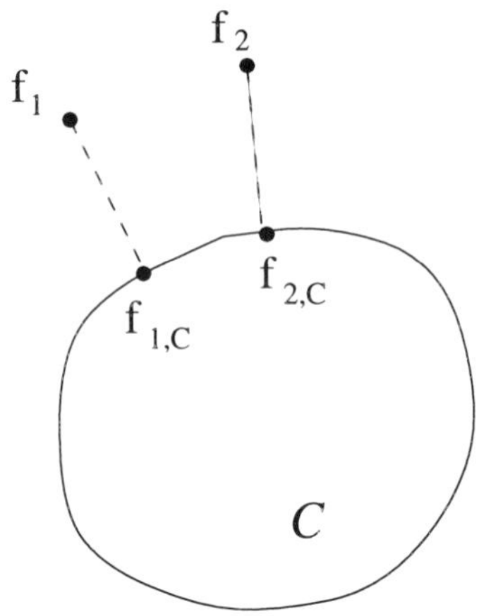

Figure F.1. Illustrating the inequality (F.15) for convex projections.

This result is rather intuitive, as suggested by figure F.1. The equality in equation (F.15) holds true if both f_1 and f_2 are elements of C. This property of the convex projections just implies that the operator P_C, defined in equation (F.14) is non-expansive.

Appendix G

The EM method

In this appendix we show that the iterative method of section 7.3 is a particular case of the EM method which is used for the computation of ML estimates in cases where the observations can be viewed as incomplete data.

This statement roughly means that we do not observe directly the random variables, whose values would constitute a complete set of data, but some functions of them, so that we have sort of indirect and incomplete observations.

According to the approach of Shepp and Vardi, in the particular example we are considering, a complete set of data is defined as a set of values of the r.v. $\eta_{m,n}$ which have the following meaning: a value of $\eta_{m,n}$ is the number of particles which are emitted at pixel n in object domain and detected at pixel m in image domain. It is reasonable to assume that all r.v. $\eta_{m,n}$ are independent and Poisson distributed. Moreover, by taking into account the meaning of $f_n^{(0)}$ and $A_{m,n}$, we can also assume that

$$E\{\eta_{m,n}\} = A_{m,n} f_n^{(0)}. \tag{G.1}$$

Then, the observed r.v. are defined by

$$\eta_m = \sum_{n=1}^{N^2} \eta_{m,n}; \quad m = 1, 2, \cdots, N^2. \tag{G.2}$$

From well-known properties of the sum of independent Poisson r.v. it follows that the η_m are also independent and Poisson distributed, with expectation values given by

$$E\{\eta_m\} = \sum_{n=1}^{N^2} E\{\eta_{m,n}\} = \sum_{n=1}^{N^2} A_{m,n} f_n^{(0)} = \left(\mathbf{A} \boldsymbol{f}^{(0)}\right)_m . \tag{G.3}$$

A value of η_m is the total number of photons detected at pixel m (in image domain). Therefore the η_m coincide with the r.v. introduced in section 7.3 and their probability distribution is given by equation (7.25).

In addition we introduce the r.v. defined by

$$\bar{\eta}_n = \sum_{m=1}^{N^2} \eta_{m,n}; \quad n = 1, 2, \cdots, N^2 \tag{G.4}$$

which are also independent and Poisson distributed with expectation values given by

$$E\{\bar{\eta}_n\} = \sum_{m=1}^{N^2} E\{\eta_{m,n}\} = \sum_{m=1}^{N^2} A_{m,n} f_n^{(0)} = \alpha_n f_n^{(0)}, \tag{G.5}$$

α_n being defined in equation (7.34). A value of $\bar{\eta}_n$ is the number of photons which are emitted at pixel n in the object domain and are detected somewhere in the image domain.

Now, if we replace $\boldsymbol{f}^{(0)}$ by an arbitrary object $\boldsymbol{f}$ in the set of permissible objects, we obtain families of r.v. $\eta_{m,n}(\boldsymbol{f})$, $\eta_m(\boldsymbol{f})$ and $\bar{\eta}_n(\boldsymbol{f})$ which contain $\boldsymbol{f}$ as a parameter. The estimation problem consists of applying the ML method to the estimation of the parameters of the r.v. $\eta_{m,n}(\boldsymbol{f})$, being given the observed values of the r.v. η_m, equation (G.2). This is clearly an *estimation problem with incomplete data*, because the data are not the values of the r.v. $\eta_{m,n}$ (these are not observed) but the values of their functions η_m

The following simple presentation of the EM method, applied to this problem, is proposed by Shepp and Vardi. Since the method is iterative, we denote by $\tilde{\boldsymbol{f}}_k$ the estimate obtained at step k. Then the next step consists of the following substeps:

(a) *Expectation*: Given $\tilde{\boldsymbol{f}}_k$ and given $\boldsymbol{g}$, the observed value of the random vector $\boldsymbol{\eta}$, compute the conditional expectation value of $\bar{\eta}_n(\tilde{\boldsymbol{f}}_k)$ (number of photons emitted at pixel n and detected somewhere in image domain) by assuming that $\boldsymbol{g}$ is the value of the random vector $\boldsymbol{\eta}(\tilde{\boldsymbol{f}}_k)$. This is the quantity:

$$E\{\bar{\eta}_n(\tilde{\boldsymbol{f}}_k)|\boldsymbol{\eta}(\tilde{\boldsymbol{f}}_k) = \boldsymbol{g}\} = \sum_{m=1}^{N^2} E\{\eta_{m,n}(\tilde{\boldsymbol{f}}_k)|\boldsymbol{\eta}(\tilde{\boldsymbol{f}}_k) = \boldsymbol{g}\} \tag{G.6}$$

and, from the independence of the r.v. $\eta_m(\tilde{\boldsymbol{f}}_k)$, it follows that it is also given by

$$\sum_{m=1}^{N^2} E\{\eta_{m,n}(\tilde{\boldsymbol{f}}_k)|\eta_m(\tilde{\boldsymbol{f}}_k) = g_m\}. \tag{G.7}$$

Therefore the problem consists of computing the conditional probability distribution of $\eta_{m,n}(\tilde{\boldsymbol{f}}_k)$ for a given value of $\eta_m(\tilde{\boldsymbol{f}}_k)$, the relationship between these r.v. being as that given in equation (G.2). If g_m is given, the possible values of each r.v. $\eta_{m,n}(\tilde{\boldsymbol{f}}_k)$ are the integer numbers from 0 to g_m (i.e. $g_m + 1$ values); since the r.v. η_m and $\eta_{m,n}$ are Poisson distributed, it is easy to show that the conditional probability distribution of $\eta_{m,n}(\tilde{\boldsymbol{f}}_k)$ is binomial and given by

$$p(\eta_{m,n}(\tilde{\boldsymbol{f}}_k) = M|\eta_m(\tilde{\boldsymbol{f}}_k) = g_m) = \frac{g_m!}{M!(g_m - M)!} p_{m,n}^M (1 - p_{m,n})^{g_m - M} \tag{G.8}$$

with

$$p_{m,n} = \frac{E\{\eta_{m,n}(\tilde{\boldsymbol{f}}_k)\}}{\sum_{n=1}^{N^2} E\{\eta_{m,n}(\tilde{\boldsymbol{f}}_k)\}}. \tag{G.9}$$

The expectation value associated with the binomial distribution (G.8) is just $g_m p_{m,n}$; then from equations (G.6) and (G.7) and from equation (G.1) (with $\boldsymbol{f}^{(0)}$ replaced by $\tilde{\boldsymbol{f}}_k$), we obtain:

$$E\{\bar{\eta}_n(\tilde{\boldsymbol{f}}_k)|\eta(\tilde{\boldsymbol{f}}_k) = \boldsymbol{g}\} = (\tilde{\boldsymbol{f}}_k)_n \sum_{m=1}^{N^2} \frac{A_{m,n} g_m}{\sum_{n'=1}^{N^2} A_{m,n'} (\tilde{\boldsymbol{f}}_k)_{n'}}. \tag{G.10}$$

(b) *Maximization*: The conditional expectation value of $\bar{\eta}_n(\tilde{\boldsymbol{f}}_k)$ is now taken as the value of $\bar{\eta}_n(\tilde{\boldsymbol{f}}_{k+1})$ and the expectation value of this r.v. is computed by means of the ML method. From the example 7.2 of section 7.1 it follows that the ML estimate of the expectation value of $\bar{\eta}_n(\tilde{\boldsymbol{f}}_{k+1})$ coincides with its value, i.e.

$$E\{\bar{\eta}_n(\tilde{\boldsymbol{f}}_{k+1})\} = E\{\bar{\eta}_n(\tilde{\boldsymbol{f}}_k)|\eta(\tilde{\boldsymbol{f}}_k) = \boldsymbol{g}\}. \tag{G.11}$$

But, from equation (G.5) (with $\boldsymbol{f}^{(0)}$ replaced by $\tilde{\boldsymbol{f}}_{k+1}$) it follows that

$$E\{\bar{\eta}_n(\tilde{\boldsymbol{f}}_{k+1})\} = \alpha_n (\tilde{\boldsymbol{f}}_{k+1})_n \tag{G.12}$$

and, by combining equations (G.10),(G.11) and (G.12) we obtain

$$\alpha_n (\tilde{\boldsymbol{f}}_{k+1})_n = (\tilde{\boldsymbol{f}}_k)_n \sum_{m=1}^{N^2} \frac{A_{m,n} g_m}{\sum_{n'=1}^{N^2} A_{m,n'} (\tilde{\boldsymbol{f}}_k)_{n'}} \tag{G.13}$$

which is precisely the iterative algorithm (7.47).

References

[1] Groetsch C W 1993 *Inverse Problems in the Mathematical Sciences* (Wiesbaden: Vieweg)

[2] Engl H W, Hanke M and Neubauer A 1996 *Regularization of Inverse Problems* (Dordrecht: Kluwer)

[3] Balakrishnan A V 1976 *Applied Functional Analysis* (New York: Springer)

[4] Royden H L 1968 *Real Analysis* (New York: Macmillan)

[5] Press W H, Teukolsky S A, Vetterling W T and Flannery B P 1992 *Numerical Recipes* (Cambridge:Cambridge University Press)

[6] Cooley J W and Tukey J W 1965 *Math Computing* **19** 297

[7] Stark H (ed) 1987 *Image Recovery: Theory and Applications* (New York: Academic)

Index